Crop Breeding, Genetics and Genomics

Crop Breeding, Genetics and Genomics

Edited by Ayden Spears

SYRAWOOD
PUBLISHING HOUSE

New York

Published by Syrawood Publishing House,
750 Third Avenue, 9th Floor,
New York, NY 10017, USA
www.syrawoodpublishinghouse.com

Crop Breeding, Genetics and Genomics
Edited by Ayden Spears

International Standard Book Number: 978-1-68286-695-5 (Hardback)

Cataloging-in-Publication Data

Crop breeding, genetics and genomics / edited by Ayden Spears.
p. cm.
Includes bibliographical references and index.
ISBN 978-1-68286-695-5
1. Plant breeding. 2. Genetics. 3. Genomics. 4. Crops. I. Spears, Ayden.
SB123 .C76 2019
631.53--dc23

TABLE OF CONTENTS

PREFACE

The study of genes, variation and heredity explored under the fields of genetics and genomics has wide applications in crop science. Crop breeding is used for improving the quality of food for humans and animals. It is accomplished using techniques of classical plant breeding and complex molecular techniques. New varieties of crops that are more productive, have higher climatic adaptability and disease resistance are being developed to optimize agricultural productivity. Modern biotechnological tools of marker-assisted selection, double haploids, reverse breeding and genetic modification are evolving methodologies of crop breeding. This book is a compilation of chapters that discuss the most vital concepts and emerging trends in the science of crop breeding, especially from the perspective of genetic and genomic studies. Multiple approaches, evaluations and methodologies have been included in this book to cover the recent advances in this field. It will be beneficial to experts and students seeking detailed information in this domain.

This book unites the global concepts and researches in an organized manner for a comprehensive understanding of the subject. It is a ripe text for all researchers, students, scientists or anyone else who is interested in acquiring a better knowledge of this dynamic field.

I extend my sincere thanks to the contributors for such eloquent research chapters. Finally, I thank my family for being a source of support and help.

Editor

Micropropagation of *Cyrtopodium paludicolum (Orchidaceae) from root tip explants*

Dayana Rotili Nunes Picolotto[1], Vespasiano Borges de Paiva Neto[2*], Fábio de Barros[3], Daly Roxana Castro Padilha[1], Ana Cláudia Ferreira da Cruz[4] and Wagner Campos Otoni[4]

Abstract: *An efficient protocol for in vitro plant propagation of Cyrtopodium paludicolum has been developed using root tips dissected from well-developed seedlings. Root tips were cultured on Knudson medium supplemented with α-naphthaleneacetic acid (NAA), and/or thidiazuron (TDZ). TDZ did not induce protocorm-like bodies (PLBs) in the NAA absence, indicating phytoregulators synergistic effect. Medium supplemented with 1.34 µM NAA and 2.27 µM TDZ resulted in better response on PLBs, and subsequent shoot differentiation (55.25 shoots per explant), and in better rooting number and root length responses, favoring acclimatization with 90% of survived plants. However, the medium supplemented with only NAA (1.34 µM) resulted in 33.50 shoots per explant. Histological sections confirmed that only one PLB was induced per responsive root tip, and it showed numerous dispersed and extended meristemoids, or division centers that originated new PLBs. Additionally, this protocol could be* an excellent model to study molecular aspects of root to shoot conversion.

Key words: *Protocorm-like bodies, thidiazuron, α-naphthaleneacetic acid, germination.*

*Corresponding author:
E-mail: vespasiano.paiva@univasf.edu.br

[1] Universidade Federal de Mato Grosso do Sul, Campus de Chapadão do Sul, CP 112, 79.560-000, Chapadão do Sul, MS, Brazil
[2] Universidade Federal do Vale do São Francisco, Campus de Ciências Agrárias, Rodovia BR 407, km 119, Lote 543, PSNC, s/n, C1, 56.300-990, Petrolina, PE, Brazil
[3] Instituto de Botânica, Núcleo de Pesquisa Orquidário do Estado, CP 68041, 04.045-972, São Paulo, SP, Brasil
[4] Universidade Federal de Viçosa, Instituto de Biotecnologia Aplicada à Agropecuária (BIO-AGRO), Laboratório de Cultura de Tecidos, Campus Universitário, Avenida Peter Henry Rolfs, s/n, 36.570-900, Viçosa, MG, Brazil,

INTRODUCTION

The genus *Cyrtopodium* (Orchidaceae) is widely distributed in tropical and subtropical countries of Central and South America (Menezes 2000), and there are approximately 39 species of *Cyrtopodium* distributed in Brazil (Barros et al. 2013). The generic name *Cyrtopodium* means "curved little foot" due to the shape of the column in the center of the flower (Guo et al. 2010). *Cyrtopodium paludicolum* is a terrestrial orchid species usually found in permanently wet soils, with thick and elongated pseudobulbs, resulting in considerable size of the species (Barros et al. 2013). Due its vegetative beauty and exuberant inflorescences (large yellow flowers on long stems, which can easily exceed two meters in height), *C. paludicolum* has been a frequent target for the local orchids collectors, as observed in visits to private collections and in conversations with farmers who cultivate the plants. Its natural propagation occurs mainly by means of seed germination. However, the natural propagation of the native species is hindered by the intensive collection from the wild, and is severely threatened by agricultural expansion in the Brazilian Savanna, which consequently reduces the appropriated areas of natural occurrence of the species. Zeng et al. (2013) and Rodrigues et al. (2015) claim that asymbiotic germination and tissue culture can provide useful means for conservation and commercial propagation of tropical orchid species critically endangered.

No report on the *in vitro* culture of *C. paludiculum* has been found in the literature. However, it was possible to find information on other species of the *Cyrtopodium* genus, such as *C. paranaense* (Guo et al. 2010) and *C. brandonianum* (Flachsland et al. 2011). Since orchids are outbreeders, seeds germination as propagation strategy results in obtaining of highly heterozygous plants. Thus, protocols that provide clonal multiplication from different vegetative explants are necessary (Chugh et al. 2009). Several explants have been used aiming at clonal propagation of orchids, such as root and stem meristems, inflorescence axis, flower bud and leaf apex (Colli and Kerbauy 1993, Chen and Chang 2006, Martin and Madassery 2006, Chugh et al. 2009, Mulgund et al. 2011). Although orchid micropropagation has shown great advances in the recent years, the widespread use of in vitro propagation is still limited by problems such as exudation of phenolic compounds from explants isolated from mature plants, somaclonal variation, and transplantation to the field (Chugh et al. 2009).

The use of root has not been considered as explant source for a great number of species, and according to Peterson (1975), root meristem consists of highly determinate cells that have limited morphogenic competence for bud formation. However, bud regeneration in root apices has been reported for some orchid species in *in vitro* conditions. Ever since, root tips have been effectively used for the induction of shoot buds and protocorm-like bodies (PBLs) of many orchids genera, such as *Vanda* (Park et al. 2003), *Oncidium* (Wu et al. 2004), *Doritaenopsis* (Lang and Hang 2006), *Catasetum* (Kerbauy 1984a), and *Cyrtopodium* (Guo et al. 2010, Flachsland et al. 2011). In all cases, the addition of at least one plant growth regulator to the induction medium was necessary.

Plant growth regulators are widely used in micropropagation of orchids, as well as the concentrations of these regulators, which may vary, for instance, according to the explants and plant species. Guo et al. (2010) and Flachsland et al. (2011) reported great success in the root-to-shoot conversion, resulting in PLBs with the addition of cytokinins thidiazuron (TDZ), 6-benzyladenine (BA) and zeatin, isolated or in combination with auxin (indol-3acetic acid), to the culture medium. Lee et al. (2013) claimed that PLBs are important in orchid micropropagation and outwardly resemble to somatic embryos in form and development. Recently, PBLs have been used to obtain synthetic seeds (Mohanty et al. 2013), cryopreservation (Silva 2013, Gogoi et al. 2013) of different orchids species.

The present study aims to assess the competence of the excised root tip of *C. paludicolum* to produce PBLs, and the influence of plant growth regulators α-Naphthalene Acetic Acid (NAA) and TDZ over this morphogenic process. In addition, tissue culture is presented as a mean to face threats of genetic diversity, and therefore offer alternative strategies for conservation of native plant species and biodiversity.

MATERIAL AND METHODS

Root tip explants

C. paludicolum seedlings with 120 days obtained from asymbiotic germination in Knudson medium (Knudson 1946) were used as explants source. Root tips (average 10 mm in length) were excised and used as plant material.

Induction medium

Root tips were placed on glass flasks (250 cm³) containing 40 mL of previously autoclaved Knudson medium supplemented with 58.43 mM sucrose, NAA (0, 1.34, 2.68 µM) in combination with TDZ (0, 2.27, 4.54 µM), solidified with 0.4% agar (HiMedia®). The pH was adjusted to 5.8 before autoclaving at 1.1 kg cm⁻² and 121 °C for 20 minutes. Plant growth regulators were filter-sterilized using sterile PES Syringe Filters (0.22 µm, TPP®) after the autoclaving process of the medium. Flasks were sealed using transparent polyvinylchloride plastic film (Dispafilm do Brasil Ltda). Cultures were maintained in a growth room at 27 ± 2 °C under 16-h photoperiod and 36 µmol m⁻² s⁻¹ irradiance.

Anatomical studies

Root tip samples were collected during culture and fixed in a solution containing formalin, acetic acid, and 50% ethyl alcohol (5 : 5 : 90 parts, respectively). Fixed samples were dehydrated in a graded ethanol series and embedded in methacrylate (Historesin, Leica Instruments, Germany). Cross and longitudinal sections (5-µm thick) were obtained using an automatic rotary microtome (RM 2155, Leica Microsystems Inc., USA) and stained with toluidine blue (O'Brien and McCully 1981). Samples were mounted in Permount on glass slides. Photographs were taken using a light microscope (Olympus AX70TRF; Olympus Optical, Japan) equipped with a digital camera (Spot Insight Color 3.2.0; Diagnostic Instruments Inc., USA).

Acclimatization

After the micropropagation stage, fifty vitroplants resulted from 1.34 µM NAA and 2.27 µM TDZ treatment were used to evaluate plant acclimatization. Vitroplants were individually placed in black plastic pots (7 cm height × 7 cm diameter) containing Plantmax® substrate. Pots were placed on benches and kept for 90 days in a greenhouse, with intermittent irrigation system, comprised of microsprinklers, activated by a timer, with two daily irrigation cycles of five minutes each. Plants received foliar fertilizer application (Nipokan®) at weekly intervals according to the manufacturer's recommendation (75 mL/100 L). After the acclimatization period, plant survival percentage was evaluated.

Statistical analysis

The experiment was arranged in a 3 x 3 factorial completely randomized design with three NAA concentrations (0, 1.34, 2.68 µM) and three TDZ concentrations (0, 2.27, 4.54 µM). Each replication consisted of one flask containing three root tip explants, with four replications per each treatment.

Morphogenetic responses of root tip explants were assessed by recording the number and length of obtained shoots; number and length of roots; and percentage of survivor explants, at 120 days of culture. Count data were transformed to $x + 0.5$, and the percentage data were transformed to arcsin x. Data were analyzed using analysis of variance (ANOVA), followed by the Tukey's test for comparison of means at 5% probability level. The statistical package SISVAR (Ferreira 2011) was used for statistical analysis.

RESULTS

Throughout *in vitro* culture of *C. paludicolum* seedlings on Knudson medium, eventual direct PLB formation was observed in the apical region (Figures 1A, B), and in non-apical root regions, although the latter was less frequent (Figure 1C). In both cases, no plant growth regulators (PGR) were added to the Knudson medium; however, in these cases,

Figure 1. Different stages of protocorm-like body (PLB) development from intact root tip (A, B, C) of *Cyrtopodium paludicolum* cultivated in Knudson medium without plant growth regulators, and from segmented root tip cultivated in Knudson medium with 1.34 µM NAA and 2.27 µM TDZ (D, E, F). (A) PLB formed from the root tip of aseptically grown seedling; (B) detail of a plantlet from a root tip-derived PLB still connected to the maternal plant; (C) well-developed plantlet obtained from PLB originated in the non-apical root tissue yet connected to maternal plant; (D) segmented root tip showing initial morphological alterations; (E) multishoots resulted from root tip derived PLB; (F) elongated shoots from root tip derived PLB; (G) Acclimatized plants. Bars = 10 mm.

PLBs were originated only when root was still attached to donor plants (Figures 1A, C). Additionally, PBLs originated in the process were capable of generating plantlets morphologically similar to donor plants. However, this phenomenon occurred in very low frequencies in excised root tips. Moreover, 91.75% (Table 2) of root tips explants died when cultured in plant growth regulators (PGR) free-medium, underscoring the role of the intact *C. paludicolum* plant for PLB induction without PGR.

Different concentrations of NAA and TDZ added to the basal Knudson medium were tested for their ability to induce PLB on excised root tips. Both NAA and TDZ were efficient in forming PLB, and significant interaction was observed between the two PGRs (Table 1), revealed by analysis of variance. When NAA was used alone, root-tip explants with PLB formation were obtained in both tested concentrations (1.34 and 2.68 µM) (Table 2). Contrarily, no improvement on PLB induction was verified when TDZ was tested alone. However, when NAA and TDZ were added together, they greatly enhanced PLB formation. Treatment containing 1.34 µM NAA combined with 2.27 µM TDZ resulted in excised root tips with higher PLBs number (Table 1, Figures 1D, F). Additionally, PBLs maintained in this PGRs combination resulted in plantlets with higher shoot length, number of roots number and length. These data suggest a synergistic effect between NAA and TDZ (Table 2), as confirmed by the statistical analyses (Table 1).

After inoculation in the presence of NAA and/or TDZ, *C. paludicolum* root tip explants presented an elongation stage. After that, approximately 71.97% root tip explants survived and formed a clear globular structure at the apical root

Table 1. Summary of the analysis of variance of shoot length, number of roots, longest root length, number of shoots, and survival (%) of *Cyrtopodium paludicolum* vitroplants derived from excised root tip explants cultivated in different combinations of plant growth regulators thidiazuron (TDZ) and α-naphtaleneacetic acid (NAA)

| Source | df | Mean Square | | | | |
| | | Explants survival (%) | Shoot | | Root | |
			number	length	length	number
TDZ	2	1139.5*	884.4**	3.4**	11.9**	13.4**
NAA	2	2264.5*	3345.0**	16.9**	7.4**	67.0**
TDZ*NAA	4	5940.5**	572.3**	3.1**	3.6**	6.9*
Error	27	759.1	24.9	0.4	0.9	2.3
Total	35					

* Significantly different at $P \leq 0.05$; ** Significantly different at $P \leq 0.001$.

Table 2. Effects of concentrations of α-naphtaleneacetic acid (NAA) and thidiazuron (TDZ) on the formation of protocorm-like bodies (PLBs) from excised root tip of *Cyrtopodium paludicolum* maintained in Knudson medium for 120 days

| Parameters | NAA (µM) | TDZ (µM) | | |
		0.00	2.27	4.54
Explant survival (%)	0.00	8.25 Bb	75.25 Aa	91.75 Aa
	1.34	91.75 Aa	66.75 Aab	83.50 Aa
	2.68	91.75 Aa	25.00 Bb	50.00 ABa
Number of shoots	0.00	0.50 Ab	4.75 Ab	1.25 Ab
	1.34	33.50 Ba	55.25 Aa	16.75 Ca
	2.68	26.75 Aa	11.50 Bb	4.50 Bb
Shoot length (mm)	0.00	2.50 Ab	12.50 Ab	2.50 Ab
	1.34	17.50 Ba	42.50 Aa	22.50 Ba
	2.68	15.00 Aa	5.00 Ab	5.00 Ab
Number of roots	0.00	0.25 Aa	1.50 Ab	1.50 Ab
	1.34	2.25 Ba	7.25 Aa	4.75 ABa
	2.68	0.25 Aa	0.25 Ab	0.50 Ab
Root length (mm)	0.00	2.50 Ba	27.50 Aa	30.00 Aa
	1.34	2.50 Ba	32.50 Aa	15.00 Bab
	2.68	2.50 Aa	2.50 Ab	10.00 Ab

Means followed by the same uppercase letter in different TDZ concentrations (line) are not significantly different according to Tukey test (*p≤0.05*). Means followed by the same lowercase letter on NAA concentration (column) are not significantly different according to Tukey test (*p≤0.05*)

region (Figure 1D), and evolved into a primary PLB (Figures 2A, B). Contrarily, medium devoid of NAA and TDZ resulted in a low frequency of root tip explants survival (8.25%) (Table 2).

The histological details of conversion of root to shoot meristem, and subsequent PBLs proliferation were observed (Figure 3). A single PLB was induced per responsive root tip (Figure 2A). Histological sections showed numerous and dispersed meristemoids or division centers (Figures 2B, D) originated from intense cellular division, primarily in the procambium and ground meristem cells, resulting in intense morphological modifications that generally occur after 40 days. Thus, primary PLB resulted from disconnected root tip (Figures 1D, 2A), and many secondary PLBs resulted from the primary one (Figures 1E, 1F, 2B, 2D), mainly in the segmented root apices cultured in Knudson medium containing 1.34 µM NAA and 2.27 µM TDZ. The cells of the earliest recognizable meristemoids or division centers were small, containing densely staining cytoplasm, and large nuclei centrally located (Figure 2D). These cells stained intensely with toluidine blue, which indicated high levels of DNA and RNA, which is typical of cell division.

Vitroplants resulted from 1.34 µM NAA and 2.27 µM TDZ treatment (Figure 1) were morphologically similar to seed-derived plants, and forty-five plants (90%) survived after acclimatization stage (Figure 1G).

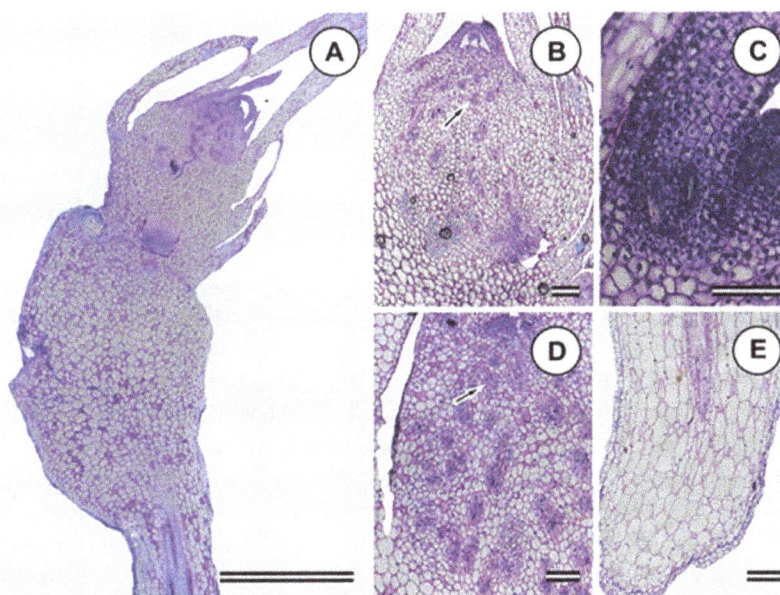

Figure 2. Longitudinal histological sections of the excised root tip of *Cyrtopodium paludicolum* during protocorm-like body (PLB) formation in Knudson medium supplemented with 1.34 µM NAA and 2.27 µM TDZ. (A) root to shoot (PLB) conversion, showing leaf primordia (arrow); (B) detail of intense cell division (arrow) in the parenchyma cells of the shoot derived from PLB root tip; (C) detail of the leaf primordia showing intense cell division; (D) detail of meristemoids resulting from cell division in the parenchyma cells (arrow); (E) Longitudinal section of root apex used as the typical explants, showing a root tip. Bars: A = 0.5 cm and B, C, D, E = 0.1 cm.

DISCUSSIONS

PLBs and their subsequent growth and development from non-apical regions of the intact root in flask-grown seedlings using free-PGR medium, as reported here with C. paludicolum, were also observed for C. paranaense by Guo et al. (2010). According to these authors, anatomical sections clearly revealed that PLBs were originated from the external cells of the root central stele, protruding through the parenchymal cortical cells and velamen.

When roots were excised and segmented, the beneficial effects of cytokinin, isolated or in combination with auxin during *in vitro* plant propagation, as reported here, are in agreement with the results obtained in other genera of orchid, such as *Cyrtopodium* (Flachsland et al. 2011), *Oncidium* (Mayer et al. 2010), *Dendrobium* (Sujjaritthurakarn and Kanchanapoom 2011), *Xenikophyton* (Mulgund et al. 2011), *Vanilla* (Giridhar and Ravishankar 2004) and *Aerides* (Devi et al. 2013). Different PGRs have been reported to enhance PLB induction from the excised root tips of several orchid species (Kerbauy 1984b, Sánchez 1988, Park et al. 2003), and depending on their concentrations, Auxins were more effective in the induction of PLBs from root tips than cytokinins. In this experiment, the medium containing only NAA was able to induce PLBs from excised root tips of *C. paludicolum*; however, the same did not happen when it contained only TDZ. Conversely, in *Catasetum fimbriatum* and *Catasetum pileatum*, excised root apices formed high percentages of PLBs in culture medium without PGRs (Kraus and Monteiro 1989, Colli and Kerbauy 1993). This indicates that the requirement for PGR in medium for root tip PLB induction is species- or genus-dependent, considering that Guo et al. (2010) and Flachsland et al. (2011) found similar results for another *Cyrtopodium* species, named *C. paranaense and C. Brandonianum,* respectively.

The best Auxin and cytokinin combination for *C. paludicolum* (1.34 μM NAA and 2.27 μM TDZ) were similar to those used by other authors in equivalent experiments (for instance, 0.5 mg L^{-1} TDZ (Flachsland et al. 2011), 0.5 mg L^{-1} NAA and 1 mg L^{-1} BAP (Bellaver et al. 2015), 10.2 mM indole-3-acetic acid (IAA) and 9.0 mM TDZ (Guo et al. 2010)). However, the number of shoots per explant obtained in this study was higher than all cited protocols, indicating that the present protocol is well adjusted in relation to PLBs induction. Differently from the results showed by Flachsland et al. (2011), which proposed the necessity to obtain root induction in the *C. brandonianum* PBLs, the present protocol obtained a great number of roots during the development of *C. paludicolum* PBLs, making unnecessary the use of root induction medium, which reinforces the efficiency of the present protocol. Therefore, the auxin and cytokinin balance showed in this study was apparently benefic to PBLs induction and development.

Despite the poor morphogenetic ability of root apical cells of higher plants, including Orchidaceae family, the usefulness of root explants for in vitro propagation purposes is due to their continuous availability, low oxidation rate, and to their ability of being explanted (Chugh et al. 2009). In relation to *C. paludicolum*, roots can be obtained during the entire year, unlike leaves, which are abscised during the drought season, a common behavior in many orchid species found in the Brazilian savannah. In addition, the leaves can be lost by herbivory.

According to Guo et al. (2010) the inhibition of root growth and meristematic cell proliferation in the root apical region is very important for conversion of root tip in PLB in C. paranaense. Further differentiation of these PLBs proceeded normally into plantlets, as observed in the present experiment with *C. paludicolum* (Figure 2B, C). The root apex region of *in vitro C. paludicolum* has similar histological traits to those of terrestrial orchids, especially to the root anatomy described for *C. paranaense* (Guo et al. 2010), presenting several cell layers (Figure 2E), with the primary meristems, including the procambium. The globular expansion increased in size as a result of division and enlargement to become a PLB, as described for *C. paranaense* by Guo et al. (2010), especially involving the parenchyma cells from root donor tissues. The PLB formation from root tips in *C. paludicolum*, particularly the existence of vascular connections between early PLBs and explants, is different from the developmental patterns of single cell-originated somatic embryo in other orchids (Kerbauy 1984b, Park et al. 2003) and similar to that in *C. paranaense* (Guo et al. 2010) and in *C. brandonianum* (Flachsland et al. 2011).

The success in the acclimatization stage assured the efficiency of this micropropagation protocol. Plants hydration was a constant care during the acclimatization stage, in order to ensure high humidity, since *C. paludicolum* is usually found in flooded areas. Rodrigues et al. (2015) obtained up to 90% of acclimatization success with *Cyrtopodium saintlegerianum*. Other authors who obtained vitroplants of the *Cyrtopodium* genus do not report acclimatization data, such as Guo et al. (2010), Flachsland et al. (2011) and Bellaver et al. (2015).

The present micropropagation protocol is an efficient method to obtain a great number of *C. paludicolum* plants using PLBs originated from excised root tips cultivated in Knudson medium added with 1.34 μM NAA and 2.27 μM TDZ. These results clearly indicate that asexual PLBs initiated from root tips resulted in regular *C. paludicolum* plants.

ACKNOWLEDGEMENTS

The authors thank the Fundação de Apoio ao Desenvolvimento do Ensino, Ciência e Tecnologia do Estado de Mato Grosso do Sul (FUNDECT), and the Conselho Nacional de Desenvolvimento Científico e Tecnológico (CNPq) for financial support.

REFERENCES

Barros F, Vinhos F, Rodrigues VT, Barberena FFVA, Fraga CN, Pessoa E M, Forster W and Menini Neto L (2013) Orchidaceae. In **Lista de espécies da flora do Brasil.** Jardim Botânico, Rio de Janeiro. Available at <http://floradobrasil.jbrj.gov.br/jabot/floradobrasil/FB11443> Accessed on April 6, 2015.

Bellaver LA, Kato KMM, Buttini S, Antonietti D, Galli S and Stefanello S (2015) Regeneração in vitro de Cyrtopodium paranaense Schltr (Orchidaceae) a partir de regiões meristemáticas. **Revista Brasileira de Energias Renováveis 4**: 100-109.

Chen JT and Chang WC (2006) Direct somatic embryogenesis and plant regeneration from leaf explants of Phalaenopsis amabilis. **Plant Biology 50**: 169-173.

Chugh S, Guha S and Rao IU (2009) Micropropagation of orchids: a review on the potential of different explants. **Scientia Horticulturae 122**: 507-520.

Colli S and Kerbauy GB (1993) Direct root tip conversion of Catasetum into protocorm-like bodies - Effects of auxin and cytokinin. **Plant Cell, Tissue and Organ Culture 33**: 39-44

Devi HS, Devi SI and Singh TD (2013) High frequency plant regeneration system of Aerides odorata Lour. through foliar and shoot tip culture. **Notulae Botanicae Horti Agrobotanici 41**: 169-176.

Ferreira DF (2011) Sisvar: a computer statistical analysis system. **Ciência e Agrotecnologia 35**: 1039-1042.

Flachsland E, Terada G, Fernández JM, Medina R, Schinini A, Rey H and Mroginski L (2011) Plant regeneration from root-tip culture of Cyrtopodium brandonianum barb. rodr. (Orchidaceae). **Propagation of Ornamental Plants 11**: 184-188.

Giridhar P and Ravishankar GA (2004) Efficient micropropagation of Vanilla planifolia Andr under influence of thidiazuron, zeatin and coconut milk. **Indian Journal of Biotecnology 3**: 113-118.

Gogoi K, Kumaria S and Tandon P (2013) Cryopreservation of Cymbidium eburneum Lindl. and C. hookerianum Rchb. f., two threatened and vulnerable orchids via encapsulation–dehydration. **In Vitro Cellular and Developmental Biology Plant 49**: 248-254.

Guo WL, Chang YCA and Kao CY (2010) Protocorm-like bodies initiation from root tips of Cyrtopodium paranaense (Orchidaceae). **Hortscience 45**: 1365-1368.

Kerbauy GB (1984a) Regeneration of protocorm-like bodies through in vitro culture of root tips of Catasetum (Orchidaceae). Zeitschrift für Pflanzenphysiologie **113**: 287-291.

Kerbauy GB (1984b) Plant regeneration of Oncidium varicosum (Orchidaceae) by means of root tip culture. **Plant Cell Reports 3**: 27-29

Knudson LA (1946) New nutrient solution for the germination of orchid seeds. **American Orchid Society Bulletin 14**: 214-217.

Kraus JE and Monteiro WR (1989) Formation of protocorm-like bodies from root apices of Catasetum pileatum (Orchidaceae) cultivated

in vitro: I. Morphological aspects. **Annals of Botany 64**: 491-498.

Lang NT and Hang NT (2006) Using biotechnological approaches for Vanda orchid improvement. **Omonrice 14**: 140-143.

Lee YI, Hsu ST and Yeung EC (2013) Orchid protocorm-like bodies are somatic embryos. **American Journal of Botanic 100**: 2121-2131

Martin KP and Madassery JP (2006) Rapid in vitro propagation of Dendrobium hybrids through direct shoot formation from foliar explants and protocorm like bodies. **Scientia Horticulturae 108**: 95-99.

Mayer JLS, Stancato GC and Appezzato-da-Glória B (2010) Direct regeneration of protocorm-like bodies (PLBs) from leaf apices of Oncidium flexuosum Sims (Orchidaceae). **Plant Cell, Tissue Organ Culture 103**: 411-416

Menezes LC (2000) **Orchids genus Cyrtopodium**: Brazilian species. IBAMA, Brasília, 208p.

Mohanty P, Nongkling P, Das MC, Kumaria S and Tandon P (2013) Short-term storage of alginate-encapsulated protocorm-like bodies of Dendrobium nobile Lindl.: an endangered medicinal orchid from North-east India. **Biotechnology 3**: 235-239.

Mulgund GS, Nataraja K, Malabadi RB and Kumar SV (2011) TDZ induced in vitro propagation of an epiphytic orchid Xenikophyton smeeanum (Reichb. f.) **Research in Plant Biology 1**: 7-15.

O'Brien TP. and McCully ME (1981) **The study of plant structure: principles and selected methods.** Termarcarphy Pty, Melburne, 352p.

Park SY, Murthy HN and Paek KY (2003) Protocorm-like body induction and subsequent plant regeneration from root tip cultures of Doritaenopsis. **Plant Science 164**: 919-923.

Peterson RL (1975) The initiation and development of root buds. In Torrey JG and Clarkson DT (eds) **The development and function of roots.** Academic Press, New York, p. 125-161.

Rodrigues LA, Paiva Neto VB, Boaretto AG, Oliveira JF, Torrezan MA, Lima SF and Otoni WC (2015) In vitro propagation of Cyrtopodium saintlegerianum rchb. f. (orchidaceae), a native orchid of the Brazilian savannah. **Crop Breeding and Applied Biotechnology 15**: 10-17.

Sánchez M (1988) Micropropagation of Cyrtopodium (Orchidaceae) through root-tip culture. **Lindleyana 3**: 93-96.

Silva JAT (2013) Cryopreservation of hybrid Cymbidium protocorm-like bodies by encapsulation–dehydration and vitrification. **In Vitro Cellular and Developmental Biology Plant 49**: 690-698.

Sujjaritthurakarn P and Kanchanapoom K (2011) Efficient direct protocorm-like bodies induction of dwarf Dendrobium using Thidiazuron. **Notulae Scientia Biologicae 3**: 88-92.

Wu IF, Chen JT and Chang WC (2004) Effect of auxins and cytokinins on embryo formation from root-derived callus of Oncidium 'Gower Ramsey'. **Plant Cell, Tissue and Organ Culture 77**: 107-109.

Zeng S, Wanga J, Wua K, Silva JAT, Zhanga J and Duan J (2013) In vitro propagation of Paphiopedilum hangianum Perner and Gruss. **Scientia Horticulturae 151**: 147-156.

2

Gene transfer utilizing pollen-tubes of *Albuca nelsonii* and *Tulbaghia violacea*

Aloka Kumari[1], Ponnusamy Baskaran[1] and Johannes Van Staden[1*]

Abstract: *Developing a tissue culture-independent genetic transformation system would be an interesting technique for gene transfer in valuable medicinal and horticultural plants. Efficient gene delivery (Agrobacterium tumefaciens strain LBA 4404: harbouring PBI121 plasmid) was achieved with Km-resistant pollen grains (pollen tube technique) and were found to be GUS-positive for Albuca nelsonii (31.3%) and Tulbaghia violacea (32.6%). The Km-resistance (95.6% for A. nelsonii and 86.7% for T. violacea) and GUS-positive (100% for A. nelsonii and 97.5% for T. violacea) putative transgenic seedlings in vitro were obtained with 200 mg L^{-1} Km. The in vitro plants were obtained from leaf explants of putative transgenic seedlings and were confirmed to be Km-resistant and GUS-positive (T. violacea, 73.7% and A. nelsonii, 80.5%). The plants were successfully acclimatized in the greenhouse. We describe a tissue culture-independent gene transfer technique with high efficiency clonal transgenic plant production for A. nelsonii and T. violacea. This can also be applied to biotechnological crop improvement of the same species and potentially to other plants.*

Key words: *Agrobacterium tumefaciens, genetic transformation, GUS-expression, in vitro regeneration, pollen transformation.*

*Corresponding author:
E-mail: rcpgd@ukzn.ac.za

[1] University of KwaZulu-Natal Pietermaritzburg, Research Centre for Plant Growth and Development, School of Life Sciences, Scottsville 3209, South Africa,

INTRODUCTION

Tulbaghia violacea (Alliaceae), commonly known as wild garlic, is cultivated for both ornamental and medicinal use (Reinten et al. 2011). It is a clump-forming perennial with narrow leaves, and produces a large cluster of fragrant, violet flowers from mid-summer to autumn (Harvey 1837). *T. violacea* produces various bioactive compounds and is used to treat earache, fever, high blood pressure, heart problems, chest complaints, high cholesterol, constipation, rheumatism and paralysis (Watt and Breyer-Brandwijk 1962). Similarly, *Albuca nelsonii* (Hyacinthaceae) is a perennial ornamental and traditional medicinal plant, which is commonly known as Nelson's slime lily (Ascough and Van Staden 2010). The species flowers during September to November. The medicinal and ornamental value of *T. violacea* and *A. nelsonii* suggests their use in studies of pollen transformation as bulbous model plants to obtain a higher efficacy of gene transfer.

Many tissue culture and genetic transformation techniques have been developed to create genetically engineered plants that can tolerate environmental stress, and improve productivity and quality of plant species (Tu et al. 2005, *García-Sogo* et al. 2010, Eapen 2011, Alikina et al. 2016; Souza et al. 2017). However, the efficiency of current gene transfer techniques are still low, judged by the difficulties in recovering fertile transgenic plants and time constraints (Saunders and Matthews 1995, Koetle et al. 2017). A need to find more efficacious and economical methods led to the development of novel alternative systems for genetic transformation that exclude tissue culture steps and rely on simple and inexpensive protocols. Pollen is an important agent for gene transfer to produce transgenic seeds directly through fertilization (Shivanna and Sawhney 2005, Eapen 2011). Pollen has been manipulated by pollen development and function, including alteration of its genome, for the production and improvement of crops and related products.

Pollen transformation is an attractive approach and helps in optimization of crop yield, hybrid seed production and crop improvement in economic plant species (Shivanna and Sawhney 2005). This is a promising research area for obtaining transgenic plants faster and easier than previous procedures (Saunders and Matthews 1995). Accordingly, the direct transformation of pollen grains could help provide an effective alternative to routine transformation methods practiced at present. The technique affords an inexpensive tissue culture-independent production of genetically uniform progeny preventing somaclonal variation. The technique is also genotype independent and applicable in both monocotyledonous and dicotyledonous plants through the regular fertilization process. Pollen transformation using pollen as a "super vector" was proposed earlier (Hess 1987). However, a routine method for pollen transformation is still lacking (Loguercio et al. 1994, Hudson et al. 2001, Eapen 2011, Han et al. 2015). Studies on selection of transformed pollen for the expression of genes are important in recombinant DNA technology, which is a major challenge for production of higher efficiency of transgenic plants. Accordingly, in the present investigation, we aimed to develop a potential alternative gene transfer technique for production of high efficiency clonal transgenic plants, *A. nelsonii* and *T. violacea*, by selection of transformed pollen via the pollen tube route. In addition, *in vitro* transgenic plantlets through leaf explants of putative transgenic seedlings were also established to confirm putative transgenics *in vitro*, as well as for independent *Agrobacterium*-mediated transformation with large-scale transgenic plant production. The techniques would contribute to modern agriculture systems including breeding and agronomic interests, as well as commercial production to improve specific plant traits with the ability to control biosafety and limit time constraints.

MATERIAL AND METHODS

Plant material and pollen transformation

Flowers (before anthesis) of *Albuca nelsonii* and *Tulbaghia violacea* were collected from the Botanical Garden of the University of KwaZulu-Natal (lat 29° 37.500 S, long 30° 24.230 E), Pietermaritzburg, South Africa, between 07:00 h and 07:30 h in the morning. Inflorescences were placed in conical flasks containing tap water and kept at growth room temperature (25 ± 2 °C) for 30 min. Fresh pollen grains were collected from each inflorescence after dehiscence of the anther. The BWK (Brewbaker and Kwack's 1963) medium containing 10% sucrose solution, 100 mg L^{-1} boric acid, 300 mg L^{-1} calcium nitrate, 100 mg L^{-1} potassium nitrate and 200 mg L^{-1} magnesium sulphate (Shivanna and Rangaswamy 1992) was used for control and transformation studies. The fresh pollen grains (1 mg) were placed in BWK medium (1 mL) containing different concentrations (100, 200, 300, 400, 500, 600 and 700 mg mL^{-1}) of kanamycin (Km) to select optimal Km concentration for inhibition of pollen tube growth for selection medium in transformation studies. *Agrobacterium tumefaciens* [LBA4404, harbouring the binary plasmid vector pCAMBIA1301, with the T-DNA region consisting of *GUS* gene driven by the Cauliflower Mosaic Virus 35S (CaMV35S) promoter] was grown on a shaker (100 rpm at 25 ± 2 °C in the dark) overnight in Luria-Bertani (LB) medium consisting of 0.1 mg mL^{-1} kanamycin, 0.15 mg mL^{-1} rifampicin and 50 µM acetosyringone. The pellet was collected at 5000 rpm by centrifugation for 20 min and resuspended in BWK medium consisting of 50 µM acetosyringone with OD600 = 0.5 density. For transformation study, 1 mg fresh pollen grains were inoculated onto 1 mL of BWK medium containing *A. tumefaciens* for 30 min under dark condition at 25 ± 2 °C for germination before being transferred to selection medium (BWK plus 200 mg L^{-1} Km for *T. violacea* and 600 mg L^{-1} Km for *A. nelsonii*) for 15 min. Fresh pollen grains (1 mg) cultured with BWK medium or BWK medium plus *A. tumefaciens* (1 mL) were used as controls to compare selection treatments. The GUS-positive pollen grains were verified under a light microscope. The pollen slurry (5 - 10 µL) was dropped evenly with the help of a micropipette (5 ml tip used without damaging the pollen tubes) onto the stigmata of emasculated flowers. Flowers were then rebagged for seed setting. The seeds were set in flowers of bagged natural (non-treated) and treated pollen grains after 8 weeks. Seeds were then collected to determine percentage of seed setting and assays for confirmation of putative transgenic plants. Seeds from both natural and treated pollen grains were sterilized with 0.2% HgCl$_2$ for 10 min, followed by five washes with sterile distilled water. The seeds were inoculated onto petri dishes containing two layers of Whatman No. 1 filter paper and moistened with 200 mg L^{-1} Km. The cultures were maintained at a temperature of 25 ± 2 ºC and light intensity of 40 µmoL m^{-2}s^{-1} provided by cool white fluorescent light (OSRAM L 58 W/740, South Africa) with a 16 h photoperiod. The efficiency (%) of Km-resistant seedling was calculated as the number of Km-resistant/total number of seeds inoculated × 100. The Km-resistant plants were used for histochemical GUS assay.

Development and confirmation of putative transgenic plantlet *in vitro*

Leaf explants were excised from 20-day-old *in vitro* germinated putative transgenic and non-transformed (control) seedlings for *in vitro* regeneration studies. Explants were cultured on MS (Murashige and Skoog 1962) medium containing 30 g L^{-1} sucrose, 8 g L^{-1} agar and supplemented with different combinations and concentrations of plant growth regulators [PGRs: benzyladenine (BA), meta-topolin riboside (*m*TR), thidiazuron (TDZ) and naphthaleneacetic acid (NAA)] for direct shoot regeneration over a period of 8 weeks. Media devoid of PGRs and shoots regenerated with 10 µM BA plus 0.5 µM NAA from leaf explants of non-transformed seedlings were used as controls. The shoots from respective shoot regeneration medium were cultured onto MS medium with 5 µM indole-3-butyricacid (IBA) or indole-3-acetic acid (IAA) plus 25 mg L^{-1} Km for rooting and selection of putative transformed shoots for 6 weeks. The combinations of PGRs are indicated in Table 2. The Km-resistant plantlets were used for histochemical GUS assay. The chemicals used in the preparation of the MS medium and PGRs were of analytical grade (Biolab, South Africa; Oxoid, England and Sigma, USA). All media were adjusted to pH 5.8 with 0.1 N NaOH before gelling with 8 g L^{-1} agar and autoclaved at 121 °C for 20 min. The cultures were maintained under 16 h photoperiod supplied by cool white fluorescent light [40 mol m^{-2}s^{-1} photosynthetic photon flux (PPF), OSRAM L 58 W/740, South Africa] at 25 ± 2 °C.

The putative transgenic plantlets (Km-resistant plus GUS-positive) were harvested and then transferred to terracotta pots (95 x 120 mm) containing a 1:1 (v/v) vermiculite:soil mixture and irrigated with tap water every third day. These plantlets were maintained in the greenhouse (25 ± 2 °C under natural photoperiod conditions and midday PPF of 950 ± 50 mol m^{-2}s^{-1}) for acclimatization.

Frequency of gene transfer by GUS assay

The plant materials (leaf tissues and plantlets) from Km-resistant and controls from *in vitro* germinated seedlings and plantlets, and one-year-old greenhouse-grown plants were used for GUS assay. The X-Gluc solutions and procedure for histochemical GUS assay was performed using β-Glucuronidase Reporter Gene Staining Kit (Sigma-Aldrich®, St. Louis, USA). The X-Gluc solutions containing plant materials were incubated at 37 °C for 24 h, followed by destaining with ethanol for 3 h to remove chlorophyll. The plant materials were then stored in 70% ethanol. The efficiency (%) of GUS expression was calculated as number of plants expressing GUS-positive/total number of plants evaluated × 100.

Statistical analysis

All experiments were conducted three times with 25 replicates for each treatment for pollen transformation, seed germination, shoot regeneration, rooting and GUS assay. Data were analyzed using a one-way analysis of variance (ANOVA), and are presented as means ± standard error. Treatment means were separated using Duncan's multiple range tests at the 5% probability level and analyzed using IBM SPSS for Windows version 23 (SPSS Inc., Chicago, IL, USA). Percentage data were arcsine square root transformed before using an analysis of variance.

RESULTS AND DISCUSSION

Gene delivery by pollen transformation

Pollen grains were inoculated on BWK medium with different concentrations of Km to investigate the optimal concentration of Km for transformation studies. Pollen tube growth was completely inhibited at 200 mg L^{-1} Km for *T. violacea* and 600 mg L^{-1} Km for *A. nelsonii* except the controls (Figure 1A, B, C, D). This would allow for efficient gene delivery via the pollen tube pathway. Optimization of Km concentration is essential for effective genetic transformation and has been reported with different methods in other plant species (Bino et al. 1987, Baskaran and Dasgupta 2012). A selection of Km concentration (200 - 400 mg L^{-1}) has been reported with less sensitive and longer pollen tube growth for chimaeric tomato (Bino et al. 1987). In this study, pollen germination was observed in all treatments in both plant species (Table 1). However, GUS-positive pollen tube growth varied significantly between BWK medium plus *A. tumefaciens* and BWK medium plus *A. tumefaciens* and Km (selection) treatments (Table 1). Frequency of GUS-positive pollen germination was 32.6% for *T. violacea* and 31.3% for *A. nelsonii* during the selection treatment (Table 1 and Figure 1E, F). The control (pollen germination with BWK medium) treatment did not show GUS-positive germination of pollen (Table 1). This indicates that successful gene delivery to the pollen could be assessed by histochemical GUS assay. Germinated pollen from all treatments was applied gently on the stigmata of emasculated flowers for seed setting. The

Figure 1. Pollen tube growth and GUS-positive expression in pollen and seedlings of *T. violacea* and *A. nelsonii*: Pollen tube growth in BWK medium for *T. violacea* (**A**) and *A. nelsonii*(**B**); Inhibition of pollen tube growth at Km for *T. violacea* (**C**) and *A. nelsonii* (**D**); GUS-positive pollen germination of *T. violacea*(**E**) and *A. nelsonii*(**F**);GUS-positive seedlings of *T. violacea*(**G**) and *A. nelsonii* (**H**); The GUS-expression of leaf in one-year-old greenhouse-grown plants of *T. violacea*(**I**) and *A. nelsonii* (**J**).*Bar A-F* 100μm; *H* 1mm and 5 mm; and *I, J* 5 mm.

Table 1. Development of putative transgenic pollen and seedlings by incubation of *Agrobacterium tumefaciens* for *T. violacea* and *A. nelsonii*

Plant name	Treatments	GUS +ve Pollen (%)	Seed setting (%)	Km-resistant Seedlings (%)	GUS +ve Km-resistant Seedlings (%)	GUS +ve greenhouse-grown plants (%)
Tulbaghia violacea	Natural plant (control)	0	85.5 ± 0.24 a	0	0	0
	BWK (control)	0	62.5 ± 0.67 b	0	0	0
	BWK + *Agro* (control)	35.9± 1.06 a	28.2 ± 1.28 c	63.1 ± 0.56 b	90.0 ± 0.28 b	25.5 ± 0.12 b
	BWK + *Agro* + Km	32.6± 1.48 b	24.8 ± 1.02 d	86.7 ± 0.27 a	97.5 ± 0.12 a	41.6 ± 0.18 a
Albuca nelsonii	Natural plant (control)	0	82.6 ± 0.27 a	0	0	0
	BWK (control)	0	80.0 ± 0.42 b	0	0	0
	BWK + *Agro* (control)	42.3± 0.87 a	77.2 ± 1.08 c	83.3 ± 0.32 b	91.7 ± 0.46 b	33.3 ± 0.20 b
	BWK + *Agro* + Km	31.3± 0.42 b	74.7 ± 0.72 d	95.6 ± 0.24 a	100 a	50.0 ± 0.26 a

Values are mean ± standard error (SE). Values followed by different letters indicate significant difference between means (*p*= 0.05); comparison by DMRT. Greenhouse grown 12-months-old plants. Km, kanamycin.

seed setting varied between natural and treated pollen grains. However, setting was 24.8% for *T. violacea* and 74.7% for *A. nelsonii* in the selection treatment, signifying that Km-resistant pollen effected seed setting, while the efficiency varied between species (Table 1). Direct application of plasmid solution or *A. tumefaciens* containing reporter genes on stigmas or styles has been observed with lack of seed setting in other plant species (Shou et al. 2002, Han et al. 2015). Therefore, Km selection in pollen tube growth is important for transformation when using pollen tubes. The results of the present study revealed that pollen transformation by Km selection is an effective technique for improved and rapid gene transfer in *T. violacea* and *A. nelsonii*.

Selection of Km-resistant and GUS confirmation

Km-resistant seedlings were screened under *in vitro* conditions. Germination of the seeds produced from control experiments (natural plant and pollen grains germinated in BWK medium) was completely inhibited after 3 weeks of culture. Km-resistant seedlings were observed in both *A. tumefaciens* alone or *A. tumefaciens* plus Km-treated pollen produced seeds. However, seeds produced from *A. tumefaciens* plus Km-treated pollen exhibited significantly improved Km-resistance (86.7% for *T. violacea* and 95.6% for *A. nelsonii*) and GUS-positive (97.5% for *T. violacea* and 100% in *A. nelsonii*) seedlings (Table 1 and Figure 1G, H). Effectiveness of Km selection and GUS expression for production of putative transgenic plants in different methods has been reported for other plant species (Bino et al. 1987, Twell et al. 1990, Baskaran and Dasgupta 2012, Baskaran et al. 2016, Koetle et al. 2017, Souza et al. 2017). GUS-positive seedlings were successfully acclimatized with 1:1 (v/v) vermiculite:soil mixture in the greenhouse. The GUS-expression with different frequency was exhibited from leaf tissues of one-year-old greenhouse-grown plants (Table 1 and Figure 1I, J). This result suggests that high efficiency of gene transfer via the pollen tube pathway is possible for the improvement of ornamental and medicinal *T. violacea* and *A. nelsonii*.

Transgenic plant production *in vitro*

Shoots were induced at the cut edges of leaf explants from all treatments, except the control (Table 2). Among the different PGR combinations, MS medium containing 10 µM BA plus 0.5 µM NAA and 10 µM TDZ plus 0.5 µM NAA combinations produced significantly higher numbers of shoots for *T. violacea* and *A. nelsonii* respectively (Table 2 and Figure 2A, B). The results strongly indicated that PGR combination for shoot regeneration is genotype/species-dependent as reported earlier (Narasimhulu and Chopra 1988, Steinitz et al. 2006, Ascough and Van Staden 2010). The shoots were rooted with IBA or IAA and Km treatments (Table 2 and Figure 2C, D). Rooting was lower in controls (Table 2). The rooted plantlets from control treatments died after 8 weeks of culture (Figure 2E, F). This indicates that the plantlets are susceptible to

Table 2. GUS-positive expression in *in vitro* regenerated plantlets of *T. violacea* and *A. nelsonii*

Plant name	PGR (µM) for shoot regeneration	Number of Shoots explant⁻¹	Shoot length (cm)	PGR (µM) + 25mg L⁻¹ Km for rooting	Number of roots shoot⁻¹	Root length (cm)	GUS⁺ᵛᵉ Plantlet (%)
Tulbaghia violacea							
	Control	0	0	0	0	0	0
	10 BA + 0.5 NAA (Control)	6.4 ± 0.78 a	17.2 ± 1.19 a	5 IBA	1.2 ± 0.12 c	2.0 ± 0.16 c	0
				5 IAA	1.6 ± 0.17 c	2.6 ± 0.20 c	0
	10 BA + 0.5 NAA	6.2 ± 0.83 a	17.6 ± 1.23 a	5 IBA	3.2 ± 0.28 b	4.8 ± 0.24 b	73.0 a
				5 IAA	5.6 ± 0.32 a	6.2 ± 0.48 a	73.3 a
	10 mTR + 0.5 NAA	3.8 ± 0.34 b	15.4 ± 0.96 b	5 IBA	3.0 ± 0.19 b	5.1 ± 0.26 ab	72.6 ab
				5 IAA	4.8 ± 0.20 ab	5.7 ± 0.48 ab	71.8 ab
	10 TDZ + 0.5 NAA	4.6 ± 0.52 b	12.0 ± 0.72 c	5 IBA	4.0 ± 0.19 b	5.9 ± 0.36 ab	73.7 a
				5 IAA	5.2 ± 0.24 a	6.9 ± 0.52 a	72.5 ab
Albuca nelsonii							
	Control	0	0	0	0	0	0
	10 BA + 0.5 NAA (Control)	3.6 ± 0.34 b	13.4 ± 1.09 a	5 IBA	1.0 ± 0.18 e	1.6 ± 0.10 d	0
				5 IAA	1.4 ± 0.12 e	1.9 ± 0.14 d	0
	10 BA + 0.5 NAA	3.2 ± 0.37 b	13.0 ± 1.06 a	5 IBA	3.4 ± 0.12 bc	3.0 ± 0.32 d	78.9 ab
				5 IAA	4.2 ± 0.24 b	7.8 ± 0.48 b	80.5 a
	10 mTR + 0.5 NAA	2.6 ± 0.24 b	9.2 ± 0.83 c	5 IBA	2.8 ± 0.18 c	3.4 ± 0.36 d	79.0 ab
				5 IAA	3.7 ± 0.14 bc	6.7 ± 0.54 c	79.7 ab
	10 TDZ + 0.5 NAA	4.8 ± 0.41 a	11.8 ± 0.96 b	5 IBA	4.0 ± 0.16 b	3.6 ± 0.48 d	79.6 ab
				5 IAA	5.8 ± 0.38 a	9.2 ± 0.32 a	80.0 a

The data were recorded after 8 weeks of culture for shoot regeneration and 6 weeks of culture for rooting. Values are mean ± standard error (SE). Values followed by different letters indicate significant difference between means (p= 0.05); comparison by DMRT. PGR, plant growth regulator. Km, kanamycin.

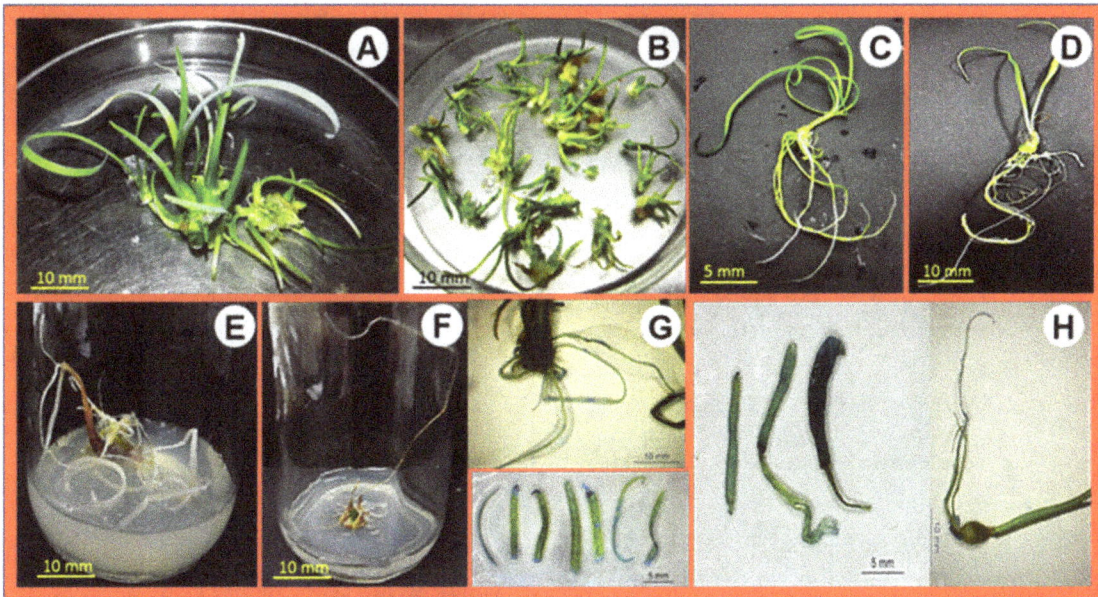

Figure 2. *In vitro* regeneration and GUS-positive expression of *T. violacea* and *A. nelsonii*. Formation of multiple shoots from 10 μM BA + 0.5 μM NAA for *T. violacea* (**A**) and 10 μM TDZ + 0.5 μM NAA for *A. nelsonii* (**B**); Rooting of shoots in *T. violacea* (**C**) and *A. nelsonii* (**D**);Necrosed plantlets of *T. violacea* (**E**) and *A. nelsonii* (**F**) in control treatments; The GUS-expression from Km-resistant plantlets of *T. violacea* (**G**) and *A. nelsonii* (**H**).Bar A, B, D, E, F 10mm; C, 5 mm; G 10 mm and 5 mm; and H 5 mm and 10 mm.

Km. Rooting rate was dependent on auxin and respective shoot regeneration treatments (Table 2). The Km-resistant plantlets showed transient GUS expression in all treatments, except the control (Table 2 and Figure 2G, H). The frequency of GUS expression varied between *T. violacea* and *A. nelsonii*, however, frequency of GUS-positive plantlets were not significantly different within the same species. Therefore, the results confirmed that *in vitro* Km-resistant putative transgenic plantlets could be achieved from leaf explants of putative transgenic seedlings for large-scale transgenics production without *Agrobacterium*-mediated transformation.

CONCLUSIONS

The present investigation was successful with higher efficiency of pollen gene transfer using kanamycin selection in valuable ornamental and medicinal plants, *T. violacea* and *A. nelsonii*. This method reduces false positives and chimeric plants by careful pollen selection, infection conditions and optimal Km concentration. This protocol has great potential for large-scale clonal transgenic plant production and commercial application of production of novel genotypes. The presented system could be adopted for temperature sensitivity and self-incompatibility of other plant species; however, the progeny needs to be screened in order to confirm genetic stability of transgenics. The *in vitro* putative transgenic plantlets through direct organogenesis were established with leaf explants of transformed seedlings and were confirmed by GUS-positive expression. Developed *in vitro* regeneration protocol has advantageous for independent-*Agrobacterium*-mediated transformation with large-scale transgenic plant production.

ACKNOWLEDGEMENTS

Financial support by the University of KwaZulu-Natal (UKZN), Pietermaritzburg is gratefully acknowledged. The authors are grateful to the Microscopy and Microanalysis Unit (MMU), UKZN, Pietermaritzburg for microscopic assistance.

REFERENCES

Alikina O, Chernobrovkina M, Dolgov S and Miroshnichenko D (2016) Tissue culture efficiency of wheat species with different genomic formulas. **Crop Breeding and Applied Biotechnology 16**: 307-314.

Ascough GD and Van Staden J (2010) Micropropagation of *Albuca bracteata* and *A. nelsonni* - Indigenous ornamentals with medicinal value. **South African Journal of Botany 76**: 579-584.

Baskaran P and Dasgupta I (2012) Gene delivery using microinjection of agrobacterium to embryonic shoot apical meristem of elite *indica* rice cultivars. **Journal of Plant Biochemistry and Biotechnology 21**: 268-274.

Baskaran P, Soós V, Balázs E and Van Staden J (2016) Shoot apical meristem injection: a novel and efficient method to obtain transformed cucumber plants. **South African Journal of Botany 103**: 210-215.

Bino RJ, Hille J and Franken J (1987) Kanamycin resistance during *in vitro* development of pollen from transgenic tomato plants. **Plant Cell Reports 6**: 333-336.

Brewbaker JL and Kwack BH (1963) The essential role of calcium ions in pollen germination and pollen tube growth. **American Journal of Botany 50**: 859-865.

Eapen S (2011) Pollen grains as a target for introduction of foreign genes into plants: an assessment. **Physiology and Molecular Biology of Plants 17**: 1-8.

García-Sogo B, Pineda B, Castelblanque L, Antón T, Medina M, Roque E, Torresi C, Beltrán JP, Moreno V and Cañas LA (2010) Efficient transformation of *Kalanchoe blossfeldiana* and production of male-sterile plants by engineered anther ablation. **Plant Cell Reports 29**: 61-77.

Han LW, Ge Y, Yao L, Cui CS and Qu SP (2015) Transformation of soft rot resistant *aiiA* gene into Chinese cabbage (*Brassica rapa* ssp. *pekinensis*) by the pollen-tube pathway. **International Journal of Agriculture and Biology 17**: 407-409.

Harvey WH (1837) *Tulbaghia violacea*. **Botanical Magazine 64**: t. 3555

Hess D (1987) Pollen-based techniques in genetic manipulation. **International Review of Cytology 107**: 367-395.

Hudson LC, Chamberlain D and Stewart JR (2001) GFP-tagged pollen to monitor pollen flow of transgenic plants. **Molecular Ecology Notes 1**: 321-324.

Koetle MJ, Baskaran P, Finnie JF, Soos V, Balázs E and Van Stadena J (2017) Optimization of transient GUS expression of *Agrobacterium*-mediated transformation in *Dierama erectum* Hilliard using sonication and Agrobacterium. **South African Journal of Botany 111**: 307-312.

Loguercio LL, Termignoni RR and Ozaki LS (1994) The reversible inhibition of pollen germination of *Nicotiana tabacum* L. - An entry into a transformation system. **Plant Cell Reports 13**: 231-236.

Murashige T and Skoog F (1962) A revised medium for rapid growth and bio assays with tobacco tissue cultures. **Physiologia Plantarum 15**: 473-497.

Narasimhulu SB and Chopra VL (1988) Species specific shoot regeneration response of cotyledonary explants of Brassicas. **Plant Cell Reports 7**: 104-106.

Reinten EY, Coetzee JH and Van Wyk B-E (2011) The potential of South African indigenous plants for the international cut flower trade. **South African Journal of Botany 77**: 934-946.

Saunders JA and Matthews BF (1995) **Pollen transformation in tobacco: Pollen electro- transformation methods in molecular biology, plant cell electroporation and electrofusion protocols**. JA Nickoloff Humana Press Inc., Totowa, 81p.

Shivanna KR and Rangaswamy NS (1992) **Pollen biology: a laboratory manual**. Springer Berlin, Germany, 119p.

Shivanna KR and Sawhney VK (2005) **Pollen biotechnology for crop production and improvement**. Cambridge University Press, New York, 261p.

Shou H, Palmer RG and Wang K (2002) Irreproducibility of the soybean pollen-tube pathway transformation procedure. **Plant Molecular Biology Reporter 20**: 325-334.

Souza RAV, Alves MC, Carneiro NP, Barros BA, Borém A and Carneiro AA (2017) Agrobacterium-mediated genetic transformation of a tropical elite maize line. **Crop Breeding and Applied Biotechnology 17**: 133-140.

Steinitz B, Amitay A, Gaba V, Tabib Y, Keller M and Levin I (2006) A simple plant regeneration-ability assay in a range of *Lycopersicon* species. **Plant Cell Tissue and Organ Culture 84**: 269-278.

Tu S, Sangwan RS, Raghavan V, Verma DPS and Sangwan-Norreel BS (2005) Transformation of pollen embryo-derived explants by *Agrobacterium tumefaciens* in *Hyoscyamus niger*. **Plant Cell Tissue and Organ Culture 81**: 139-148.

Twell D, Yamaguchi J and McCormick S (1990) Pollen-specific gene expression in transgenic plants: coordinate regulation of two different tomato gene promoters during microsporogenesis. **Development 109**: 705-713.

Watt JM and Breyer-Brandwijk MG (1962) **The medicinal and poisonous plants of southern and eastern Africa**. 2nd edn., Livingstone, London, 1457p.

3

Oleic acid variation and marker-assisted detection of Pervenets mutation in high- and low-oleic sunflower cross

Aleksandra Dimitrijević[1], Ivana Imerovski[1], Dragana Miladinović[1]*, Sandra Cvejić[1], Siniša Jocić[1], Tijana Zeremski[1] and Zvonimir Sakač[1]

Abstract: *High-oleic sunflower oil is in high demand on the market due to its heart-healthy properties and richness in monounsaturated fatty acids that makes it more stable in processing than standard sunflower oil. Consequently, one of sunflower breeder's tasks is to develop stable high-oleic sunflower genotypes that will produce high quality oil. We analyzed variability and inheritance of oleic acid content (OAC) in sunflower, developed at the Institute of Field and Vegetable Crops, by analyzing F_1 and F_2 progeny obtained by crossing a standard linoleic and high-oleic inbred line. F_2 individuals were classified in two groups: low-oleic with OAC of 15.24-31.28% and high-oleic with OAC of 62.49-93.82%. Monogenic dominant inheritance was observed. Additionally, several molecular markers were tested for the use in marker-assisted selection in order to shorten the period of detecting high-oleic genotypes. Marker F4-R1 was proven to be the most efficient in detection of genotypes with Pervenets (high-oleic acid) mutation.*

Key words: *Helianthus annuus L., fatty acids, molecular marker, breeding.*

*Corresponding author:
E-mail: dragana.miladinovic@ifvcns.ns.ac.rs

[1] Institute of Field and Vegetable Crops, Maksima Gorkog 30, 21000 Novi Sad, Serbia,

INTRODUCTION

Sunflower (*Helianthus annuus* L.) is the fourth most important source of edible vegetable oil in the world after palm, soybean and canola oil, contributing up to 12% of the edible oil produced globally (Rauf et al. 2017). Sunflower oil has multiple uses in both food and non-food industries (biofuel, lubricants, surfactants, polymer synthesis). Standard sunflower oil is naturally rich in polyunsaturated linoleic acid that makes up about 70% of the total sunflower oil content, and the second most abundant is monounsaturated oleic acid contributing with 20% (Cvejić et al. 2014a). The expanding and ever growing market demands not only standard oils, but also vegetable oils with altered properties. Creation of the first sunflower cultivar Pervenets (Soldatov 1976) with elevated oleic acid content (OAC) in the late 20th century allowed the expansion of the sunflower breeding programs and, ultimately, enabled sunflower breeders to comply with the market demands. Pervenets was obtained by treating the seed of VNIMK 8931 variety with 0.5% DMS solution. In the M3 generation, Soldatov selected single plants containing over 70% oleic acid, developing the Pervenets variety with 80-90% oleic acid in oil (Soldatov 1976, Lacombe et al. 2004, Cvejić et al. 2014b). Later on, more cultivars with altered fatty acid content were developed

(Andrich et al. 1992, Osorio et al. 1995, Fernández-Martinez et al. 1997, Škorić et al. 2007, Velasco et al. 2008, Leon et al. 2013a, b, Alberio et al. 2016, Cvejić et al. 2016). However, when it comes to the development of sunflower genotypes with the elevated OAC, the Pervenets remain the most commonly used source of the increased OAC.

The advantages of high-oleic oils are numerous, not only for the human consumption, but also in food processing. Due to its high oxidative stability, high-oleic sunflower oil is more stable in the frying process and exposure to high temperatures. Lack of trans fatty acids renders it healthier. In addition, high-oleic oil has reduced rancidity and extended shelf life. In processing, high-oleic oil has reduced cleaning costs and it is easier to transport and store (Vannozzi 2006). Furthermore, a diet rich in high-oleic sunflower oil and margarine has positive effects on blood lipids and factor VII coagulant activity (factor VIIc) (Allman-Farinelli et al. 2005). All the benefits mentioned make high-oleic sunflower oil in high demand by both the industry and the consumers. Consequently, significant attention is paid not only to the creation of high-oleic sunflower lines and hybrids, but also to shedding light on the mechanism(s) that cause the significant OAC increase.

Inheritance of OAC has widely been investigated. The results reported so far differ, but there is a common consensus that OAC is greatly affected by the genetic background of the recipient genotype (Schuppert et al. 2006, van der Merwe et al. 2013, Ferfuia et al. 2015), and by the environment (Izquierdo and Aguirrezábal 2008, Hlisnikovský et al. 2015, Regitano Neto et al. 2016). Urie (1984) reported the dominant mode of inheritance for high OAC, while Fick (1984) reported the partially dominant mode of inheritance. Joksimović et al. (2006) showed that additive gene action plays a more important role in OAC than the non-additive gene action. In addition, there were different reports describing the presence of a modifier gene and one or more genes influencing OAC (Fernández et al. 1999, Lacombe et al. 2004, Bervillé 2010). Ferfuia and Vannozzi (2015) reported that the high-oleic trait is influenced by at least three loci and reported a significant maternal effect on OAC. Since the high-oleic trait is such a complex one, these authors expressed the need for conducting further testing in different sunflower crossings and in different locations under different temperature and field conditions, in order to get a better insight into the genetic system controlling oleic acid levels and the effect of environment on OAC.

On the molecular level, it was reported that the increased OAC in Pervenets was caused by the partial duplication of the *FAD2-1* allele, which had caused silencing of the *FAD2-1* gene (Lacombe et al. 2002, Schuppert et al. 2006). Therefore, both the standard and the high-oleic genotypes contain the *FAD2-1* sequence. However, in the high-oleic genotype there is an addition of the intergenic region (IGR) separating common *FAD2-1* sequence and the truncated intron and exon that make up the duplicated sequence designated as *FAD2-1D* (Schuppert et al. 2006). This *FAD2-1* duplication is termed *Ol* mutation.

FAD2-1 encodes FAD2 (oleoyl-phosphatidyl choline desaturase), an enzyme that catalyzes synthesis of linoleic acid from oleic acid. In sunflower genome, there are three FAD genes: *FAD2-1*, *FAD2-2*, *FAD2-3* (Hongtrakul et al. 1998, Martínez-Rivas et al. 2001). While *FAD2-2* and *FAD2-3* are weakly expressed, *FAD2-1* is strongly expressed in developing seeds of standard type sunflower (Martínez-Rivas et al. 2001). The seed-specific *FAD2-1* gene has unequivocally been associated with the Ol_1 gene (Hongtrakul et al. 1998). Different authors mapped Ol_1-*FAD2-1* locus on LG14 (Lacombe and Berville 2001, Pérez-Vich et al. 2002, Schuppert et al. 2006). Some of the first molecular studies conducted by Dehmer and Friedt (1998) found RAPD PCR fragments AC10-765 and F15-690 at 7.2 cM and 7.0 cM from Ol_1, respectively. Later on, Pérez-Vich et al. (2002) reported a QTL describing 84.5% of the phenotypic variation in C18:1 content. These authors concluded that the detected QTL was most likely one of the *Ol* genes controlling high OAC. Schuppert et al. (2006) developed dominant INDEL markers for detection of *Ol* mutation (presence or absence of tandem *FAD2-1* repeats) and identified almost 50 SNPs and several INDELs downstream of *FAD2-1*. Lacombe et al. (2009) developed two types of markers; the first was a codominant SSR marker located in the intron of the putative *FAD2-1* gene and tightly linked to the mutation, and the second group comprised the dominant markers specific for the mutation. In the last several years, identification of the appropriate and easy-to-use molecular markers for the detection of high-oleic genotypes has intensified owing to an increasing interest of breeding companies to offer a greater variety of sunflower high-oleic genotypes. Several authors have worked on the development of the methods for *Ol* gene detection with molecular markers (Nagarathna et al. 2011, Singchai et al. 2013, Bilgen 2016, Dimitrijević et al. 2016), but the obtained methods and results showed the need for further validation in different sunflower populations and genetic backgrounds (Singchai et al. 2013, Bilgen 2016).

In the present work, the high-oleic line Ha-26-Ol was crossed with the standard linoleic line RAJ-SIN-IMI, both developed at the Institute of Field and Vegetable Crops. The goal was twofold: (1) to analyze the variability of OAC in these genotypes and (2) to analyze and identify easy-to-use molecular markers that will be the most efficient in the detection of *FAD2-1D* in sunflower, distinguishing sunflower genotypes with elevated OAC.

MATERIAL AND METHODS

Plant material

High-oleic B line Ha-26-Ol and low-oleic line SIN-RAJ-IMI were used as the parental lines. Ha-26-Ol is a near isogenic line developed from Ha-26 (Škorić and Jocić 1998), with average OAC exceeding 80%. SIN-RAJ-IMI is a standard linoleic imidazoline-tolerant line created at the Institute of Field and Vegetable Crops with 20% OAC on average. Ha-26-Ol and SIN-RAJ-IMI were crossed to produce F_1 progeny. F_2 generation, consisting of 86 individuals, was produced by self-fertilization of a single F_1 plant.

Parental lines Ha-26-Ol and SIN-RAJ-IMI, F_1 and F_2 were grown in the field. Leaf samples for molecular analysis were collected at two leaf-stage, and seeds for oleic acid content analysis at physiological maturity on R9 (Schneiter and Miller 1981) from ten plants/heads of each parental line and F_1 progeny, as well as from all of 83 F_2 plants.

Oleic acid content

Fatty acid composition was analyzed by gas chromatography (GC). Samples for GC analysis were prepared in a hydraulic press (Sirio, Mikodental 10-tons strength, cc 400 bars). Two grams of seeds were pressed to yield approximately 0.5 ml of oil that was used for GC analysis. In the reaction vial 270 µL of TMSH (transesterification agent) was added to exactly 30 µL of oil, well shook in the vortex, and kept at room temperature for an hour.

Oleic acid was identified by use of a reference mixture of fatty acids methyl esters (FAME). Multi-standard (FAME RM-1, Cat. no. O7006, Supelco) containing the methyl esters of palmitic, stearic, oleic, linoleic, linolenic and arachidic fatty acids was used to confirm the retention times, as well as to confirm that the peak areas reflected the actual composition of these mixtures.

Agilent 5890 gas chromatograph equipped with flame ionization detector (FID) was used for the analysis of fatty acid methyl esters. Fused silica capillary column with polyethylene stationary phase (HP-INNOWAX, 30 m × 0.25 mm i.d. and 0.25 µm film thickness) was used for the separation. The sample volume injected was 1 µL and split ratio was 1:50. Helium was used as a carrier gas at a constant pressure of 53 kPa at 50 °C. The injector and detector temperatures were set at 250 °C and 280 °C, respectively, and the initial temperature of 50 °C was held for 1 min, then increased to 200 °C at a rate of 25 °C min^{-1}, followed by another increase to 230 °C at a rate of 3°C/min, and then maintained for 18 min. The results were processed by the ChemStation software and expressed as the percentage of individual fatty acids in the oil sample. Oleic acid content (OAC) is shown as the percentage of the content of the total fatty acids.

Molecular analysis

Leaves of parental lines (bulk sample – 10 plants per sample), F_1 (bulk sample – 10 plants), and individual samples of 86 F_2 plants were used for the molecular analysis. Samples were immediately put in liquid nitrogen and later stored at -70°C until DNA extraction.

DNA was extracted from leaves using the modified CTAB protocol (Permingeat et al. 1998). The following markers were used for the detection of *FAD2-1D* or *FAD2-1* sequence: F13-R5 and F4-R1 (F 13 – 5'-TCAACAGCCTCTTCCTCCTCAG-3'; R5 – 5'-GTAGTTTTGGAAAGCTAGAGACC-3'; F4 – 5'-GTAACGTCTGCGCGCTTGCAGACATCA-3'; R1 – 5'-GGTTTTGCATGAGGGACTCGATCGAGTG-3') (Schuppert et al. 2006) and Fsp-b-R1 (Fsp-b 5'-AGAAGAGGGAGGTGTGAAG-3'; R1 – 5'-AGCGGTTATGGTGAGGTCAG-3') (Lancombe et al. 2009). PCR with primers F13-R5 and F4-R1 was performed as described by Schuppert et al. (2006), and with Fsp-b-R1 as described by Lancombe et al. (2009) in the PCR reaction described by Dimitrijević et al. (2010). Products of PCR amplification were run on 2% agarose gels stained with ethidium-bromide and visualized with the BIO-Print system (Vilber Lourmat, Marne-La-Vallée, France).

RESULTS AND DISCUSSION

Oleic acid content

GC analysis of parental lines showed that high-oleic parental line Ha-26-Ol contained 86% of oleic acid, and the standard linoleic line SIN-RAJ-IMI contained OAC 17.89%. The average OAC of the F_1 progeny was 88.44% indicating the dominant mode of inheritance of OAC, while the OAC of the F_2 progeny ranged between 15.24% and 93.82%. Analysis of distribution of OAC in the tested F_2 individuals showed two groups: the first group in which OAC varied between 15.24% and 31.28% (low-oleic group), and the second group where OAC ranged between 62.49% and 93.82% (high-oleic group) (Figure 1). None of the F_2 individuals had OAC in between these two groups. Therefore, on the

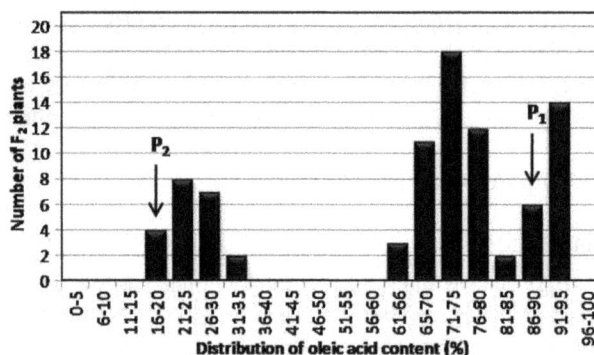

Figure 1. Distribution of oleic acid content (%) in examined F_2 sunflower plants, P_1 is high-oleic parental line – Ha-26-Ol and P_2 is standard parental line RAJ-SIN-IMI

basis of phenotype, the threshold for high OAC between the two groups could be set at 60%. High OAC threshold is an important parameter in breeding and seed production since oils with OAC above a certain threshold receive a prime over the regular price in today's market (Angeloni et al. 2016). Furthermore, the distribution of OAC of F_2 individuals and threshold set allows breeders to classify the individuals in low-, normal-, mid-, or high-oleic group and evaluate whether the parental lines used for crossing should be used in future breeding. This is especially important since OAC is highly dependent on both the genotype and its interaction with the environment (van der Merwe et al. 2013), and *Ol* locus exhibits genetically unstable expression (Demurin and Škorić 1996).

There is no general consensus on the threshold value, and it is set for each cross individually due to the influence of genetic background and conditions in which the plants were grown. The threshold set in our research is within the range of previously reported thresholds for high OAC, which were set between 55% and 70% (Lacombe and Bervillé 2001, Lacombe et al. 2002, Bilgen 2016). The analysis of F_2 generation showed grouping of the individuals in two groups: 65 plants were in the high-oleic group and 21 plants were in the low-oleic group. The ratio 65:21 fits 3:1 ratio, which is in agreement with monogenetic dominant inheritance (2 alleles at one locus) (χ^2, p>0.05). Monogenic dominant mode of inheritance was also reported by Urie (1984), Lacombe et al. (2000), Lacombe and Bervillé (2001). In addition, OAC of F_1 progeny was similar to the OAC of the high-oleic parent, thus further supporting the dominant monogenetic inheritance in the cross performed in our study. OAC of F_1 progeny can be a good initial parameter for determination of OAC inheritance in sunflower. This is in agreement with the results of Varès et al. (2002) who detected that OAC of F_1 obtained by crossing high-oleic and linoleic lines had similar OAC as high-oleic parental line in the dominant inheritance, while in intermediate inheritance OAC of F_1 was lower than OAC of high-oleic parental line.

Within the high-oleic group, two sub-groups were observed: 43 plants in the 61-80% OAC and 22 plants in the 80-95% OAC sub-group. The reason for this sub-grouping could be the number of copies of *FAD2-1D* alleles: 22 plants had two *FAD2-1D* alleles (homozygous plants; *OlOl*), and 43 plants had one *FAD2-1D* (heterozygous plants; *Olol*). If observed through the segregation ratio of F_2 progeny, it fits 1:2:1 ratio (22(*OlOl*):43(*Olol*):21(*olol*)) for the dominant monogenic inheritance of high-oleic acid content trait (χ^2, p>0.01). Bearing in mind that 2 g of seeds of each F_2 plant was used for GC analysis, when homozygous F_2 plant was analyzed all the seeds within one head were homozygous for high-oleic trait, while heterozygous F_2 plants contained two alleles in three different combinations: homozygous dominant (*OlOl*), homozygous recessive (*olol*) and heterozygous (*Olol*) which influenced OAC content. This means that heterozygous F_2 plants had a certain variation in OAC in the seeds from a single head, which is supported by the results reported by Demurin et al. (2000) who found great variation in OAC in F_2 progeny plants that varied from 24% to 68% in a single sunflower head due to the segregation in the two parent phenotypic classes.

Molecular analysis

Two markers used in this study enabled the discrimination between high-oleic parental line Ha-26-Ol and standard linoleic line SIN-RAJ-IMI, while one marker was monomorphic (Figure 2). Marker F13-R5 (Schuppert et al. 2006) was developed as codominant INDEL for *FAD2-1* detecting minimum 8 alleles in several confectionary and oilseed sunflower lines, open-pollinated cultivars and sunflower wild populations. It was shown to be monomorphic between the parental lines used in this study (Figure 2A). This marker amplified a band of approximately 340 bp in both parental lines and F_1 progeny.

Markers F4-R1 (Schuppert et al. 2006) and Fsp-b-R1 (Lacombe et al. 2009) enabled the discrimination between parental lines; however, neither of them enabled discrimination between homozygous and heterozygous genotype (Figure 2 B, C, respectively). Molecular marker Fsp-b-R1, also known as N1-3F/N2-1R (Berville et al. 2009), was created to amplify high-oleic specific element and it amplifies a band of approximately 891 bp. In our study, Fst-b-R1 amplified a band approximately 900 bp in high-oleic line and F_1 progeny, while this band was absent from the standard linoleic line. In addition, this primer amplified several common bands between parental lines and F_1. The forward primer of the F4-R1 marker is located in intergenic DNA sequences upstream of *FAD2-1D* making it a *FAD2-1D* specific marker. In our study, F4-R1 amplified a single band approximately 650 bp in high-oleic line Ha-26-Ol and in F_1 progeny. The absence of this band was recorded in SIN-RAJ-IMI line. Similarly, Schuppert et al. (2006) amplified a 653 bp band in mutant lines (high-oleic lines), while this band was absent in standard sunflower lines.

Due to clearer profiles obtained by using F4-R1, this primer was used for molecular analysis of *FAD2-1D* in F_2 individuals (Figure 2D). It amplified a band of the expected length, approximately 650 bp, in 65 F_2 plants (that had OAC higher than 61%), while this band was not amplified in 21 F_2 plants (in which OAC was lower than 32%). Similarly, Bilgen (2016) was able to discriminate between high-oleic (60-92%) and low-oleic (below 60%) sunflower F_3 individuals by the use of two types of molecular markers developed by Berville et al. (2009). The first marker was HO PCR specific fragment-marker (N1-3F/N2-1R) that amplified a 870 bp fragment in genotypes carrying Pervenets mutation and codominant SSR (N1-1F/N1-1R) that amplified a 249 bp fragment in homozygous low-oleic genotype and 246 bp fragment in homozygous high-oleic genotype. Nagarathna et al. (2011) and Singchai et al. (2013) successfully discriminated between the high-oleic and low-oleic genotypes by using N1-3F/N2-1R. This primer amplified a band 800 to 900 bp in high-oleic genotypes, which was absent in low- and mid-oleic genotypes. In addition, Singchai et al. (2013) identified 9 additional polymorphic SSRs between high- and low-oleic sunflower lines, while 27 SSRs were monomorphic. Lacombe et al. (2002) used probes (molecular hybridization) for discrimination between genotypes with OAC higher than 65 and 70% (higher OAC for both homozygous and heterozygous plants) and lower OAC (homozygous for *ol*). The probe revealed 8 kb and 15 kb *HindIII* fragment in low-oleic and high-oleic genotype, respectively. Lacombe et al. (2002) found 2 exceptions in which two genotypes with low-oleic phenotype carried the mutation that should have led to the increase in OAC.

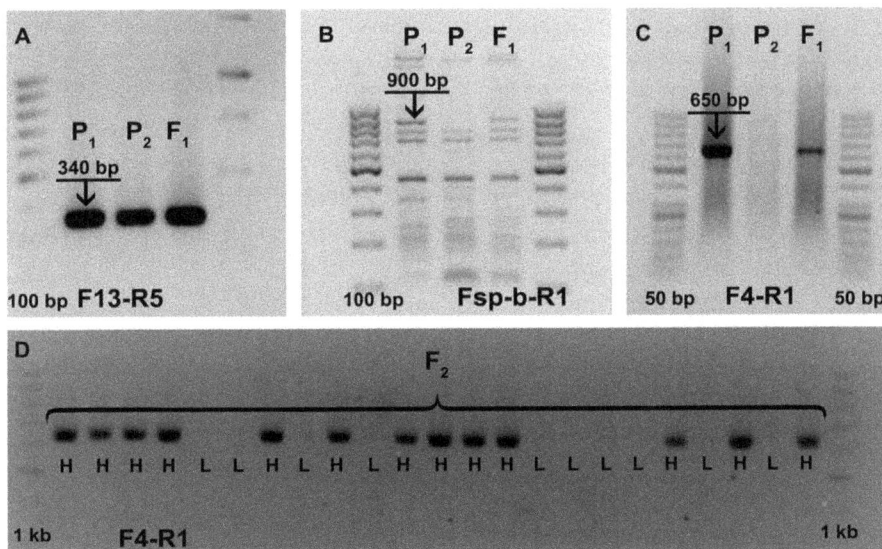

Figure 2. Molecular profiles of parental lines – high-oleic Ha-26-Ol (P_1), standard line (RAJ-SIN-IMI) (P_2), F_1 obtained by use of different molecular markers: A) F13-R5 (Schuppert et al. 2006); B) Fsp-b-R1 (Lacombe et al. 2009); C) F4-R1 (Schuppert et al. 2006); D) profiles of individuals of F_2 population obtained by amplification with F4-R1 (Schuppert et al. 2006) (DNA ladder 50 bp, 100 bp and 1 kb, Thermo Scientific)

Due to the well-known influence of genetic background on OAC (van der Merwe et al. 2013, Ferfuia et al. 2015,) certain markers in different sunflower populations should be validated (Bilgen 2016, Singchai et al. 2013). Similarly, Imerovski et al. (2014) validated SSR markers for downy mildew resistance in sunflower. Marker F4-R1 was previously tested in numerous sunflower genotypes (Schuppert et al. 2006, Dimitrijević et al. 2016) and now it was validated in F_2 sunflower population. It is easy to use and requires use of only agarose gels. The advantage of using molecular markers is that the genotyping results can be obtained before the sunflower growth is completed. Consequently, breeders can discard genotypes without Pervenets mutation before flowering saving both time and money on maintaining the unwanted genotypes.

ACKNOWLEDGMENTS

This work is a part of the project TR31025, supported by Ministry of Education, Science and Technological Development, Republic of Serbia.

REFERENCES

Alberio C, Izquierdo NG, Galella T, Zuil S, Reid R, Zambelli A and Aguirrezábal LA (2016) A new sunflower high oleic mutation confers stable oil grain fatty acid composition across environments. **European Journal of Agronomy 73**: 25-33.

Allman-Farinelli MA, Gomes K, Favaloro EJ and Petocz P (2005) A diet rich in high-oleic-acid sunflower oil favorably alters low-density lipoprotein cholesterol, triglycerides, and factor VII coagulant activity. **Journal of the American Dietetic Association 105**: 1071-1079.

Andrich G, Balzini S, Zinnai A, Fiorentini R, Baroncelli S and Pugliesi C (1992) The oleic/linoleic ratio in achenes coming from sunflower lines treated with hard X-rays. In **Proceedings of the 13th international sunflower conference**. ISA, Pisa, p. 1544-1549.

Angeloni P, Echarte MM, Irujo GP, Izquierdo N and Aguirrezábal L (2016) Fatty acid composition of high oleic sunflower hybrids in a changing environment. **Field Crops Research 202**: 146-157.

Berville A, Lacombe S, Veillet S, Granier C, Leger S and Jouve P (2009) Method of selecting sunflower genotypes with high oleic acid content in seed oil. **United States patent application** US 11/587,956.

Berville A (2010) Oil composition variations. In Hu J, Seiler G and Kole C (eds) **Genetics, genomics and breeding of sunflower**. Science Publisher, Boca Raton, FL, p. 253-277.

Bilgen BB (2016) Characterization of sunflower inbred lines with high oleic acid content by DNA markers. In Kaya Y and Hasancebi S (eds) **Proceedings of the 19th international sunflower conference**. ISA, Edirne, p. 662-668.

Cvejić S, Jocić S, Dimitrijević A, Imerovski I, Miladinović D, Jocković M and Miklič V (2016) An EMS mutation altering oil quality in sunflower inbred line. In Kaya Y and Hasancebi S (eds) **Proceedings of the 19th international sunflower conference**. ISA, Edirne, p. 422-430.

Cvejić S, Jocić S, Miladinović D, Jocković M, Imerovski I, Sakač Z and Miklič V (2014b) Development and utilization of sunflower genotypes with altered oil quality. **Journal of Processing and Energy in Agriculture 18**: 191-195.

Cvejić S, Miladinović D and Jocić S (2014a) Mutation breeding for changed oil quality in sunflower. In Tomlekova NB, Kozgar MI, Wani MR (eds) **Mutagenesis: exploring genetic diversity of crops**. Wageningen Academic Publishers, Wageningen, p. 77-96.

Dehmer KJ and Friedt W (1998) Development of molecular markers for high oleic acid content in sunflower (*Helianthus annuus* L.). **Industrial Crops and Products 7**: 311-315.

Demurin Y and Škorić D (1996) Unstable expression of *Ol* gene for high oleic acid content in sunflower seeds. In **Proceedings of the 14th international sunflower conference**. ISA, Beiging/Shenyang, p. 12-20.

Demurin YA, Škorić D, Verešbaranji I and Jocić S (2000) Inheritance of increased oleic acid content in sunflower seed oil. **Helia 23**: 87-92.

Dimitrijević A, Imerovski I, Miladinović D, Jocković M, Cvejić S, Jocić S, Zeremski T and Sakač Z (2016) Screening of the presence of *ol* gene in NS sunflower collection. In Kaya Y and Hasancebi S (eds) **Proceedings of the 19th international sunflower conference**. ISA, Edirne, p. 661-667.

Dimitrijević A, Imerovski I, Miladinović D, Tancić S, Dusanić N, Jocić S and Miklič V (2010) Use of SSR markers in identification of sunflower isogenic lines in late generations of backcrossing. **Helia 33**: 191-198.

Ferfuia C and Vannozzi GP (2015) Maternal effect on seed fatty acid composition in a reciprocal cross of high oleic sunflower (*Helianthus annuus* L.). **Euphytica 205**: 325-36.

Ferfuia C, Turi M and Vannozzi GP (2015) Variability of seed fatty acid composition to growing degree-days in high oleic acid sunflower genotypes. **Helia 38**: 61-78.

Fernández H, Baldini M and Olivieri AM (1999) Inheritance of high oleic acid content in sunflower oil. **Journal of Plant Breeding and Genetics 53**: 99-103.

Fernández-Martinez JM, Osorio J, Mancha M and Garces R (1997) Isolation of high palmitic mutants on high oleic background. **Euphytica 97**: 113-116.

Fick GN (1984) Inheritance of high oleic acid in the seed oil of sunflower. In **Proceedings of the 6th sunflower research workshop**. Sunflower Association, Bismarck, p. 9.

Hlisnikovský L, Kunzová E, Hejcman M, Škarpa P, Zukalová H and Menšík L

(2015) The effect of climate, nitrogen and micronutrients application on oiliness and fatty acid composition of sunflower achenes. **Helia** 38: 221-39.

Hongtrakul V, Slabaugh MB and Knapp SJ (1998) A seed specific D12 oleate desaturase gene is duplicated, rearranged, and weakly expressed in high oleic acid sunflower lines. **Crop Science 38**: 1245-1249.

Imerovski I, Dimitrijević A, Miladinović D, Jocić S, Dedić B, Cvejić S and Šurlan-Momirović G (2014) Identification and validation of breeder-friendly DNA markers for Pl arg gene in sunflower. **Molecular Breeding 34**: 779-788.

Izquierdo NG and Aguirrezábal LA (2008) Genetic variability in the response of fatty acid composition to minimum night temperature during grain filling in sunflower. **Field Crops Research 106**: 116-25.

Joksimović J, Atlagić J, Marinković R and Jovanović D (2006) Genetic control of oleic and linoleic acid contents in sunflower. **Helia 29**: 33-40.

Lacombe S and Berville A (2001) A dominant mutation for high oleic acid content in sunflower (*Helianthus annuus* L.) seed oil is genetically linked to a single oleate-desaturase RFLP locus. **Molecular Breeding 8**: 129-137.

Lacombe S, Guillot H, Kaan F, Millet C and Berville A (2000) Genetic and molecular characterization of the high oleic content of sunflower oil in Pervenets. In **Proceedings of the 15th international sunflower conference**. ISA, Toulouse, p. 12-15.

Lacombe S, Kaan F, Griveau Y and Berville A (2004) The pervenets high oleic mutation: methodological studies. **Helia 27**: 41-54.

Lacombe S, Leger S, Kaan F, Berville A and Sas M (2002) Genetic, molecular and expression features of the Pervenets mutant leading to high oleic acid content of seed oil in sunflower. **Oléagineux, Corps gras, Lipides 9**: 17-23.

Lacombe S, Souyris I and Berville AJ (2009) An insertion of oleate desaturase homologous sequence silences via siRNA the functional gene leading to high oleic acid content in sunflower seed oil. **Molecular Genetics and Genomics 281**: 43-54.

Leon AJ, Zambelli AD, Reid RJ, Morata MM and Kaspar M (2013a) Nucleotide sequences mutated by insertion that encode a truncated oleate desaturase protein, proteins, methods and uses. **WIPO Patent** WO/2013/004281, Jan 10.

Leon AJ, Zambelli AD, Reid RJ, Morata MM, Kaspar M, Martinez-Force E, Garcés R, Salas JJ and Venegas-Caleron M (2013b) Isolated mutated nucleotide sequences that encode a modified oleate destaurase sunflower protein, modified protein, methods and uses. **WIPO Patent** WO/2013/004280, Jan 10.

Martínez-Rivas JM, Sperling P, Luehs W and Heinz E (2001) Spatial and temporal regulation of three different microsomal oleate desaturase genes (FAD2) from normal-type and higholeic varieties of sunflower (*Helianthus annuus* L.). **Molecular Breeding 8**: 159-168.

Nagarathna TK, Shadakshari YG and Ramanappa TM (2011) Molecular analysis of sunflower (*Helianthus annuus* L.) genotypes for high oleic acid using microsatellite markers. **Helia 34**: 63-68.

Osorio J, Fernández-Martinez JM, Mancha M and Garces R (1995) Mutant sunflower with high concentration in saturated fatty acid in the oil. **Crop Science 35**: 739-742.

Pérez-Vich B, Fernández-Martínez JM, Grondona M, Knapp SJ and Berry ST (2002) Stearoyl-ACP and oleoyl-PC desaturase genes cosegregate with quantitative trait loci underlying high stearic and high oleic acid mutant phenotypes in sunflower. **Theoretical and Applied Genetics 104**: 338-349.

Permingeat HR, Romagnoli MV and Vallejos RH (1998) A simple method for isolating high yield and quality DNA from cotton (*Gossypium hirsutum* L.) leaves. **Plant Molecular Biology Reported 16**: 1-6.

Rauf S, Jamil N, Ali Tariq S, Khan M and Kausar M (2017) Progress in modification of sunflower oil to expand its industrial value. **Journal of the Science of Food and Agriculture**. Doi: 10.1002/jsfa.8214.

Regitano Neto A, Miguel AMRDO, Mourad AL, Henriques EA and Alves RMV (2016) Environmental effect on sunflower oil quality. **Crop Breeding and Applied Biotechnology** 16: 197-204.

Schneiter A and Miller JF (1981) Description of sunflower growth stages. **Crop Science 21**: 901-903.

Schuppert GF, Tang S, Slabaugh MB and Knapp SJ (2006) The sunflower high-oleic mutant Ol carries variable tandem repeats of FAD2-1, a seed-specific oleoyl-phosphatidyl choline desaturase. **Molecular Breeding 17**: 241-256.

Singchai A, Muangsan N and Machikowa T (2013) Evaluation of SSR markers associated with high oleic acid in sunflower. **International Journal of Biological, Food, Veterinary and Agricultural Engineering** 7: 631-634.

Škorić D and Jocić S (1998) Good combining ability for productivity of line Ha-26. In Turkulov J (ed) **Proceedings of the 39th oil industry conference**. Faculty of Techology, Novi Sad, p. 53-59.

Škorić D, Jocić S, Lečić N and Sakač Z (2007) Development of sunflower hybrids with different oil quality. **Helia 30**: 205-212.

Soldatov KI (1976) Chemical mutagenesis in sunflower breeding. In **Proceedings of the 7th international sunflower conference**. ISA, Krasnorar, p. 352-357.

Urie AL (1984) Inheritance of very high oleic acid content in sunflower. In **Proceedings of the 6th sunflower research workshop**. Sunflower Association, Bismarck, p. 8-9.

van der Merwe R, Labuschagne MT, Herselman L and Hugo A (2013) Stability of seed oil quality traits in high and mid-oleic acid sunflower hybrids. **Euphytica 193**: 157-68.

Vannozzi GP (2006) The perspectives of use of high oleic sunflower for oleochemistry and energy raws. **Helia 29**: 1-24.

Varès D, Lacombe S, Griveau Y, Berville A and Kaan F (2002) Inheritance of oleic acid content of F$_1$ seed in a complete diallel cross between seven sunflower lines. **Helia 25**: 105-12.

Velasco L, Pérez-Vich B and Fernández-Martínez JM (2008) A new sunflower mutant with increased levels of palmitic acid in seed oil. **Helia 31**: 55-60.

4

Introgression of the RI$_{adg}$ allele of resistance to potato leafroll virus in Solanum tuberosum L.

Otávio Luiz Gomes Carneiro[1], Silvia Regina Rodrigues de Paula Ribeiro[2], Carolina Mariane Moreira[3], Marcio Lisboa Guedes[2], Danilo Hottis Lyra1 and César Augusto Brasil Pereira Pinto2*

Abstract: *Genetic resistance to Potato Leafroll Virus (PLRV) is polygenic, which hinders the obtainment of resistant cultivars. However, works carried out at the International Potato Center have identified an andigena accession, LOP-868, with high resistance level and low accumulation of PLRV due to the gene of major effect RI$_{adg}$. We verify the transfer of the RI$_{adg}$ allele to clones of the cross between LOP-868 and UFLA clones, by using the SCAR RGASC850 molecular marker; to evaluate the reaction of these clones to PLRV by inoculating the virus using aphids; and to analyze their agronomic performance of clones. Among the clones inoculated with viruliferous aphids, 49.3% were negative to the serological test, indicating possible resistance. Clones containing the RI$_{adg}$ allele were identified by the RGASC850 molecular marker, which demonstrates the possibility of transferring the RI$_{adg}$ allele of resistance to PLRV from LOP-868 to Solanum tuberosum. Some clones that presented the RI$_{adg}$ allele are also promising for agronomic performance.*

Key words: *Potato, disease resistance, viral disease.*

*Corresponding author:
E-mail: cesarbrasil@dbi.ufla.br

[1] Universidade de São Paulo (USP), Escola Superior de Agricultura "Luiz de Queiroz" (ESALQ), Departamento de Genética, Avenida Pádua Dias, 11, 13.418-900, Piracicaba, SP, Brazil
[2] Universidade Federal de Lavras (UFLA), Campus UFLA, Departamento de Biologia, CP 3037, 37.200-000, Lavras, MG, Brazil
[3] Instituto Federal de Educação, Ciência e Tecnologia do Sul de Minas Gerais, Campus Poços de Caldas, Avenida Dirce Pereira Rosa, 300, 37.713-100, Poços de Caldas, MG, Brazil,

INTRODUCTION

Potato (*Solanum tuberosum* L.) is a plant that is vegetatively propagated for commercial purposes, allowing the dissemination of diseases, especially viruses. These pathogens sharply decrease crop yield and quality, known as degeneration (Figueira 1995). In Brazil, viral diseases are more severe than in temperate countries due to the high potential for dissemination by insect vectors, especially the green peach aphid (*Myzus persicae*), since it occurs in every region and season (Gallotti et al. 1992). Despite the possibility of controlling this virus by applying systemic insecticides to reduce the vector population (Lowery and Boiteau 1988), environmental implications and high costs have led to the search for other control strategies, such as genetic resistance, in the context of integrated management.

Potato leafroll virus (PLRV) is considered as one of the main viruses in potato crops in Brazil. It is responsible for more than 50% loss in tuber yield during secondary infection, being historically mentioned as one of the main phytosanitary problems in the production of potato seed in Brazil (Souza Dias and Iamauti 2005).

The obtainment of national cultivars adapted to the Brazilian growing conditions and resistant to the main diseases is the most viable alternative to make the crop more yielding and profitable to the producer (Zaag 1987). However, the development of PLRV-resistant cultivars is laborious, since this resistance is governed by several genes, which hinders the selection of resistant genotypes (Cockerham 1945). Another problem relates the interaction of PLRV with other viruses, especially PVX and PVY, which can drastically reduce the level of resistance to PLRV (Brandolini et al. 1992).

However, even with the polygenic control, Velásquez et al. (2007) have demonstrated the existence of a major gene or a genomic region, RI_{adg}, which controls a single mechanism of resistance to the infection and to PLRV accumulation. The cultivar LOP-868 (*S. tuberosum* ssp. *andigena*) has been mentioned in countless studies as an interesting source of resistance to PLRV, presenting the RI_{adg} gene in the duplex condition (Mihovilovich et al. 2007, Velásquez et al. 2007, Mihovilovich et al. 2014). Nevertheless, due to the binding of genes that confer resistance to PLRV to undesirable traits, such as irregular shape, small tubers, and low yield (Barker and Harrison 1985), the commercialization of the tubers of this cultivar becomes impracticable.

The objective of this work was to evaluate the reaction of clones obtained from the crosses of a PLRV resistance source (LOP-868), by inoculating them with viruliferous aphids, in order to verify the transfer of the RI_{adg} allele from the resistance source to *S. tuberosum,* and to detect the presence of this allele by Sequence Characterized Amplified Region (SCAR), and to evaluate the agronomic traits of the experimental clones.

MATERIAL AND METHODS

Trials were carried out at the Plant Virology Laboratory of the Phytopathology Department, at the Laboratory of Molecular Genetics, and in the greenhouse of the Department of Biology, of the Federal University of Lavras (UFLA). In addition, an experiment was carried out in the experimental area of the Agricultural Research Company of Minas Gerais (EPAMIG), in Lambari, MG.

A total of 220 potato clones from six families obtained by biparental crosses between the PLRV-resistant clone LOP-868 and clones from the UFLA Potato Breeding and Genetics Program were evaluated.

Inoculation with viruliferous aphids

In order to evaluate the reaction of the experimental clones, inoculations were carried out using viruliferous aphids (*Myzus persicae*), reared in *Datura stramonium* L. plants in a greenhouse with natural temperature and luminosity. Wingless adults were removed from the leaves and transferred with a fine brush to 15 cm diameter Petri dishes containing leaves of *D. stramonium* L., placed on 1 cm of 1% agar-water. Each Petri dish was sealed with perforated polyvinyl chloride film (PVC), allowing aeration, and was identified and maintained in an air-conditioned chamber at 25 ± 2 ºC, with a relative humidity of 70 ± 10%, with 12-hour photoperiod.

Potato tubers (*S. tuberosum*) contaminated with PLRV, previously evaluated by the *Double Antibody Sandwich - Enzyme Linked Immunono Sorbent Assay* (DAS-ELISA) (Clark and Adams 1977) were planted in 5.0 L pots, and maintained in a greenhouse. Approximately 30 days after seedlings emergence, these plants served as food for healthy aphids, which were placed in contact with these plants for 96 hours for the acquisition access period (AAP). For this, aphids were collected from the leaves of *D. stramonium* L. by using a brush, and transferred to the adaxial part of the leaf of the potato plants. This procedure was carried out under temperature ranging from 20 to 25ºC, and relative humidity between 65 and 90%.

The experimental clones were planted in 0.5 L plastic pots with commercial substrate, and as soon as the seedlings reached between 15 and 20 cm in height, they were inoculated with 50 viruliferous aphids at nymph stage, characterizing a high aphid pressure (CIP 2010). Two plants of each clone were inoculated, characterizing two replications per treatment. The number of replications was ideal for this study; if more plants of each clone had been used, it would have required an increase in the number of aphids, and it would have been more laborious, making the experiment unfeasible. In addition, since the treatments consisted of potato clones, no variation was observed between individuals.

Aphids were placed in contact with each plant for 96 hours for the inoculation access period (IAP), and were collected from the leaves of each inoculum source with the aid of a brush, and transferred to the adaxial part of the leaves of the clones. In order to avoid breaking the stylet of the aphids when they were captured on the leaf, the insects were touched with the brush to stimulate their movement, and were subsequently captured. This procedure was also carried out in the greenhouse, under cages with anti-aphid screen, with temperature ranging between 20 and 25 °C, and relative humidity between 65 and 90%. After 96 hours, the number of aphids was visualized in the abaxial part of each leaf of the plant, and the total number of aphids per plant was obtained. Afterwards, aphids were eliminated with insecticide.

After two and three weeks aphids had been removed from potato plants, three to four young leaves of each plant were collected in order to perform the DAS-ELISA serological test (Clark and Adams 1977), using polyclonal antisera (Agdia) for PLRV diagnosis.

The cultivars Ágata, Caesar, Markies and Perricholi were used as susceptible control, and LOP-868 was used as resistant control.

Detection of the presence of the RI_{adg} allele

At approximately 20 to 30 days after planting, young leaves of the plants were collected for DNA extraction by using the method described by Ferreira and Grattapaglia (1998) with modifications.

PCR reaction was performed for a total volume of 25 µL, containing: 5.0 µL 5X Buffer Colorless Go; 1.5 µL $MgCl_2$ (25 mM); 1.0 µL dNTP (10 mM); 0.75 µL of each primer (0.4 µM); 0.125 µL of the enzyme Go Taq Flexi DNA polymerase; 0.25 µL genomic DNA (10 ng); and 15.625 µL ultrapure water. The pair of SCAR primers was used, designated RGASC850, developed by Mihovilovich et al. (2014), which produces an 850 bp fragment, typical of the RI_{adg} allele.

Amplification reactions were carried out in a thermocycler with the following programming: 35 cycles at 95 °C for 1 minute for denaturation; primer annealing at 46 °C for 1 minute, followed by a final extension for 1 minute at 72 °C. The products obtained from this DNA amplification reaction were then analyzed on 1% agarose gel stained with Gel Red Nucleic Acid Gel Stain.

Cultivar Perricholi was used as susceptible control, and LOP-868 was used as resistant control.

Agronomic evaluation of clones

The field trial was installed in the experimental area of the EPAMIG in Lambari, MG, from October 2013 to January 2014 (lat 21º 58' S, long 45º 22' W, at alt 845 m asl). The region presents high-altitude tropical climate, with annual rainfall of 1642 mm, and average annual temperature of 20.8 ºC. The soil of the experiment is characterized as dystrophic dark red latosol.

An augmented block design with 134 clones was used as regular treatments, distributed in eight blocks with 24 treatments each. Common treatments were the cultivars Cupido, Voyager, Caesar, and clone CBM 16-16. Each plot consisted of a row with five plants spaced 0.35 m x 0.75 m apart.

At planting, fertilization was carried out with the formulation 4-14-8 (N, P_2O_5 e K_2O), at a dose of 3.0 t ha^{-1}, and with soil insecticide (aldicarb) at a dosage of 10.0 kg ha^{-1}. At approximately 40 days after planting, topdressing fertilization was carried out, with 300 kg ha^{-1} ammonium sulphate and 160 kg ha^{-1} potassium chloride, followed by hilling. Phytosanitary treatments were applied throughout the experiment, in order to prevent the competition with weed and the damage caused by pests and diseases, according to the standard of the commercial plantation in the region.

The following agronomic traits were evaluated: i. total tuber yield per plant (g planta^{-1}); ii. tubers specific weight, obtained by the formula: SW = air weight (air weight - water weight)$^{-1}$, having the weights determined in a hydrostatic scale; iii. grade of general appearance of tubers, ranging from 1 (poor appearance) to 5 (great appearance), taking into account the shape, skin roughness, and eye depth; the grades were given by three evaluators, being: 1. (round), 2 (round-oval), 3 (oval), 4 (oval-elongated), and 5 (elongated).

Data were subjected to analysis of variance for each trait separately, by using the augmented blocks design (Federer 1956). Analysis of variance was carried out by the mixed procedure of the statistical package SAS (SAS Institute 2008).

RESULTS AND DISCUSSION

Inoculation of PLRV by means of viruliferous aphids

The mean number of aphids plant^{-1} was 77.3. Of the total plants evaluated in the study, 82.3% had mean above 50 aphids plant^{-1}, which can be considered as a high inoculum pressure (CIP 2010).

The results of the serological test for the susceptible controls Ágata, Caesar, Markies and Perricholi were positive, while LOP-868 showed negative results. The cultivars Perricholi and LOP-868 have been used as controls to evaluate resistance to PLRV due to the absence or presence of the RI_{adg} allele, respectively (Velásquez et al. 2007, Mihovilovich et al. 2014).

Of the experimental clones evaluated in the study, 49.3% presented negative results in the DAS-ELISA serological test at 30 and 40 days after inoculation (DAI), which could be an indication that these clones have resistance level to PLRV. Other works, such as that of Velásquez et al. (2007) and Mihovilovich et al. (2014), who used PLRV inoculation by means of viruliferous aphids, adopted the DAS-ELISA serological test only at 40 DAI, which is enough time for the test to accurately detect the presence or absence of PLRV in the plant.

Beekman (1987) states that resistance to infection is highly dependent on inoculum pressure, which is governed by the population levels of vector aphids. In the present study, of the plants with negative results at 30 and 40 DAI, 77.7% had more than 50 aphids plant^{-1}. On the other hand, of the plants with positive results at 30 and 40 DAI, 81.4% had more than 50 aphid plant^{-1}. In this way, the inoculum pressure was adequate for most treatments, and the application of 50 viruliferous aphids plant^{-1} was ideal for the infection, since under this pressure, 100% of the inoculated susceptible cultivars were infected. According to Mihovilovich et al. (2007), differences between the resistance levels of treatments were defined according to the application of different aphid pressures, and the application of 50 aphids plant^{-1} was considered adequate, since a pressure of aphids per plant higher than this could result in greater damage caused by the pest than by the virus infection.

Identification of the RI_{adg} allele

The evaluation of the plants by using the RGASC850 marker identified the RI_{adg} allele. Figure 1 shows two androgenic cultivars, LOP-868 and Perricholi, with PLRV reaction known, and 14 clones with the 850bp band, which identifies the presence of the RI_{adg} allele. The cultivar LOP-868 is resistant to PLRV and presents the band, whereas the cultivar Perricholi is susceptible and does not present the band, as verified by Mihovilovich et al. (2014).

Of the total number of clones evaluated with the molecular marker, 31% presented the band (presence of the RI_{adg} allele), and 69% did not present the band (absence of the RI_{adg} allele). Since the parent LOP-868 presents the RI_{adg} allele in the duplex condition (Mihovilovich et al. 2014), approximately 83% of the progenies containing the allele were expected.

Of the clones evaluated by both the RGASC850 marker and by the inoculation of PLRV by means of viruliferous aphids, 22.2% presented the band and possible resistance to PLRV at 30 and 40 DAI, i.e., they had a negative result in the DAS-ELISA test; conversely, 29% of the clones did not present the band and had a positive result in the serological

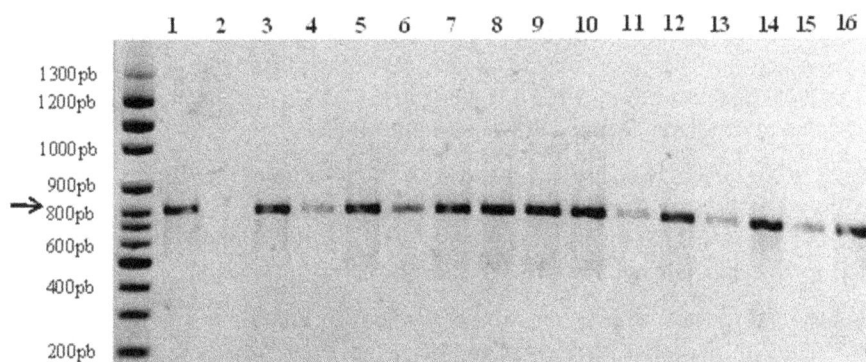

Figure 1. Presence or absence of bands (indicated by the arrow) amplified by the RGASC850 pair of primers by means of PCR, in the resistant [LOP-868 (1)] and susceptible [Perricholi (1)] controls, and in the resistant clones OGC 1-02 (3), OGC 1-12 (4), OGC 2-18 (5), OGC 2-26 (6), OGC 5-07 (7), OGC 5-51 (8), OGC 5-79 (9), OGC 6-44 (10), OGC 6-45 (11), OGC 6-46 (12), OGC 6-52 (13), OGC 6-69 (14), OGC 6-89(15), and OGC 6-105 (16).

test. Both results were expected, since in the presence or absence of the RI_{adg} allele, the absence or presence of PLRV is expected, respectively.

Approximately 40.3% of the clones did not present the band, nor possible resistance to PLRV at 30 and 40 DAI. These results show that despite the adequate number of aphids per plant, escapes may have occurred, which is a relatively common situation in artificial inoculations, reinforcing the usefulness of the marker for the selection of clones that are resistant to PLRV. In addition, Chuquillanqui and Jones (1980) commented that the technique of transferring aphids from inoculum plants to the treatment with a brush disturbs the aphids, and/or leads to the fatigue of the operator. In the present work, about 4000 aphids were manipulated at each inoculation (totaling approximately 17000 aphids). The acetone technique, developed by Souza Dias et al. (1993), could have avoided operator fatigue; nevertheless, the authors do not confirm that the technique has no effect on the results of the virus transmission.

In addition to escapes, these clones, which are denoted as possible resistant by the inoculation, but do not present the band, can be considered as false positives, due to the recombination of the marker gene with the RI_{adg} allele. This hypothesis should not be discarded, since even if the marker is at 1.2 cM from the RI_{adg} resistance allele, a small probability of exchange between the resistance gene and the marker can still be expected. This was the case of some treatments identified as possible false positives in the work of Mihovilovich et al. (2014). Even with the RI_{adg} allele, PLRV infection can occur and reach levels of up to 20% in LOP 868 when virus pressure is high (Velásquez et al. 2007).

Problems in the detection of PLRV by the DAS-ELISA test may also occur, resulting in false positives or false negatives. In the case of low concentration virus in the plant, as in the case of PLRV (family *Luteoviridae*), inconstant results are described in the literature (Guedes 2000). This fact can be explained by differences in the amount of antigen and the development of each plant. When comparing three techniques to detect *Barley yellow dwarf virus*–PAV-IL (family *Luteoviridae*), Figueira et al. (1997) concluded that the results of the DAS-ELISA test in barley plants were variable, especially when different coatings and/or conjugated antibody preparations were used.

Mihovilovich et al. (2014) also found unexpected results in their study, such as the presence of homomorphic band of the SCAR RGASC850 marker in a highly susceptible cultivar, Flor Blanca. According to these authors, this fact is not surprising, since the potato genome presents high frequency of RGA genomic regions, in addition to homologous sequences (Bakker et al. 2011). Therefore, the presence of homomorphic bands in the clones of the present study also deserves attention, since they were evaluated as possible susceptible clones by the inoculation, and presented the band.

Agronomic performance

Wide variation was observed for the total tuber yield, from clones that produced few grams to clones with production superior to 1000 g plant^{-1}. The means of the clones were low; however, approximately 10% of the clone population had a mean above 400 g plant^{-1} (Table 1). The low total tuber yield of the clones is due to the parent LOP-868, which is not adapted and presents several other undesirable traits, such as a great number of small tubers. In addition, the mean temperature throughout the experiment in the field was very high, of approximately 23.10 ºC, hindering the performance of the clones. The ideal temperatures for potato cultivation are between 15 and 21 ºC (Haverkort and Verhagen 2008). The effects of high temperatures have been reported in several works carried out in the south of the state of Minas Gerais (Menezes et al. 2001, Benites and Pinto 2011, Figueiredo et al. 2015). In the present study, the control CBM 16-16 stood out probably due to its tolerance to heat (Menezes et al. 2001).

The specific weight of the tubers is highly correlated with starch content and tubers dry matter, influencing the absorption of oil in the frying process. This trait presented means ranging from 1.0260 to 1.0920. The overall mean of the clones for this trait was low (1.0509), also due to high temperatures. Even the heat-tolerant control CBM 16-16 (Menezes et al. 2001) had a low mean, but it was higher than that of the other treatments (Table 1).

In relation to the shape and appearance of tubers, most of the clones presented round-oval tubers and poor appearance (Table 1). In general, tubers of the clones were small and numerous, which are traits inherited from LOP-868. These traits are probably related to the gene that confers resistance to PLRV (Barker and Harrison 1985). In addition, the high temperatures must have been responsible, in part, for these undesirable traits.

Of the clones tested with the molecular marker RGASC850 and evaluated in the field, 40% presented the RI_{adg} allele.

Table 1. Adjusted means of the clones with the RI_{adg} allele, and of the controls for the traits total tuber yield, tuber specific weight, shape grades, and tubers appearance

Clone	Total yield (g planta^{-1})	Specific weight	Shape	Appearence
OGC 1-02	174.3	1.0420	1.1	3.1
OGC 1-05	81.8	1.0467	3.3	2.1
OGC 1-10	141.8	1.0443	3.3	1.8
OGC 1-12	314.3	1.0473	1.4	2.3
OGC 1-22	64.3	1.0381	0.7	2.4
OGC 2-01	174.3	1.0414	2.0	2.3
OGC 2-11	36.8	1.0642	2.0	2.8
OGC 2-14	111.8	1.0394	1.8	2.1
OGC 2-16	349.3	1.0551	2.3	2.6
OGC 2-18	216.8	1.0792	2.3	3.0
OGC 2-26	308.1	1.0477	1.4	2.9
OGC 4-02	291.5	1.0508	1.4	2.4
OGC 4-16	253.1	1.0510	2.0	2.2
OGC 5-02	210.2	1.0410	1.3	1.9
OGC 5-07	332.5	1.0589	1.6	3.1
OGC 5-11	95.0	1.0478	3.1	2.6
OGC 5-22	136.4	1.0554	2.0	2.7
OGC 5-37	336.4	1.0487	2.0	2.4
OGC 5-46	182.5	1.0780	1.1	1.9
OGC 5-51	82.5	1.0426	2.1	2.4
OGC 5-58	141.5	1.0455	2.4	1.9
OGC 5-69	121.5	1.0372	2.7	2.9
OGC 5-72	291.5	1.0521	2.7	2.9
OGC 5-75	52.5	1.0767	2.1	2.3
OGC 5-79	114.4	1.0304	2.3	3.1
OGC 6-13	448.7	1.0676	1.9	2.8
OGC 6-16	82.5	1.0916	1.7	2.1
OGC 6-20	261.5	1.0389	2.4	3.1
OGC 6-24	307.5	1.0494	2.1	1.9
OGC 6-33	151.6	1.0654	1.3	2.6
OGC 6-34	289.4	1.0432	1.8	3.1
OGC 6-43	216.4	1.0427	2.0	1.9
OGC 6-44	351.5	1.0422	2.1	3.1
OGC 6-45	495.3	1.0611	2.0	3.1
OGC 6-46	270.3	1.0588	3.5	3.2
OGC 6-52	145.0	1.0673	1.1	2.6
OGC 6-68	332.5	1.0503	2.1	3.6
OGC 6-69	270.3	1.0551	1.7	3.2
OGC 6-82	185.2	1.0367	2.3	2.2
OGC 6-89	315.3	1.0430	3.7	2.6
OGC 6-104	285.2	1.0265	2.3	3.4
OGC 6-105	161.2	1.0407	2.9	2.8
OGC 6-107	252.7	1.0406	2.4	2.9
OGC 6-119	165.8	1.0689	2.1	2.6
General mean of clones	217.1	1.0509	1.9	2.6
CBM 16-16	724.7	1.0592	3.2	3.1
Caesar	359.8	1.0560	3.7	3.1
Cupido	330.8	1.0478	2.9	2.8
Voyager	462.0	1.0525	3.1	3.1
CV (%)	67.8	0.68	32.5	18.1
LSD (5%)[a]	432.45	0.0172	1.70	1.18

[a] Minimum significant difference at 5% probability for means comparison between a control and a clone.

Some clones are worth mentioning, such as clone OGC 6-45, which presented total tuber yield similar to that of the cultivars; tuber specific weight higher than that of the heat tolerant control (CBM 16-16); and a reasonable general appearance of tubers. The clones OGC 2-18, OGC 6-68, OGC 6-34, and OGC 6-44 should also be mentioned (Table 1).

Since some of the parents of the experimental clones also have the Ry_{adg} alleles and/or the Rx_1 allele, which confer extreme resistance to Potato virus Y (PVY) and Potato virus X (PVX), respectively, further trials should be performed on these clones in order to verify the presence of these alleles, adding multiple resistance to these viral diseases.

The inoculation of clones by means of viruliferous aphids with PLRV is laborious and time-consuming, and is not effective in proving resistance or susceptibility of the clones. In order to confirm the resistance or susceptibility, several replications are required. Nevertheless, the association the RGASC850 marker with the inoculation of the clones with aphids leads to greater safety in proving resistance or susceptibility of the clones.

Clones containing the Rl_{adg} allele were identified by means of the molecular marker SCAR RGASC850, showing that the PLRV-resistant Rl_{adg} allele can be transferred from LOP-868 to *Solanum tuberosum*. These clones should be used in backcross programs with commercial potato clones.

ACKNOWLEDGMENTS

The authors thank CAPES, CNPq and FAPEMIG for the financial support.

REFERENCES

Bakker E, Borm T, Prins P, Vossen E, Uenk G, Arens M, Boer J, Eck H, Muskens M, Vossen J, Linden G, Ham R, Klein-Lankhorst R, Visser R, Smant G, Bakker J and Goverse A (2011) A genome wide genetic map of NB-LRR disease resistance loci in potato. **Theoretical and Applied Genetics 123**: 493-508.

Barker H and Harrison BD (1985) Restricted multiplication of *Potato leafroll virus* in resistance potato genotypes. **Annals of Applied Biology 107**: 205-212.

Beekman AGB (1987) Breeding for resistance. In Bokx JA and Want JPH (eds) **Viruses of potatoes and seed-potato production**. Center for Agricultural Publishing and Documentation, Wageningen, p. 162-170.

Benites FRG and Pinto CABP (2011) Genetic gains for heat tolerance in potato in three cycles of recurrent selection. **Crop Breeding and Applied Biotechnology 11**: 133-140.

Brandolini A, Caligari PDS and Mendonza HA (1992) Combining resistance to *Potato leafroll virus* (PLRV) with immunity to potato viruses X and Y (PVX and PVY). **Euphytica 61**: 37-42.

Centro Internacional de la Papa (2010) **Procedimientos para pruebas de evaluaciones estándar de clones avanzados de papa**. CIP, Lima, 96p.

Chuquillanqui C and Jones RAC (1980) A rapid technique for assessing the resistance of families of potato seedlings to potato leaf roll virus. **Potato Research 23**: 121-128.

Clark MF and Adams AN (1977) Characteristics of the microplate method for enzyme-linked immunosorbent assay for the detection of plant viruses. **Journal of General Virology 34**: 475-483.

Cockerham G (1945) Some genetical aspects of resistance to potato viruses. **Annals of Applied Biology 32**: 280-281.

Federer WT (1956) Augmented (orhoonuiaku) designs. **Hawaiian Planters Record 55**: 191-208.

Ferreira ME and Grattapaglia D (1998) **Introdução ao uso de marcadores moleculares em análise genética**. Editora Biosystems, Brasília, 220p.

Figueira AR, Domier LL and D'arcy CJ (1997) Comparison of techniques for detection of *Barley yellow dwarf virus*–PAV-IL. **Plant Disease 81**: 1236-1240.

Figueira AR (1995) Viroses da batata e suas implicações na produção de batata-semente no estado de Minas Gerais: histórico do problema e soluções. **Summa Phytopathologica 21**: 268-269.

Figueiredo ICR, Pinto CABP, Ribeiro GHMR, Lino LO, Lyra DH and Moreira CM (2015) Efficiency of selection in early generations of potato families with a view toward heat tolerance. **Crop Breeding and Applied Biotechnology 15**: 210-217.

Gallotti GJM, Hirano E and Bertocini O (1992) Virose da batata: principais causas de degenerescência. **Agropecuário Catarinense 5**: 47-48.

Guedes MV (2000) Eficiência na detecção dos vírus PLRV, PVX E PVY em tecidos vegetais de gema apical dormente mais estolão, brotos e folhas da batata, pelo método sorológico enzyme-linked immunosorbent assay-elisa. **Scientia Agraria 1**: 80.

Haverkort AJ and Verhagen A (2008) Climate change and its repercussions for the potato supply chain. **Potato Research 51**: 223-237.

Lowery DT and Boiteau G (1988) Effects of five insecticides on the probing, walking, and settling behavior of the green peach aphid and the buckthorn aphid (Homoptera: Aphididae) on potato. **Journal of Economic Entomology 81**: 208-214.

Menezes CB, Pinto CABP and Lambert ES (2001) Combining ability of potato genotypes for cool and warm season in Brazil. **Crop Breeding and Applied Biotechnology 1**: 145-157.

Mihovilovich E, Aponte M, Lindqvist-Kreuze H and Bonierbale M (2014)

An RGA-derived SCAR marker linked to PLRV resistance from *Solanum tuberosum* ssp. *andigena*. **Plant Molecular Biology Reporter 32**: 117-128.

Mihovilovich E, Alarcon L, Perez AL, Alvarado J, Arellano C and Bonierbale M (2007) High levels of heritable resistance to *Potato leafroll virus* (PLRV) in *Solanum tuberosum* subsp. *andigena*. **Crop Science 47**: 1091-1103.

SAS Institute (2008) **Statistical analysis software**: SAS 9.2.

Souza Dias JAC and Iamauti MT (2005) Doenças da batateira (*Solanum tuberosum* L.). In Kimati H, Amorim L, Rezende JAM, Bergamin-Filho A and Camargo LEA (eds) **Manual de fitopatologia**. Agronômica Ceres, São Paulo, p. 119-142.

Souza Dias JAC, Slack SA, Yuki VA and Rezende JAM (1993) Use of acetone to facilitate aphid harvesting for plant vírus transmission assays. **Plant Disease 77**: 744-746.

Velásquez AC, Mihovilovich E and Bonierbale M (2007) Genetic characterization and mapping of major gene resistance to *Potato leafroll virus* in *Solanum tuberosum* ssp. *andigena*. **Theoretical and Applied Genetics 114**: 1051-1058.

Zaag DE (1987) Yield reduction in relation to virus infection. In Bokx JA and Want JPH (eds) **Viruses of potatoes and seed-potato production**. Pudoc, Wageningen, p. 146-150.

5

Diallel analysis in agronomic traits of Jatropha

Paulo Eduardo Teodoro[1], Erina Vitório Rodrigues[2], Leonardo de Azevedo Peixoto[1], Bruno Galvêas Laviola[2] and Leonardo Lopes Bhering[1*]

Abstract: *The objectives of this study were to estimate the general and specific combining ability of the parents, and to verify the existence of maternal effect and inbreeding depression in Jatropha. The experiment was carried out from 2010 to 2015, in the municipality of Planaltina, Distrito Federal. The following traits were evaluated: plant height, stem diameter, canopy projection between the row, canopy projection on the row, number of branches, mass of one hundred grains, and grain yield. Cytoplasmic effects and effects of female parent nuclear genes were observed for all traits. Dominance effects were predominant in the genetic control of all traits. Genotypes 107 and 190 were the superior parents for the reduction of the size, and for the increase of grain yield. No inbreeding depression was observed for grain yield. The most promising crosses for the conduction of segregating populations and increment in grain yield were 190x107, 107x190 and 259x107.*

Key words: *Jatropha curcas L, reciprocal effect, inbreeding depression, quantitative genetics.*

*Corresponding author:
E-mail: leonardo.bhering@ufv.br

[1] Universidade Federal de Viçosa, Departamento de Biologia Geral, 36.571-000, Viçosa, MG, Brazil
[2] Embrapa Agroenergia, 70297-400, Brasília, DF, Brazil,

INTRODUCTION

Jatropha curcas L. has unisexual, insect-pollinated monoeceious flowers, and shows protandry, however, hermaphroditic, self-pollinated flowers may occasionally occur (Kumar and Sharma 2008). It is a species of great importance due to its multiple use (Laviola et al. 2013). It was first introduced in Brazil to be used as hedges, and for small-scale oil production (Rosado et al. 2010). However, faced with the need for alternative sources of cleaner energy, and to the high energy demand required by the population, the interest in Jatropha cultivation has increased, since it is a promising culture to be used in biofuel production (Ong et al. 2013, Spinelli et al. 2014, Laosatit et al. 2017). Despite this, there are few studies on the breeding of this culture, since this species is under domestication.

Thus, aiming at the implementation of a breeding program of Jatropha in Brazil, researches have focused on the implementation of germplasm bank (Laviola et al. 2012a); on studies on genetic diversity (Rosado et al. 2010, Wen et al. 2010); and on estimates of genetic parameters (Laviola et al. 2012a, Santana et al. 2013). Estimates of genetic parameters are important in the orientation of breeding programs, since they support the selection process, and serve as reference for the understanding of the genetic structure of the population, in order to use it in a more accurate way in the breeding program (Rosado et al. 2010).

Concomitantly with these studies, parents selection is crucial. This step must be carried out very carefully, since it does not ensure the obtainment of progenies with high genetic potential when it is based only on desirable agronomic traits. Thus, diallel crosses is a technique that assists the choice of parents, aiming at improving accuracy in selection. This methodology selects superior parents based on the genotypic values, on the general combining ability (GCA), and on the specific combining ability (Cruz et al. 2012).

Diallel analyses estimate parameters that are useful in parents selection for hybridization and in the understanding of the genetic effects involved in the determination of the traits. Among the commonly used methods, the model of Griffing (1956) stands out for estimating GCA and SCA. The first refers to the performance of a parent in a series of crosses, and is associated with additive genetic effects; the second refers to the performance of specific hybrid combinations in relation to the parents means, and is related to the dominance and epistasis effects and several types of gene interactions. Moreover, they inform about the importance of reciprocal and maternal effects, and about the capitalization of heterosis (Cruz et al. 2012).

Although diallel analyses are useful in the understanding of the genetic basis of traits and in decision-making regarding the parents and the promising hybrid combinations, there are few reports related to this technique in Jatropha (Biabani et al. 2012, Santana et al. 2013, Islam et al. 2013). Therefore, the objectives of this study were to estimate the general combining ability (GCA) and the specific combining ability (SCA) of the parents used in the crosses, and to verify the existence of maternal effect and of inbreeding depression.

MATERIAL AND METHODS

Experimental area

The experiment was carried out from 2010 to 2015, in the municipality of Planaltina, Distrito Federal (lat 15° 35' 30'' S, long 47° 42' 30'' W, alt 1007 m asl). The climate is tropical with dry winter and rainy summer (Aw), with average annual temperature of 21 °C, relative humidity of 68%, and average rainfall of 1,100 mm/year. The soil is classified as Oxisol, with high clay content. The soil of the experimental area was corrected with limestone to raise the base saturation to 60%. At planting, 400 g per plant of superphosphate were applied to the soil. After planting, 200 g per plant of 20:00:20 fertilizer formulation were applied, which were divided into three applications at 30, 60 and 90 days. The experiment was carried out without irrigation. Management practices were based on Rosado et al. (2010).

Experimental design and traits evaluated

Crosses were carried out in March, 2010, in a complete diallel scheme, totaling three selfings and six hybridizations. The experiment consisted of randomized blocks design, with five replications, and three plants per plot, spaced 4x2 m apart. Three genotypes were used in the crosses, which were selected in the germplasm bank, with the following characteristics: small size (CNPAE-107), high grain yield (CNPAE-190), and resistance to powdery mildew (CNPAE-259). Only three parents were chosen, since they represent all the genetic variability for the most important agronomic traits in jatropha in Brazil. The seeds of the crosses obtained (F_1s) were sown in January 2011.

The following traits were evaluated: plant height (PH, m), stem diameter (SD, mm), canopy projection between rows (CPB, m), canopy projection on the row (CPR, m), number of branches (NB), mass of a hundred grains (MHG, g), and grain yield (GY, g plant^{-1}). The traits PH, CPB, CPR, and NB were evaluated in the third year after sowing (2013); SD was measured in the first year (2011); MHG was measured in the fourth year (2014); and GY was evaluated in the fifth year (2015).

Genetic-statistics analysis

Initially, analyses of variance were carried out for each trait, according to the model: $Y_{ij} = \mu + g_i + b_j + \varepsilon_{ij}$, in which Y_{ij} is the valued measured in the plot, analyzed for the i-th genotype, in the j-th block; μ is the overall mean; g_i is the effect of the i-th genotype, considered as fixed (i = 1, 2, ..., 9); b_j is the effect of the j-th block, considered as fixed (j = 1, 2, ..., 5); e_{ij} is the effect of the random error associated with the ij observation with ~NID $(0, \sigma^2)$.

The quadratic component associated with the genotypic effects (ϕ_g), with the environmental variance ($\hat{\sigma}_e^2$); as well as the quadratic component associated with phenotypic effects (ϕ_f), coefficient of genotypic determination (H²), selective accuracy (Ac), coefficient of genotypic variation (CV$_g$), coefficient of environmental variation (CVe), coefficient of relative variation (CVr), given by the ratio CV$_g$/CV$_e$ were obtained by analysis of variance. Subsequently, means clustering was carried out for each trait by using the Scott-Knott test.

Diallel analyses were carried out according to the model 1 of Griffing (1956):

$Y_{ij} = \mu + g_i + g_j + s_{ij} + r_{ij} + \varepsilon_{ij}$, in which Y_{ij} is the mean value of the hybrid combination (i ≠ j) or of the parent (i = j); μ is the general mean; g_i and g_j are the effects of the general combining ability of the i-th genotype or of the j-th parent (i, j = 1, 2 and 3); s_{ij} is the effect of the specific combining ability for crosses between parents i and j; r_{ij} is the reciprocal effect that measures the difference provided by parent i or j when used as male or female in the cross ij; and e_{ij} is the effect of the mean error associated with the ij observation with ~ NID (0, σ²).

In this model, the g_i effect was defined as: i ranging from 1 to number of genotypes (nine), since the diallel analysis was performed using the experimental unit mean (three plants per experimental unit). Also, it was considered $s_{ij} = s_{ji}$, $r_{ij} = - r_{ji}$ and $r_{ii} = 0$. From this model, estimates of general combining ability (GCA) and of specific combining ability (SCA), reciprocal effects, and quadratic components (Φ) were obtained. All statistical analyses were carried out using the Genes software (Cruz 2013).

RESULTS AND DISCUSSION

Genetic parameters

The estimates of genetic variability (ϕ_g) were higher than the estimates of environmental variance ($\hat{\sigma}_e^2$) for all traits (Table 1), which reveals predominance of genetic effect. Estimates of coefficient of genotypic determination (H²) obtained for all traits can be considered high (> 70%), since they result from polygenic traits, which are governed by several genes of little effect on the phenotype. The square root of H² reflects in the selective accuracy statistics (Ac), which indicates the quality of the information and of the procedures used in the prediction of genetic values. This parameter is associated with the precision of the selection, and refers to the correlation between predicted genetic values and actual genetic values of the individuals (Resende and Duarte 2007). Thus, the genetic values predicted for the population are reliable due to the high estimates of Ac for all traits.

For Cruz et al. (2012), the experimental precision is high when the estimates of the coefficient of environmental variation (CV$_e$) are lower than 20%, as in the present study, except for GY (CV$_e$ = 28%). Borges et al. (2014) found CV$_e$ of 29, 39 and 39% for GY in the first, second and third year of Jatropha production, which was similar to the values observed in this study. Estimates of CV$_e$ obtained in this work are in accordance with those reported in other studies on Jatropha (Laviola et al. 2012b, Laviola et al. 2013, Bhering et al. 2013, Borges et al. 2014, Ramos et al. 2014, Spinelli et al. 2015).

The coefficient of genetic variation (CV$_g$) quantifies the proportion of the genetic variability available for selection (Cruz et al. 2012). The ratio between this parameter and CV$_e$ results in the coefficient of relative variation (CV$_r$). Thus, the values obtained from CV$_g$ provided CV$_r$ superior to 1 for all traits indicates favorable situation for the selection of

Table 1. Estimates of genetic parameters for the traits plant height (PH), stem diameter (SD), canopy projection between rows (CPB), canopy projection on the row (CPR), mass of one hundred grains (MHG), and grain yield (GY) evaluated in nine Jatropha genotypes

Parameters	PH	SD	CPB	CPR	NB	MHG	GY
ϕ_g	0.137	17.71	0.22	0.05	11.13	34.81	186751.17
$\hat{\sigma}_e^2$	0.003	1.88	0.01	0.01	3.81	3.35	17624.38
ϕ_f	0.140	19.59	0.23	0.06	14.94	38.16	204375.55
H² (%)	98	90	97	87	75	91	91
Ac (%)	99	95	98	93	86	96	96
CV$_g$ (%)	15	6	18	11	15	9	41
CV$_e$ (%)	5	4	7	9	20	6	28
CV$_r$ (%)	3	1	3	1	1	1	1

ϕ_g: quadratic component associated with genotypic effects; $\hat{\sigma}_e^2$ environmental variance; ϕ_f: quadratic component associated with phenotypic effects; H²: coefficient of genotypic determination; Ac: selective accuracy; CV$_g$: coefficient of genotypic variation; CV$_e$: coefficient of environmental variation; CV$_r$ coefficient of relative variation.

superior genotypes in the population to be obtained from crosses.

The genetic variability reported in this study indicates the possibility of obtaining gains with selection in populations, and proves that the chosen genotypes are contrasting among each other, as observed by Rosado et al. (2010), at molecular level.

Diallel analysis

Diallel analysis made it possible to know the performance of the parents among each other, their hybrid combinations, and the nature of the reciprocal effect (Table 2). Estimates of general and specific combining ability, besides the reciprocal effect, were significant by the F-test for all traits, except for GCA for canopy projection on the row (CPR). These results indicate the existence of additive genetic effects, non-additive genetic effects, and/or cytoplasmic effects of female genitor involved in the control of the traits.

However, when evaluating the quadratic components (Φ), dominance effects (Φ_{sc}) were superior in the genetic control of all traits, which indicates the existence of differences in the gene compositions of the parents used in the study. When higher estimates of Φsc are observed in relation to the others, the parents are divergent for most of the loci in dominance, and the heterotic manifestations exhibited by these crosses may result from genetic complementation between the loci that control the evaluated traits.

Therefore, these results suggest that methods that prioritize the capitalization of heterosis must be used in order to

Table 2. Mean squares and quadratic components (Φ) obtained by diallel analysis for the traits plant height (PH), stem diameter (SD), canopy projection between rows (CPB), canopy projection on the row (CPR), mass of one hundred grains (MHG), and grain yield (GY) evaluated in nine Jatropha genotypes

SV	df	PH	SD	CPB	CPR	NB	MHG	GY
Treatment	8	0.55[+]	147.11[+]	0.70[+]	0.637[+]	176.32[+]	462.84[+]	1549525.65[+]
GCA	2	0.40[+]	373.17[+]	0.04[+]	0.003	117.20[+]	756.98[+]	2193597.26[+]
SCA	3	0.94[+]	90.03[+]	1.23[+]	0.799[+]	210.57[+]	546.34[+]	2154437.05[+]
Reciprocal	3	0.25[+]	53.49[+]	0.61[+]	0.898[+]	181.49[+]	183.26[+]	515233.17[+]
Residue	32	0.02	9.39	0.03	0.039	19.03	16.73	88121.92
Φ_{gc}	-	0.01	12.13	0.00	0.0	3.27	24.67	70182.51
Φ_{sc}	-	0.19	16.13	0.24	0.15	38.31	105.92	413263.02
Φ_{rc}	-	0.02	4.41	0.06	0.09	16.25	16.65	42711.12

[+]: significant at 5% probability by the F test, respectively; SV: source of variation; GCA: general combining ability; SCA: specific combining ability; Φgc, Φsc and Φrc: quadratic components of GCA, SCA and reciprocal effects, respectively.

Table 3. Estimates of general combining ability (GCA) and specific combining ability (SCA) obtained by diallel analysis for the traits plant height (PH), stem diameter (SD), canopy projection between rows (CPB), canopy projection on the row (CPR), mass of one hundred grains (MHG), and grain yield (GY) in Jatropha

Genotypes	PH	SD	CPB	CPR	NB	MHG	GY
				GCA estimates			
107	-0.08	0.25	-0.01	0.002	-1.00	2.63	194.82
190	-0.06	-3.65	-0.03	-0.008	-1.28	-5.79	113.90
259	0.13	3.39	0.04	0.011	2.28	3.16	-308.72
♂ x ♀				SCA estimates			
107x107	-0.42	-3.05	-0.42	-0.34	5.33	-9.41	-2.66
107x190	0.37	1.81	0.44	0.36	-5.89	2.67	228.45
107x259	0.05	1.25	-0.03	-0.02	0.56	6.74	-225.80
190x107	0.04	-2.22	0.05	-0.45	3.17	4.25	132.02
190x190	-0.31	-3.90	-0.41	-0.26	4.56	4.78	267.98
190x259	-0.06	2.09	-0.03	-0.10	1.33	-7.44	-496.43
259x107	-0.27	-1.05	-0.30	-0.23	-3.50	4.19	333.39
259x190	-0.07	3.17	-0.30	-0.15	-5.67	-4.40	161.23
259x259	0.01	-3.34	0.06	0.12	-1.89	0.70	722.23

improve the population for these traits (genetic dominance and divergence). Moreover, reciprocal recurrent selection should be used to improve the intrapopulational hybrids.

Genotypes 107 and 190 presented positive GCA estimates for GY, and negative estimates for PH (Table 3), and can be considered good alleles donor to increase GY and reduce the size of Jatropha plants, which are two of the main objectives of the breeding program of this culture. Genotype 259 stood out with positive GCA estimates for all traits, except for GY, whose value was negative. GCA estimates provide information on the concentration of predominantly additive genes, and their effects allowed the identification of parents to be used in the obtainment of the population for selection. By analyzing these estimates, the favorable alleles for the evaluated traits are dispersed among the parents, and some reciprocal recurrent selection cycles may be necessary.

Another strategy to be used in the populations is the indirect selection, due to high estimates of R^2 and Ac for all traits. Similar results were reported by Santana et al. (2013), who identified two genotypes with favorable alleles for six and four of the eight evaluated traits in diallel crosses of Jatropha. Moreover, Biabani et al. (2012) obtained GCA estimates that allowed identifying a parent effective for vegetative traits (PH, SD, and number of leaves), probably due to the high genotypic correlation between them.

SCA refers to the deviation of the hybrid's performance in relation to what would be expected based on the GCA of their parents. The SCA effects suggest the presence of non-additive interactions resulting from allelic complementation between the parents, enabling the improvement in the estimate of the genetic gain by exploring the heterosis with the use of different crosses. However, by observing the four most promising crosses, based on GY, two of them are derivative from selfing (259x259 and 190x190), which suggests the absence of inbreeding depression for this trait.

The crosses 259x107 and 107x190 also proved to be promising for the obtainment of populations based on SCA estimates for GY. These results were expected, since Rosado et al. (2010) had already observed genetic diversity at molecular level between these genotypes, which were clustered separately. The conduction of segregating populations from the cross 259x107 can be a favorable strategy for Jatropha breeding, since it enables the selection of genotypes of small size, with high grain yield, and which are resistant to powdery mildew, a disease that frequently occurs in the Brazilian Savanna. In addition, this cross combines parents with high GCA estimates for all traits, making it easier the selection of superior populations.

Comparison of means

Plant height is an important trait in the choice of the most appropriate spatial arrangement for planting, since Jatropha can exceed 5 m height. Thus, the breeder's efforts focus on selecting plants with small size for easier harvesting. Therefore, the selfings 107x107 and 190x190 were the most promising for PH reduction. Moreover, the hybrid formed by crossing the parents 107 and 190 and the selfings presented high mean for GY by the Scott-Knott test; thus, they formed the cluster with high GY (Table 4). Teodoro et al. (2016) observed that PH presents negative cause-effect relation with GY, suggesting that self-pollination of these genotypes may have contributed to the increase in the frequency of favorable alleles for the increment of GY and to the reduction of the size.

Table 4. Mean values for the traits plant height (PH), stem diameter (SD), canopy projection between rows (CPB), canopy projection on the row (CPR), mass of one hundred grains (MHG), and grain yield (GY) evaluated in nine Jatropha genotypes

Genotypes	PH		SD		CPB		CPR		NB		MHG		GY	
107x107	1.72	e	71.54	c	1.97	b	1.83	b	27.87	a	61.44	d	1286.94	a
107x190	2.90	a	72.95	c	3.16	a	2.33	a	17.74	b	67.07	c	1451.12	a
107x259	2.32	c	76.22	b	2.12	b	1.87	b	24.80	a	79.49	a	670.63	b
190x107	2.63	b	75.26	b	2.95	a	2.38	a	15.73	b	63.76	d	1556.09	a
190x190	2.16	d	71.08	c	2.14	b	1.92	b	19.67	b	61.67	d	1432.44	a
190x259	2.36	c	83.22	a	2.23	b	1.89	b	21.33	a	69.36	c	588.52	b
259x107	2.77	a	79.98	a	2.93	a	2.29	a	23.07	a	73.86	b	1067.66	a
259x190	2.63	b	76.01	b	2.91	a	2.30	a	25.54	a	68.15	c	239.48	c
259x259	2.79	a	82.08	a	3.06	a	2.37	a	22.47	a	74.38	b	1093.51	a

Means followed by the same letter in the same column do not differ by the Scott-Knott test.

SD, MHG and CPB have positive cause-effect relation with GY (Teodoro et al. 2016). Thus, indirect selection based on these traits can be carried out with the cross 259x107, which was also present in the cluster of higher GY. Also, as previously stated, the genotypes obtained from this cross presented small size, resistance to powdery mildew, and high grain yield. The crosses 190x107 and 107x190 may also be considered promising for the obtainment of superior populations, since they had high means for GY.

Therefore, by observing the cluster of genotypes with higher GY, almost all of the genotypes come from crosses with genotypes of high GCA for this trait (107 and 190), which reinforces the need of including these genotypes in future recurrent selection cycles for the obtainment of populations. Other relevant information for Jatropha breeding program was the high GY obtained by crosses derivative from selfings, which indicates the absence of inbreeding depression for this trait in Jatropha. However, this trait must be evaluated during the advancement of selfing generations before generalizing these results.

Therefore, since parents were chosen by molecular diversity previously reported (Rosado et al. 2010, Bhering et al. 2015), and based on all the results found in this study, the next step in the Brazilian jatropha breeding is to select superior individuals formed by the crosses 190x107, 107x190 and 259x107, and use them for the next breeding cycle, aiming to improve grain yield and reduce plant height.

REFERENCES

Bhering LL, Barrera CF, Ortega D, Laviola BG, Alves AA, Rosado TB and Cruz CD (2013) Differential response of Jatropha genotypes to different selection methods indicates that combined selection is more suited than other methods for rapid improvement of the species. **Industrial Crops and Products 41**: 260-265.

Bhering LL, Peixoto LA, Leite NLSF and Laviola BG (2015) Molecular analysis reveals new strategy for data collection in order to explore variability in Jatropha. **Industrial Crops and Products 74**: 898-902.

Biabani A, Rafii M, Saleh G, Shabanimofrad M and Latif M (2012) Combining ability analysis and evaluation of heterosis in *Jatropha curcas* (L). **Australian Journal of Crop Science 6**: 1030-1036.

Borges CV, Ferreira FM, Rocha RB, Santos AR and Laviola BG (2014) Productive capacity and genetic progress of physic nut. **Ciencia Rural 44**: 64-70

Cruz CD (2013) GENES - a software package for analysis in experimental statistics and quantitative genetics. **Acta Scientiarum Agronomy 35**: 271-276.

Cruz CD, Regazzi AJ and Carneiro PCS (2012) **Modelos biométricos aplicados ao melhoramento genético**. Editora UFV, Viçosa, 514p.

Griffing B (1956) Concept of general and specific combining ability in relation to diallel crossing systems. **Australian Journal of Biological Sciences 9**: 463-93.

Islam A, Anuar N, Yaakob Z, Ghani JA and Osman M (2013) Combining ability for germination traits in *Jatropha curcas* L. **The Scientific World Journal 13**: 1-6.

Kumar A and Sharma S (2008) An evaluation of multipurpose oil seed crop for industrial uses (*Jatropha curcas* L.): a review. **Industrial crops and products 28**: 1-10.

Laosatit K, Mokrong N, Tanya P and Srinives P (2017). Overcoming crossing barriers between jatropha (*Jatropha curcas* L.) and castor bean (*Ricinus communis* L.) **Crop Breeding and Applied Biotechnology 17**: 164-167.

Laviola BG, Alves AA, Gurgel FdL, Rosado TB, Costa RD and Rocha RB (2012a) Estimate of genetic parameters and predicted gains with early selection of physic nut families. **Ciência e Agrotecnologia 36**: 163-70

Laviola BG, Alves AA, Gurgel FL, Rosado TB, Rocha RB and Albrecht JC (2012b) Estimates of genetic parameters for physic nut traits based in the germplasm two years evaluation. **Ciência Rural 42**: 429-435

Laviola BG, Oliveira AMC, Bhering LL, Alves AA, Rocha RB and Gomes BEL (2013) Estimates of repeatability coefficients and selection gains in Jatropha indicate that higher cumulative genetic gains can be obtained by relaxing the degree of certainty in predicting the best families. **Industrial Crops and Products 51**: 70-76

Ong H, Silitonga A, Masjuki H, Mahlia T, Chong W and Boosroh M (2013) Production and comparative fuel properties of biodiesel from non-edible oils: Jatropha curcas, Sterculia foetida and Ceiba. **Energy Conversion and Management 73**: 245-255.

Resende MVD and Duarte JB (2007) Precisão e controle de qualidade em experimentos de avaliação de cultivares. **Pesquisa Agropecuária Tropical 37**: 182-194.

Ramos HCC, Pereira MG, Viana AP, Luz LN, Cardoso DL and Ferreguetti GA (2014) Combined selection in backcross population of papaya (*Carica papaya* L.) by the mixed model methodology. **American Journal of Plant Sciences 5**: 2973-2983.

Rosado TB, Laviola BG, Faria DA, Pappas MR, Bhering LL and Quirino B (2010) Molecular markers reveal limited genetic diversity in a large germplasm collection of the biofuel crop L. in Brazil. **Crop Science 50**: 2372-82

Santana UA, Carvalho Filho JLS, Blank AF and Silva-Mann R (2013) Combining ability and genetic parameters of physic nut genotypes for morphoagronomic traits. **Pesquisa Agropecuaria Brasileira 48**: 1149-1156.

Spinelli VM, Dias LAS, Rocha RB and Resende MDV (2014) Yield performance of half-sib families of physic nut (*Jatropha curcas* L.). **Crop Breeding and Applied Biotechnology 14**: 49-53.

Spinelli VM, Dias LAS, Rocha RB and Resende MDV (2015) Estimates of genetic parameters with selection within and between half-sib families of *Jatropha curcas* L. **Industrial Crops and Products 69**: 355-61.

Teodoro PE, Costa RD, Rocha RB and Laviola BG (2016) Contribution of agronomic traits for grain yield in physic nut. **Bragantia 75**: 51-6

Wen M, Wang H, Xia Z, Zou M, Lu C and Wang W (2010) Developmenrt of EST-SSR and genomic-SSR markers to assess genetic diversity in *Jatropha curcas* L. **BMC Researches Notes 3**: 1-5

6

SCS122 Miura - New Rice Cultivar

Rubens Marschalek[1], Jose Alberto Noldin[1], Ester Wickert[1], Klaus Konrad Scheuermann[1], Moacir Antonio Schiocchet[1], Domingos Savio Eberhardt[1], Ronaldir Knoblauch[1], Eduardo Hickel[1], Gabriela Neves Martins[1], Juliana Vieira Raimondi[2] and Alexander de Andrade[1*]

Abstract: *The rice cultivar SCS122 Miura has modern architecture, lodging resistance, late maturity cycle, moderate resistance to blast, high yield potential, long grains and adequate cooking quality. Industrial tests have demonstrated that the grains are suitable for parboiled rice. SCS122 Miura is recommended for all rice-producer regions of the state of Santa Catarina, Brazil.*

Key words: *Oryza sativa, breeding, variety.*

*Corresponding author:
E-mail: alexanderandrade@epagri.sc.gov.br

[1] Epagri, EEI, CP 277, 88301-970, Itajaí, SC, Brazil
[2] Avantis, Av. Marginal Leste, n. 3600, 88339-125, BC, SC, Brazil

INTRODUCTION

The state of Santa Catarina is the second largest irrigated rice producer in Brazil, with a cultivated area of 147,400 ha, and an annual production of 1.052.300 tons of paddy rice (CONAB 2016). The activity provides high generation of foreign exchange and has a great socio-economic role in the state. The development of cultivars adapted to different soil and climatic regions is fundamental to maintain Santa Catarina rice industry competitive, and consequently the local farmers. SCS122 Miura is a new irrigated cultivar developed by Epagri Rice Breeding Program, and it is recommended for all rice-producer regions of Santa Catarina. Industrial tests carried out with this cultivar showed that the grains are suitable for parboiled rice.

PEDIGREE AND BREEDING METHOD

SCS122 Miura was derived from a single cross between the line PR122 (IAPAR) and SCSBRS Tio Taka (Rangel et al. 2007), in 2003. In the following year (2004), F_1 plants were crossed with Epagri's inbreed line SC 354, which is a multispike line. The seeds obtained were sown to form the F_2 population. F_2 generation plants showed genetic variability, as expected. The selection process started with phenotypic traits of interest, such as plant type, height, yield, grain type, number of grain per panicle, and panicle type. Plants individual selection formed F_3 families, which, in turn, originated the F_4 generation. Agronomical traits, such as yield, were evaluated, starting with F_3 families. Seeds of the plants selected in the F_4 generation formed the lines of the "preliminary trial", and of the F_5 generation. During these stages, populations were grown following Epagri's technological recommendations for rice (Eberhardt and Schiocchet 2015). At every stage, the occurrence of blast (*Pyricularia oryzae*) and other diseases were carefully registered, in order to select genotypes with disease tolerance. In the F_6 generation (2011/2012), the new inbreed line was named as SC 681, and was again evaluated for yield potential, lodging resistance, plant height, tolerance to iron toxicity, blast resistance, and grain quality (Marschalek et al. 2017). Moreover, the line SC 681 was included in the trials to evaluate the Value for Cultivation and Use (VCU), carried out for two seasons (2012 and 2013) in Itajaí, Pouso Redondo and Massaranduba. Statistical analysis was carried out using the Scott-Knott test at 5%. The line SC 681 showed uniformity, good milling yield, and good agronomic traits, being suitable for milled and parboiled rice. SC 681 presented lodging resistance, moderate resistance to blast, and susceptibility to rice leaf scald. Based on the results of the VCU trials, the line SC 681 was released in 2017, as SCS122 Miura. The cultivar's name was a tribute to the phytopathologist Lucas Miura, who spent many years as a dedicated scientist at the Experiment Station of Itajaí/Epagri.

TRAITS PERFORMANCE

The agronomic traits of the cultivar SCS122 Miura are listed in Table 1. These evaluations were based on the Handbook of Research Methods in Rice (Embrapa 1977). SCS122 Miura has modern architecture, with late maturity life cycle (137-144 days), high tillering capacity, and erect and hairy leaves. The cultivar is resistant to lodging, which is considered as essential trait for the pre-germinated system, the most commonly used in Santa Catarina. The cultivar also presents intermediate shattering and intermediate resistance to iron toxicity and blast disease. SCS122 Miura presented mean yield of 9.388 kg ha^{-1} in VCU trials (Itajaí, Pouso Redondo and Massaranduba counties), which is higher than those of the controls (Table 2). SCS122 Miura also has good processing performance for parboiling rice. Therefore, as a result of the good agronomic traits, and due to acceptable industrial and sensory performance, this cultivar is recommended to rice-producers in all rice production areas in the state of Santa Catarina.

SCS122 Miura has long and translucent grains, acceptable cooking qualities, and milling yield of 67.5% (Table 3). The evaluations for the industrial grain traits showed that the cultivar is suitable for the parboiling process, and both milled and parboiled grains present glassy appearance.

Table 1. Agronomic traits of cultivar SCS122 Miura in VCU trials (Itajaí, Pouso Redondo and Massaranduba), in 2012/13 and 2013/14

Plant trait	Description
Leaf color	Green
Leaf pubescence	Medium
Flag leaf angle	Upright
Tillering	High
Cycle to maturity	137-144 days
Plant height	74.6 cm
Lodging	Resistant
Disease resistance:	
Leaf blast	Moderately resistant
Panicle blast	Moderately resistant
Brown spot	Moderately resistant
Rice leaf scald	Susceptible
Iron toxicity tolerance	Susceptible
Glumella color	Golden
Apex color at maturity	White
Awns	Absent
Shattering	Intermediate

Table 2. Mean grain yield (kg ha^{-1}) of SCS122 Miura, Epagri 109, and SCS116 Satoru, in VCU trials (Itajaí, Pouso Redondo and Massaranduba), in 2012/2013 and 2013/2014 seasons

Cultivars	Itajaí		Pouso Redondo		Massaranduba		Means
	2012	2013	2012	2013	2012	2013	
			kg ha^{-1}				
SCS122 Miura	9.900	8.196	11.100	9.984	9.900	7.249	9.388 a
Epagri 109	8.800	5.383	9.200	8.274	8.900	7.121	7.946 b
SCS116 Satoru	9.600	6.962	7.600	9.381	8.900	6.207	8.108 b

Means followed by the same letter are not significantly different by the Scott-Knott's test at 5% probability.

Table 3. Physical characteristics of grains of SCS122 Miura rice cultivar

Cultivar	Traits*				Grain size (mm)**				
	Total	AC	GT	WB	L	W	T	L/W	Class
SCS122 Miura	67.5%	I	H	2	7.23	2.08	1.72	3.48	Long-thin

Total: Total percentage of grains milled; AC: Percentage of amylose content; GT: Gelatinization temperature (I: intermediate, H: high); WB: White belly grade; L: Grain length; W: Grain width; T: Thickness; and L/W: Length width ratio. *Embrapa/CNPAF, ** 2,129 grains analyzed by the Image Rice Grain Scanner (Marschalek et al. 2017).

Cooking tests confirmed the good quality of the cultivar, since grains remain loose with soft texture, good aroma, and normal taste. Sensory evaluation of parboiled rice resulted in good consumer acceptance.

FOUNDATION SEED PRODUCTION

SCS122 Miura is protected by the Ministério da Agricultura, Pecuária e Abastecimento (Ministry of Agriculture, Livestock and Supply) under the registration number 36176. Genetic seed stock is kept at the Experiment Station of Itajaí/Epagri, located at Rodovia Antônio Heil, n.6800, Itaipava, CEP 88318-112, Itajaí, SC, Brazil. Certified seeds of the cultivar SCS122 Miura are produced by the Santa Catarina Rice Seed Producers Association (Acapsa).

ACKNOWLEDGEMENTS

This technology was financially supported by the National Council for Scientific and Technological Development (CNPq), Santa Catarina Research Foundation (Fapesc), and by the Brazilian Innovation Agency (Finep). The authors are grateful to the Santa Catarina Rice Seed Producers Association (Acapsa), to the Union of Industries of Rice in the State of Santa Catarina (Sindarroz-SC), and to the Brazilian Agricultural Research Corporation (Embrapa Rice & Beans) for the sensorial and chemical analysis of grains.

REFERENCES

CONAB - Companhia Nacional de Abastecimento (2016) **Acompanhamento da safra brasileira de grãos 2015/2016 – segundo levantamento.** Conab, Brasilia, 156p. Available at < http://www.conab.br >. Accessed on December 05, 2016.

Embrapa (1977) **Manual de métodos de pesquisa de arroz: 1ª aproximação.** Embrapa Arroz e Feijão, Goiânia, 106p.

Eberhardt DS and Schiocchet MA (2015) **Recomendações para a produção de arroz irrigado em Santa Catarina (Sistema pré-germinado).** Epagri, Florianópolis, 92p.

Marschalek R, Vieira J, Schiocchet MA, Ishiy T and Bacha RE (2008) Melhoramento genético de arroz irrigado em Santa Catarina. **Agropecuária Catarinense 21:** 54-56.

Marschalek R, Silva MC, Santos SB, Manke JR, Bieging C, Porto G, Wickert E and Andrade A (2017) Image - Rice Grain Scanner: a three-dimensional fully automated assessment of grain size and quality traits. **Crop Breeding and Applied Biotechnology 17:** 89-97.

Rangel PHN, Brondani C, Morais OPDE, Schiocchet MA, Borba TCO, Rangel PN, Brondani RPV, Yokoyama S, Bacha RE and Ishiy T (2007) Establishment of the irrigated rice cultivar SCSBRS Tio Taka by recurrent selection. **Crop Breeding and Applied Biotechnology 7:**103-110.

7

MG Travessia: a coffee arabica cultivar productive and responsive to pruning

Gladyston Rodrigues Carvalho[1]*, Gabriel Ferreira Bartholo[2], Antônio Alves Pereira[3], Juliana Costa de Rezende[1], Cesar Elias Botelho[1], Antônio Carlos Baião de Oliveira[2] and Felipe Lopes da Silva[4]

Abstract: *This paper presents the results of progeny 1190-1170-2, which was recorded as 'MGS Travessia' and selected based on its performance in the state of Minas Gerais. The cultivar has short size, cylindrical canopy, high yield capacity, high vegetative vigor, very satisfactory husk/bean ratio, grain quality compatible to traditional cultivars, and is very responsive to skeleton pruning.*

Key words: Coffea arabica, *selection, plant breeding.*

*Corresponding author:
E-mail: carvalho@epamig.br

[1] Empresa de Pesquisa Agropecuária de Minas Gerais (EPAMIG), Unidade Sul, Campus da Universidade Federal de Lavras (UFLA), CP 176, 37.200-000, Lavras, MG, Brazil
[2] Embrapa Café, Parque Estação Biológica - PqEB., 70.770-901, Brasília, DF, Brazil
[3] EPAMIG, Unidade Sudeste, Vila Gianetti, 46/47, Campus da UFV, 36.570-000, Viçosa, MG, Brazil
[4] Universidade Federal de Viçosa (UFV), Departamento de Fitotecnia, Avenida P.H. Rolfs, 36.570-000, Viçosa, MG, Brazil

INTRODUCTION

In order to recover important alleles from 'Mundo Novo' cultivar; to diversify the traits of 'Catuaí' cultivar; and to select earlier and more vigorous, productive, uniform fruit maturation forms, artificial crosses were carried out by the Agronomic Institute of Campinas (Instituto Agronômico de Campinas - IAC), in the 60s, between 'Catuaí' and 'Mundo Novo' cultivar (Carvalho et al. 2008). In the early 70s, with the introduction of these genetic materials in the state of Minas Gerais, by the State System of Agricultural Research System (Empresa de Pesquisa Agropecuária de Minas Gerais- EPAMIG/ Universidade Federal de Lavras- UFLA/ Universidade Federal de Viçosa- UFV), new backcrosses were carried out, and selections were intensified, resulting in 'Rubi' and 'Topázio' cultivars.

Subsequently, evaluations and selection of segregating progenies were carried out; these progenies resulted from a cross between 'Catuaí Amarelo IAC H 2077-2-12-70' and 'Mundo Novo IAC 515-20', which presented potential to generate progenies with traits of interest. Thus, this study aimed to obtain an Arabica coffee cultivar carrying traits demanded by coffee farmers and coffee drinkers, which were selected based on the performance, especially yield, after nine harvests, being the last one after skeleton pruning.

MATERIAL AND METHODS

Experiments were carried out in EPAMIG's Experimental Farms located in the municipalities of Três Pontas and São Sebastião do Paraíso, and on private farms in the municipalities of Capelinha (Fazenda Resplendor) and Campos Altos (Fazenda Ouro Verde). Thus, the main coffee regions of the state were represented in the experiment (South and Southwest of Minas Gerais, Vale do Jequitinhonha and Alto Paranaíba). Table 1 shows the edaphoclimatic characteristics of the experiments sites.

The performance of the commercial cultivars 'Catuaí Vermelho IAC 99', Rubi MG 1192', 'Acaiá Cerrado 1474', 'Catuaí Vermelho IAC 144' and 'Catuaí Amarelo IAC 62' was assessed together with other progenies developed by EPAMIG's coffee breeding program. This progenies are in the fourth generation by self-fertilization after the second backcross between 'Catuaí Amarelo IAC H 2077-2-12-70' and 'Mundo Novo IAC 515-20'. This paper presents the results of '1190-1170-2' progeny, which was registered at the National Register of Cultivars of the Ministry of Agriculture, Livestock and Supply (RNC/MAPA), as 'MGS Travessia'. This cultivar was selected among other tested progenies since it excelled in all the analyzed traits.

Experiments were carried out in a randomized block design with four replications and six plants per plot, all of which were considered useful, spaced 2.50 x 0.70 m apart (Três Pontas), 3.50 x 1.00 m (São Sebastião do Paraíso), and 3.50 x 0.50 m (Campos Altos and Capelinha). Experiments were carried out according to the technical recommendations for Arabica coffee culture for the state of Minas Gerais (Reis and Cunha 2010). Preventive or curative phytosanitary treatment was performed by chemicals, according to seasonality of pests and diseases.

The yield was analyzed annually in liters of freshly harvested coffee per plot. Coffee plants were harvested between the months of May and August each year. Subsequently, the yield was converted into yield of 60 kg bags of green coffee per hectare (bags ha^{-1}). This conversion was carried out by approximating the values, considering mean yield of 480 liters of freshly harvested coffee for each bag of 60 kg processed coffee (Carvalho et al. 2009). During analyses, mean yield (bags ha^{-1}) per biennium was considered, which was obtained by the mean of two consecutive harvests. The experiments in Três Pontas and São Sebastião do Paraíso were assessed in eight harvests and those in Campos Altos and Capelinha were assessed in six harvests.

After the eighth harvest, skeleton pruning was carried out in the coffee from Três Pontas. There, some traits, such as yield, vegetative vigor, percentage of floating grains and high sieve grains were assessed before and after pruning. Husk/bean ratio and percentage of peaberry grains after pruning were also assessed. Percentage of floating grains fruits was measured by the method proposed by Antunes and Carvalho (1957), by placing 100 cherry coffee fruits in water, and those which remained on the surface were considered floating. Husk/bean ratio was obtained by gallons of coffee harvested in the Field divided by the number of processed coffee bags per hectare. Samples of 500 g of coffee from each plot were used to classify the grains by size, in interleaved screens of different sizes. High sieve grains classification was carried out by the sum of screen 16 and above, and the peaberry grains were classified by the sum of grains retained in screen 10 and below. The screens used in the experiment were specific for the classification of these grains (Brazil 2003). Statistical analyses were carried out using the software Sisvar (Ferreira 2008). The means of the assessed traits were grouped by the Scott-Knott test, at a 5% probability (p <0.05).

RESULTS AND DISCUSSION

Significant effect (p <0.05) was found for yield for all cultivars in the four study sites. Cultivar 'MGS Travessia' stood out regarding grain yield in São Sebastião do Paraíso. It formed an isolated group in the mean comparison by the Scott-Knott test (Table 2). In the experiment installed in Campos Altos, two medium-sized groups were observed, and the top position was occupied by 'MGS Travessia', 'Rubi MG 1192', 'Acaiá Cerrado 1474' and 'Catuaí Amarelo IAC 62', with yields between 60.06 and 67.91 bags ha^{-1}. Similarly, in Capelinha, 'MGS Travessia', 'Rubi MG 1192', 'Acaiá Cerrado 1474', 'Catuaí Vermelho IAC 144' and 'Catuaí Amarelo IAC 62' stood out with the highest yield. Mean yield evaluation in Três Pontas formed three groups: 'Rubi MG 1192' occupied the top position, bottom position was occupied by 'Catuaí Vermelho IAC 99' and 'Acaiá Cerrado 1474', leaving 'MGS Travessia' in an intermediate position, with a mean yield of 41.32 bags ha^{-1} (Table 2).

One of the objectives of the cross that originated 'MGS Travessia' cultivar was to give 'Catuaí' cultivar greater vegetative vigor (Fazuoli et al. 2002). By the results observed in the assessments for this trait, this objective was achieved, since 'MGS Travessia' cultivar had a high mean for vegetative vigor (Table 3). These results are similar to those obtained in the evaluation of 'Rubi MG 1192', which is considered a very strong genetic material (Carvalho et al. 2008) and can be attributed to the high vegetative vigor presented by 'Mundo Novo IAC 515-20', one of the parents used in the artificial cross that originated 'MGS Travessia'.

Table 1. Edaphoclimatic characteristics of the experimental sites

Traits	Três Pontas	São Sebastião do Paraíso	Campos Altos	Capelinha
Local relief	wavy	slightly wavy	flat	wavy
Altitude	900 m	890 m	1.230 m	820 m
Latitude	21° 22′ 01″ S	20° 55′ S	19° 41′ 47″ S	21° 40′ S
Longitude	45° 30′ 45″ W	46° 55′ W	46° 10′ 17″ W	45° 55′ W
Annual Precipitation*	1670 mm	1470 mm	1830 mm	1450 mm
Annual Temperature*	20.1 °C	20.8 °C	17.6 °C	21.3 °C

* Annual average data.

Table 2. Mean yield of processed coffee in bags ha^{-1} of the studied cultivars in the first eight harvests in Três Pontas and São Sebastião do Paraíso and in the first six harvests in Capelinha and Campos Altos

Cultivars	São Sebastião do Paraíso	Campos Altos	Capelinha	Três Pontas
MGS Travessia	22.86a	60.06a	35.86a	41.32 b
Catuaí vermelho IAC 99	18.70b	53.63b	19.41b	34.35 c
Rubi MG 1192	20.90b	61.67a	25.29a	47.00 a
Acaiá Cerrado 1474	15.90b	60.88a	31.86a	31.83 c
Catuaí vermelho IAC 144	20.86b	53.61b	35.23a	-
Catuaí amarelo IAC 62	-	67.91a	25.74a	-

Means followed by the same letter do not differ by the Scott-Knott test (P <0.05).

Table 3. Mean coffee vegetative vigor; mean percentage of floating grains and mean percentage of high sieve grains assessed in the eighth harvest (before pruning) of four Arabica coffee cultivars in experiment installed in the EPAMIG's Experimental Farm in Três Pontas

Cultivars	Vigor	Floating grains	Sieve
MGS Travessia	7.33 a	5.00 a	49.00 b
Catuaí vermelho IAC 99	5.33 b	4.67 a	49.00 b
Rubi MG 1192	7.33 a	3.00 a	45.00 c
Acaiá Cerrado 1474	6.33 b	4.33 a	66.67 a

Means followed by the same letter do not differ by the Scott-Knott test (P <0.05).

There were no significant differences between the percentages of floating grains for the cultivars (Table 3). According to Carvalho et al. (2006), less than 10% of floating grains is considered satisfactory by breeders during evaluation and selection of coffee in breeding programs, since most of the commercial lines present percentage around that level. In this study, all cultivars had floating grains percentages ranging from 3-5% (Table 3). This trait again shows the potential of these genetic materials, since the smaller the percentage of floating grains, the greater is the yield of cherry coffee when compared with processed coffee, and consequently the greater is the yield potential of the cultivar.

The classification of grain in high sieve grains detected the formation of three distinct groups by the Scott-Knott test (p <0.05). 'Acaiá Cerrado 1474' stood out, since it presents well-developed grains. 'Rubi MG 1192' cultivar was the one with the lowest percentage of grains retained in high screens, while 'MGS Travessia' and 'Catuaí Vermelho IAC 99' cultivars are in intermediate position (Table 3).

Cultivar behavior after pruning

The traits yield, husk/bean ratio, percentage of floating grains, high sieve grains, and peaberry grains were assessed after skeleton pruning, in Três Pontas (Table 4). Regarding yield, 'MGS Travessia' presented best post-pruning performance. Although data refer to the first harvest after pruning, it can be inferred that 'MGS Travessia' had excellent response to skeleton pruning, which rises great interest, especially for farmers who opt by production system "Safra Zero".

'Acaiá Cerrado MG 1474' stood out for requiring only 380.12 liters of freshly harvested coffee for each 60 kg bag of processed coffee. On the other hand, the other cultivars showed lower husk/bean ratio, requiring larger volumes of freshly harvested coffee to fill a 60 kg bag of processed coffee (Table 4). Higher percentage of well-developed grains was observed, when compared with floating ones, for all cultivars, ranging from 94.70% to 97.57%. 'Rubi MG 1192' and

Table 4. Mean yield (bags ha⁻¹); husk/bean ratio (gallons of coffee harvested in the field divided by the number of processed coffee bags per hectare); percentage of floating grains, of grains classified in high sieve grains, and peaberry grains, assessed after pruning. EPAMIG's Experimental Farm, Três Pontas

Cultivars	Yield	Husk/bean ratio	Floating grains	High sieve	Peaberry grains
MGS Travessia	99.40 a	456.19b	5.30 b	72.21 b	7.90 a
Catuaí Vermelho IAC 99	72.90 b	455.81b	4.96 b	78.61 a	7.66 a
Rubi MG 1192	81.43 b	487.98b	3.24 a	81.31 a	6.33 a
Acaiá Cerrado 1474	79.17 b	380.12a	2.43 a	84.30 a	7.24 a

Means followed by the same letter do not differ by the Scott-Knott test (P <0.05).

'Acaiá Cerrado MG 1474' stood out for presenting smaller percentages of floating grains (Table 4).

Two groups with different means were formed by high sieve grains. 'Catuaí Vermelho IAC 99', 'Rubi MG 1192' and 'Acaiá Cerrado MG 1474' (Table 4) were found at the highest position; and 'MGS Travessia' was found at the bottom position, with 72.21% of grains classified in high sieve grains. Thus, all the assessed genetic materials showed high percentage of grains classified in high sieve grains. These results are very relevant, since this is a trait of great interest in coffee business, due to its high commercial value.

No significant difference was found among cultivars in percentage of peaberry grains, since they all presented low percentage for this type of grain (Table 4). There is no requirement for maximum content of peaberry grains as a criterion for assessing the quality of an Arabica coffee plant variety. Guimarães et al. (2002) reported that for seeds, standardization criteria indicate tolerance of up to 12% peaberry grains. Therefore, results observed in this study indicate that all cultivars showed very satisfactory values for this trait.

Table 5. Morphological, physiological and agronomic traits of 'MGS Travessia' with the respective descriptions

Characteristics	Type
Plant height	Low to medium
Canopy architecture	Cylindrical
Canopy radius	Medium
Internode length	Short
Secondary plagiotropic branching	High
Young leaf colour	Green
Leaf size	Medium
Color of ripe fruits	Yellow
Fruit shape	Oblong
Grain lenght	Medium
Grain width	Short and wide
Ripening cycle	Medium
Leaf edge curling	Little curled
Rust resistance	Susceptible
Nematode resistence	Susceptible
Vegetative vigor	High

Other morpho-agronomic descriptors

Other morpho-agronomic descriptors were observed in 'MGS Travessia' (Table 5). The cultivar has low to medium size, similar to 'Catuaí Vermelho IAC 99', and cylindrical canopy architecture. The canopy radius is medium. The plagiotropic branches have short internodes, with abundant side branches. It has yellow fruits when ripe, oblong and green colored leaves with curled edge. Grains have intermediate size and short and wide shape.

Planting recommendations

'MGS Travessia' presented excellent agronomic behavior in three major regions of the state of Minas Gerais, showing wide adaptation in very diverse environments. Therefore, this cultivar of intermediate maturation cycle may be planted safely in several producing regions of Minas Gerais. Spacing suitable for planting is at least 2.50 x 0.70 m. Conditions for growing and cultivation are similar to those recommended for Topázio MG 1190 cultivar. 'MGS Travessia' has also been recommended for the production system "Safra Zero", due to its high responsiveness to skeleton pruning.

ACKNOWLEDGMENTS

To the National Science and Technology Institute (INCT-Café/CNPq), to the Coffee Research Consortium, and the Research Support Foundation of the State of Minas Gerais (Fapemig) for financial support for the project, and to the owners of Ouro Verde and Resplendor Farms for making the experimental area available. To the fellowship of research productivity (PQ) granted by the National Council for Scientific and Technological Development (CNPq) and Fapemig.

REFERENCES

Antunes Filho H and Carvalho A (1957) Melhoramento do cafeeiro, ocorrência de lojas vazias em frutos de café Mundo Novo. **Bragantia 13**: 165-179.

Brasil - Ministério da Agricultura, Pecuária e Abastecimento (2003) **Instrução Normativa n. 8, 11 de junho de 2003**. Available at <http:www.abic.com.br/arquivos/abic_nm_a1d_inst_normativa 08.pdf>. Accessed on Jan 3, 2009.

Carvalho CHS, Fazuoli LC, Carvalho GR, Guerreiro Filho O, Pereira AA, Almeida SR, Matiello JB, Bartholo GF, Sera T, Moura WM, Mendes ANG, Rezende JC, Fonseca AFA, Ferrão MAG, Ferrão RG, Nacif AP, Silvarolla MB and Braghini MT (2008) Cultivares de café arábica de porte baixo. In Carvalho CHS (ed) **Cultivares de café: origem, características e recomendações**. Embrapa Café, Brasília, p. 155-252.

Carvalho GR, Botelho CE, Bartholo GF, Pereira AA, Nogueira AM and Carvalho AM (2009) Comportamento de progênies F_4 obtidas por cruzamentos de 'Icatu' com 'Catimor'. **Ciência e Agrotecnologia 33**: 47-52.

Carvalho GR, Mendes ANG, Bartholo GF and Amaral MA (2006) Avaliação e seleção de progênies resultantes do cruzamento de cultivares de café Catuaí com Mundo Novo. **Ciência e Agrotecnologia 30**: 844-852.

Fazuoli LC, Filho HPM, Gonçalves W, Gerreiro Filho O and Silvarolla MB (2002) Melhoramento do Cafeeiro: variedades do tipo arábica obtidas no Instituto Agronômico de Campinas. In Zambolim L (org) **O estado da arte de tecnologias na produção de café**. UFV, Viçosa, p. 163-216.

Ferreira DF (2008) Sisvar: um programa para análises e ensino de estatística. **Revista Symposium 6**: 36-41.

Guimarães RJ, Mendes ANG and Souza CAS (2002) Colheita. In Guimarães RJ, Mendes ANG and Souza CAS (eds) **Cafeicultura**. UFLA/FAEPE, Lavras, p. 285-300.

Reis PR and Cunha RL (2010) **Café arábica do plantio a colheita**. Embrapa Informação Tecnológica, Brasília, 896p.

8

Recurrent selection of popcorn composites *UEM-CO1* and *UEM-CO2* based on selection indices

Rafael Augusto Vieira[1], Renato da Rocha[1], Carlos Alberto Scapim[1] and Antonio Teixeira do Amaral Junior[2]

Abstract: *Selection indices were applied to data sets of 169 half-sib families of the popcorn composites UEM-Co1 and UEM-Co2 in four cycles of recurrent selection. From 2005 to 2008, the experiments were arranged in a 13 by 13 lattice square design, with two replications per cycle and composite. Genetic gains for popping expansion (PE) and grain yield (GY) were estimated based on several selection indices and truncation selection. The magnitude and balance of gains estimated for each trait by the indices were compared by an auxiliary statistical value (C_i). This value C_i consists of an arbitrary value, resulting from differences between the gains estimated for n traits by truncation selection and by index i. The indices of Subandi and Mulamba and Mock were the most promising to estimate high and balanced genetic gains for PE and GY in recurrent selection of half-sib popcorn families.*

Key words: *Zea mays L., multiple-trait selection, popping expansion, grain yield.*

*Corresponding author:
E-mail: ?

[1] Universidade Estadual de Maringá, Departamento de Agronomia, 87.020-900, Maringá, PR, Brazil
[2] Universidade Estadual do Norte Fluminense Darcy Ribeiro, Laboratório de Melhoramento Genético Vegetal, 28.013-602, Campos dos Goytacazes, RJ, Brazil,

INTRODUCTION

Popcorn is highly appreciated by the Brazilian population and the crop acreage in the country is on the rise. However, factors such as the limited availability of hybrid seeds and high-quality varieties are pressing towards higher popcorn imports (Sawazaki et al. 2003). On the other hand, researchers are dedicated to developing superior and adapted genotypes, especially with regard to popping expansion (PE) and grain yield (GY) (Amaral Júnior et al. 2013, Ribeiro et al. 2016). In some situations, the obstacle of negative correlation between PE and GY has to be overcome, so that the developed product satisfies both producers and consumers (Zinsly and Machado 1987, Carpentieri-Pípolo et al. 2002, Broccoli and Burak 2004, Faria et al. 2008, Freitas et al. 2013).

With a view to increase the accuracy of choice of the genotypes that contain a combination of both agronomic and quality traits in a single line, selection indices are very useful breeding tools for crops such as popcorn, for allowing the choice of the target traits for improvement. In this context, Granate et al. (2002) evaluated half-sib families of the popcorn composite CMS-43, and found that the Smith and Hazel index predicted higher gains for a greater number of traits, and that the percentage gain for the combination of the two most important traits was 9.14%. In a study of the perfromance of S_1 and S_2 progenies obtained from the popcorn population Beija-Flor, Vilarinho et al. (2003) selected superior families with the Mulamba and Mock index recommended by the authors. To select progenies in a third cycle population of intrapopulation recurrent selection

(UNB-2U/UENF-14) by the method of half-sib families, Santos et al. (2007) found that the Mulamba and Mock index using arbitrary weights, resulted in higher gains for most traits, including PE and GY (7.6% and 10%). In the fourth recurrent selection cycle of the same population, Freitas Júnior et al. (2009) selected 30 superior families by the Mulamba and Mock selection index with arbitrary weights, (10.55% for PE and 8.50% for GY). In an evaluation of the fifth selection cycle of population UNB-2U, Rangel et al. (2011) found that the selection of superior families should be based on the Mulamba and Mock index, for estimating highest gains (6.01% for PE and 8.53% for GY). Analyzing the sixth recurrent selection cycle in population UENF-14, Ribeiro et al. (2012) observed, as in the previous publications, that the Mulamba and Mock index predicted the best gains (PE 10.97% and GY 15.30%), based on random economic weights.

Although the above studies demonstrate the efficiency of the tested indices, to date only part of the methodologies have been applied. Among the indices used in popcorn breeding, those of Smith (1936) and Hazel (1943), Williams (1962), Pesek and Baker (1969) and Mulamba and Mock (1978) are noteworthy. However, the indices of Tallis (1962), Cunningham et al. (1970) and Subandi et al. (1973) have not been applied in segregating popcorn populations so far. Moreover, previous studies failed to indicate a measurable form of comparing the selection indices in terms of the genetic gain estimates. In this context, this paper proposes and describes the statistical value C_i, which is promising for indicating which index estimates high and balanced genetic gains for traits. In this sense, the different selection indices for popcorn were compared, based on data collected in four recurrent selection cycles of half-sib families of the composites *UEM-Co1* and *UEM-Co2*.

MATERIAL AND METHODS

Four recurrent selection cycles (C_0, C_1, C_2, and C_3) of the popcorn composites *UEM-Co1* and *UEM-Co2* were evaluated in this study. Composite *UEM-Co1* has yellow grain, resulting from open pollination among 17 popcorn genotypes. Composite *UEM-Co2* has white grain, derived from 12 popcorn genotypes. The selection cycles of the composites *UEM-Co1* and *UEM-Co2* were performed and evaluated in the municipality of Iguatemi, Maringá, in the north of Paraná, Brazil. One hundred and sixty-nine half-sib families were tested in a 13 by 13 lattice square design, with two replications for each selection cycle of the composites. The experimental units consisted of a single 5-m row, spaced 0.90 m apart, with a total of 25 plants per plot. The experiments were carried out between 2005 and 2008. Basal and topdressing fertilization were applied according to the crop requirements, as indicated by soil analysis. Other cultural practices were applied as recommended for maize cultivation in southern Brazil.

The half-sib families were evaluated for the traits grain yield (GY) and popping expansion (PE) in each selection cycle of the two composites. Grain yield was measured by weighing the amount of grain produced per plot, adjusted to kg ha^{-1}, at 13% moisture. The PE was determined based on a grain sample of 30 g taken from the intermediate portion of the cobs, at 13% moisture (mL g^{-1}). The grain samples were popped for 2.5 minutes at 270 °C in an electric popcorn popper. The best families were selected and recombined, according to the method between and within half-sib families (Paterniani 1967).

For data analyses, analysis of variance (ANOVA) was performed for each selection cycle and composite. The intra-block analysis followed the lattice model: $Y_{ijk} = m + g_i + b_{k/j} + r^j + E_{ijk}$, where: Y_{ijk} = observation of each half-sib family i^{th}, located in the kth block in the jth replication; m = overall mean; g_i = effect of each half-sib family; $b_{k/j}$ = effect of the kth block, in the jth replication; r_j = effect of the rth replication of the experiment; E_{ijk} = experimental error associated with Y_{ijk}; and g_i and $b_{k/j}$ were adjusted by analysis of variance.

The effect of half-sib families (g_i) was considered random. Heritability and variability (CVg) were estimated from the expected mean squares of the analysis of variance. For all selection and cycles and composites, the genetic gains for GY and PE were estimated by selection indices and truncation selection (direct selection). The following indices were used: i) Smith (1936) and Hazel (1943) (classical index); ii) Williams (1962); iii) Mulamba and Mock (1978); iv) Cunningham et al. (1970); v) Tallis (1962); vi) Pesek and Baker (1969) and vii) Subandi et al. (1973). The methodologies proposed by Tallis (1962) and Cunningham et al. (1970) consist of restricted indices, unlike the others used in this study. For the indices i, ii, iii, iv and v, three sets of economic weights were used, namely: a) CVg for both traits, b) 350 for PE and 1 for GY, and c) 1 for both evaluated traits. The selection intensity of this study was 20%, selecting, consequently, a total of 34 families per cycle and composite. The statistical analyses were performed with software Genes (Cruz 2013).

The statistical value C_i was designed to compare the selection indices with regard to maximization and balance of genetic gains for GY and PE. This value is based on the difference between the genetic gain for each trait of a set, estimated by an index, and the maximum possible gain of this trait in the study population. We suggest the measurement of the maximum genetic gain by truncation selection. The C_i value should be calculated for each selection index under study. The most appropriate selection index is the one represented by the lowest C_i value, indicating effectiveness and the possibility of considerable genetic gain for several traits under the specific conditions of the study population. The reason is that in every trait, the genetic gains of an index are subtracted from the maximum possible genetic gains in the study population. It is expected that good selection indices will estimate high gains for the traits with highest chances for improvement in the population, as estimated by truncation selection. In the opposite case, the higher C_i value indicates the situation. This procedure is performed at the beginning of the calculation. The basis of the maximum possible gain is a key point of using C_i. Nevertheless, it is also expected that good selection indices will achieve relevant gains for the other target traits, a condition measured by the sum of all effects of n characters, incorporated in the general expression of C_i. Furthermore, when the genetic gain of truncation selection is reduced to one trait, the difference between the maximum gain and the gain estimated by an index is not as important in terms of increase in the C_i value. In this sense, the statistical value C_i allows an identification of the selection index that fits best in view of the possibilities of breeding and limitations of each population, i.e., this index can provide high and balanced genetic gains for the studied traits and makes better use of the opportunities for breeding intrinsic to the study population.

The general expression to calculate C_i is: $C_i = \displaystyle\sum_{j=1}^{J} \dfrac{(GG_{ij} - GGts_j)^2}{\dfrac{2}{|GGts_j|}}$

Where: GG_{ij} = is the genetic gain estimated by the ith selection index for the jth trait; $GGts_j$ = maximum possible genetic gain for the jth trait, estimated by truncation selection. If the truncation selection cannot be estimated, $GGts_j$ can be represented by the maximum genetic gain for the jth trait, for the ith selection index. In this study, the genetic gains for each index were estimated over four selection cycles of the two composites at a selection intensity of 20%. Then the statistical value C_i was applied, using Microsoft Office Excel spreadsheets.

RESULTS AND DISCUSSION

The results of the analysis of variance (ANOVA) for all cycles and composites are shown in Table 1. Significant differences were observed for the source of variation (p <0.05) for all cycles and composites for both traits (PE and GY). It is well-documented that the presence of genetic variability is imperative for good results with selection (Silva et al. 2001). In this context, the genetic variability for the composite *UEM-Co2* was generally greater, a factor that can explain the relatively higher gains of this composite than of *UEM-Co1* in the different selection cycles (Table 1).

Different conditions of variability (CVg) and heritability based on the family means were observed for the study traits of the composites throughout the selection cycles. The CVg values varied from 3.86 to 15.04% for PE and from 7.20 to 12.54% for GY. High heritability estimates were also found, indicating the possibility of selecting half-sib families with good accuracy. In fact, the heritability of composite *UEM-Co1* in C_0 was 0.638 for PE and 0.634 for GY. The heritability estimated for composite *UEM-Co2* was 0.754 and 0.606 for PE and 0.697 and 0.757 for GY, in the cycles C_0 and C_2, respectively (Table 1). High heritability values for PE and GY were also reported by other authors (Pacheco et al. 1998, Coimbra et al. 2002, Santos et al. 2008).

The results of truncation selection at a selection intensity of 20%, for the composites *UEM-Co1* and *UEM-Co2*, are shown in Tables 2 and 3. For both composites, positive genetic gains were estimated in most selection cycles on the basis of truncation selection. For composite *UEM-Co1*, the most significant gains for GY were estimated in cycles C_0 and C_3 (14.2% and 11.3% respectively), while for PE, in the cycles C_0 and C_1 (11.1% and 8.0%, respectively). The highest gains for PE and GY were estimated in the first selection cycle C_0 (Table 2). For composite *UEM-Co2*, based on truncation selection, the genetic gains estimates were highest in the cycles C_2 (16.5%) and C_0 (14.1%) for GY, while higher PE estimates were found in the cycles C_0 and C_1 (18.2 and 13.6%, respectively) (Table 3). These results indicate that, for both populations and traits, truncation selection would maximize the efficiency in the initial breeding cycle (C_0), which may be related to the greater genetic variability and heritability commonly found in this cycle (Table 1). Although the estimated values were

Table 1. Analysis of variance (ANOVA) for popping expansion (PE) and grain yield (GY) in two popcorn populations in four selection cycles of two popcorn composites *UEM-Co1* and *UEM-Co2*

Source of variation	df	Mean squares							
		PE				GY			
		C_0	C_1	C_2	C_3	C_0	C_1	C_2	C_3
UEM-Co1									
Families	168	27.3*	22.3*	19.2*	17.6*	1.90^{E+5*}	1.67^{E+5*}	1.70^{E+5*}	1.98^{E+5*}
Block/Replications	24	9.5	10.8	30.4	58.8	0.89^{E+5}	1.26^{E+5}	4.29^{E+5}	4.12^{E+5}
Replications	1	18.9	10.0	4.5	84.5	2.52^{E+5}	7.34^{E+6}	5.62^{E+6}	1.22^{E+6}
Erro	144	9.9	10.3	11.9	14.2	7.00^{E+4}	9.70^{E+4}	1.09^{E+5}	8.40^{E+4}
Mean		29.6	30.6	31.3	32.3	1.97^{E+3}	2.17^{E+3}	2.15^{E+3}	2.35^{E+3}
CVe		10.63	10.49	11.02	11.67	13.45	14.38	15.38	12.36
CVg		9.96	8.02	6.11	4.04	12.48	8.58	8.16	10.17
Heritability		0.638	0.540	0.379	0.193	0.634	0.415	0.361	0.576
UEM-Co2									
Families	168	31.2*	26.0*	22.7*	17.6*	1.98^{E+5*}	1.70^{E+5*}	1.42^{E+6*}	1.97^{E+5*}
Blocks/Replications	24	11.3	5.5	15.7	52.5	0.96^{E+5}	1.22^{E+5}	4.83^{E+5}	4.22^{E+5}
Replications	1	0.1	1.1	4.3	3.6	0.73^{E+5}	1.70^{E+5}	3.42^{E+4}	1.96^{E+5}
Error	144	7.7	6.2	9.0	14.2	6.00^{E+4}	9.84^{E+4}	8.31^{E+4}	8.78^{E+4}
Mean		22.8	28.2	33.3	33.9	2.18^{E+3}	2.63^{E+3}	2.87^{E+3}	3.10^{E+3}
CVe		12.17	8.83	9.01	11.12	11.26	11.92	10.04	9.56
CVg		15.04	11.10	7.87	3.86	12.06	7.20	12.54	7.53
Heritability		0.750	0.762	0.606	0.194	0.697	0.422	0.757	0.554

*Significant at 5% probability; C_0, C_1, C_2, and C_3, indicating the recurrent selection cycles, 0, 1, 2, and 3, respectively

Table 2. Genetic gains estimated by selection indices and truncation selection for grain yield (GY) and popping expansion (PE) in four cycles of the composite *UEM-Co1* with yellow popcorn grain

Selection indices and truncation selection[†]	Estimated genetic gains							
	C_0		C_1		C_2		C_3	
	GY	PE	GY	PE	GY	PE	GY	PE
Truncated for GY	14.2	1.9	7.9	2.1	7.1	0.4	11.3	0.1
Truncated for PE	-0.1	11.1	0.9	8.0	-0.8	5.3	-0.9	2.4
SH_{Cvg}	14.1	1.0	7.9	2.2	6.9	-1.3	11.1	-0.3
$SH_{350.1}$	4.2	10.8	3.0	7.7	1.7	5.1	8.2	2.0
$SH_{1.1}$	14.1	1.0	7.9	2.2	6.9	-1.3	11.1	-0.3
W_{Cvg}	14.2	1.9	7.9	2.1	7.1	0.4	11.3	0.1
$W_{350.1}$	5.7	9.8	3.6	7.8	2.6	5.0	5.4	2.3
$W_{1.1}$	14.2	1.9	7.9	2.1	7.1	0.4	11.3	0.1
MM_{Cvg}	11.5	6.9	6.1	5.5	5.2	3.8	10.0	1.4
$MM_{350.1}$	2.0	10.3	0.9	8.0	0.2	5.2	0.1	2.4
$MM_{1.1}$	9.9	7.9	6.0	5.5	5.0	3.9	8.2	1.8
$Cunn_{Cvg}$	13.4	4.9	7.4	4.2	1.7	5.1	11.2	0.5
$Cunn_{350.1}$	10.3	2.0	-0.4	7.8	6.9	2.0	1.9	2.4
$Cunn_{1.1}$	13.0	5.6	7.4	4.2	6.9	2.0	10.8	1.1
$Tallis_{Cvg}$	10.9	7.7	6.5	5.7	5.3	3.9	7.2	2.1
$Tallis_{350.1}$	10.9	7.7	6.5	5.7	5.3	3.9	7.2	2.1
$Tallis_{1.1}$	10.9	7.7	6.5	5.7	5.3	3.9	7.2	2.1
PB_{1-DPG}	10.9	7.3	5.5	5.7	5.3	3.9	7.2	2.1
Sub	11.7	6.9	6.7	5.5	5.6	3.7	9.6	1.6

[†]SH_{Cvg}, $SH_{350.1}$, $SH_{1.1}$: indices based on Smith (1936) and Hazel (1943); W_{Cvg}, $W_{350.1}$, $W_{1.1}$: indices of Williams (1962); MM_{Cvg}, $MM_{350.1}$, $MM_{1.1}$: indices based on the sum of "ranks" of Mulamba and Mock (1978); $Cunn_{Cvg}$, $Cunn_{350.1}$, $Cunn_{1.1}$: restricted index of Cunningham et al. (1970); $Tallis_{Cvg}$, $Tallis_{350.1}$, $Tallis_{1.1}$: restricted selection index of Tallis (1962); PB_{1-DPG}: index proposed by Pesek and Baker (1969) based on a genetic standard deviation as desired genetic gain; Sub: multiplicative index of Subandi et al. (1973); the following economic weights were tested: CVg 350 for PE; 1 for GY; and 1 for both (GY and PE).

higher for both traits, this direct selection strategy - also called truncated - is inadequate for satisfactory simultaneous genetic gains for two or more traits, as in the case of PE and GY, which are occasionally negatively correlated as well (Pacheco et al. 1998, Carpentieri-Pípolo et al. 2002, Daros et al. 2004b, Faria et al. 2008).

With regard to the applied selection indices, in a first analysis, the classical index of Smith and Hazel (1936, 1943) and that of Williams (1962) did not estimate balanced gains for both traits of the study populations (Table 2 and Table 3), suggesting limited applicability. Our results for the index of Williams (1962) confirmed those of Granate et al. (2002), who could find no advantage of using it. Other authors, however, reported different results. For example Daros et al. (2004a), when selecting families by the Smith and Hazel index, predicted relatively high genetic gains (17.8% for PE and 26.95% for GY). Similarly, Granate et al. (2002) found that this index allowed the prediction of higher gains for a larger number of characters, with joint gains of 9.14% for GY and PE. The results of the above authors differ from those in this study, since the Smith and Hazel index proved unsatisfactory for the prediction of genetic gains of the two main target traits in popcorn breeding (Tables 2 and 3). For both composites, the index of Tallis (1962), for all studied possibilities of economic weights, and that of Pesek and Baker (1969) using a genetic standard deviation as desired gain, estimated balanced genetic gains for GY and PE, except in cycle C_3 of composite *UEM-Co1*, with gains of 7.2% for GY and 2.1 for PE (Table 2). For composite *UEM-Co2* in cycle C_3, however, genetic gains of 12.8% and 6.9% for GY and PE, respectively, were estimated by the Pesek and Baker method (Table 3). Good simultaneous gains were predicted by the index of Pesek and Baker (7.99% for GY, and 10.75% for PE), according to Freitas Junior et al. (2009), in an evaluation of 200 full-sib families in cycle C_4 of the popcorn population UENF-14. Other selection indices, as that of Mulamba and Mock, also estimated balanced gains for the selection cycles of both composites. For this the economic weights were determined as CVg and 1 for PE and 1 for GY, in cycle C_1 of composite *UEM- Co1-* (Table 2) and in C_0 of *UEM-Co2* (Table 3). The use of this index was also effective in popcorn breeding of the germplasm studied by Vilarinho et al. (2003), Santos et al. (2007), Freitas Júnior et al. (2009), Amaral Júnior et al. (2010), Rangel et al. (2011), Ribeiro et al. (2012) and Freitas et al. (2014).

For the index of Subandi et al. (1973), the genetic gains estimated for composite *UEM-Co1* in cycles C_0 to C_3 for GY ranged from 5.6% to 11.7%, and from 1.6% to 6 9% for PE. For *UEM-Co2*, genetic gains between 5.2% and 14.7% were estimated for GY and from 1.6% to 12.7% for PE (Tables 2 and 3).

Table 3. Genetic gains estimated by selection indices and truncation selection for grain yield (GY) and popping expansion (PE) in four cycles of the composite *UEM-Co2* with yellow popcorn grain

Selection indices and truncation selection[†]	Estimated genetic gains							
	C_0		C_1		C_2		C_3	
	GY	PE	GY	PE	GY	PE	GY	PE
Truncada para GY	14.1	2.7	6.7	4.2	16.5	2.4	8.0	-0.1
Truncada para PE	0.2	18.2	0.2	13.6	1.4	8.6	-0.1	2.3
SH_{CVg}	14.1	2.4	6.7	4.6	16.5	2.4	7.8	-0.5
$SH_{350.1}$	4.0	16.3	0.9	13.5	11.5	7.3	4.2	2.1
$SH_{1.1}$	14.1	2.4	6.7	4.6	16.5	2.4	7.8	-0.5
W_{CVg}	14.1	3.0	6.7	4.2	16.5	2.4	8.0	0.2
$W_{350.1}$	4.0	16.3	2.4	13.0	10.3	7.6	3.7	2.2
$W_{1.1}$	14.1	3.0	6.7	4.2	16.5	2.4	8.0	0.2
MM_{CVg}	10.0	12.4	4.6	10.9	14.2	6.0	7.1	1.4
$MM_{350.1}$	0.9	16.4	0.2	13.6	8.2	4.1	-0.1	2.3
$MM_{1.1}$	10.7	11.9	5.1	9.7	12.9	6.5	5.5	1.8
$Cunn_{CVg}$	11.8	10.6	5.7	9.3	15.8	4.3	8.0	0.6
$Cunn_{350.1}$	0.6	16.4	-0.4	13.4	4.1	8.2	2.3	2.3
$Cunn_{1.1}$	12.8	9.1	6.1	7.9	14.7	5.8	7.8	1.0
$Tallis_{CVg}$	11.3	11.4	5.9	8.6	12.8	6.9	5.0	2.0
$Tallis_{350.1}$	11.3	11.4	5.9	8.6	12.8	6.9	5.0	2.0
$Tallis_{1.1}$	11.3	11.4	5.9	8.6	12.8	6.9	5.0	2.0
PB_{1-DPG}	11.3	11.4	5.9	8.6	12.3	6.9	5.0	2.0
Sub	10.1	12.7	5.2	10.1	14.7	5.8	6.7	1.6

[†]SH_{CVg}, $SH_{350.1}$, $SH_{1.1}$: indices based on Smith (1936) and Hazel (1943); W_{CVg}, $W_{350.1}$, $W_{1.1}$: indices of Williams (1962); MM_{CVg}, $MM_{350.1}$, $MM_{1.1}$: indices based on the sum of "ranks" of Mulamba and Mock (1978); $Cunn_{CVg}$, $Cunn_{350.1}$, $Cunn_{1.1}$: restricted index of Cunningham et al. (1970); $Tallis_{CVg}$, $Tallis_{350.1}$, $Tallis_{1.1}$: restricted selection index of Tallis (1962); PB_{1-DPG}: index proposed by Pesek and Baker (1969) based on a genetic standard deviation as desired genetic gain; Sub: multiplicative index of Subandi et al. (1973); the following economic weights were tested: CVg 350 for PE; 1 for GY; and 1 for both (GY and PE).

Table 4. Statistical value C_i calculated to compare the genetic gains estimated by selection indices in two popcorn populations

Indices	C_i^+		Mean
	UEM-Co1	UEM-Co2	
	GY-PE	GY-PE	
SH_{CVg}	12.33	13.78	13.05
$SH_{350.1}$	7.57	7.99	7.78
$SH_{1.1}$	12.33	13.78	13.05
W_{CVg}	9.36	12.79	11.07
$W_{350.1}$	6.77	7.49	7.13
$W_{1.1}$	9.36	12.79	11.07
MM_{CVg}	2.40	2.90	2.65
$MM_{350.1}$	17.28	16.79	17.03
$MM_{1.1}$	2.73	3.34	3.03
$Cunn_{CVg}$	5.48	4.25	4.86
$Cunn_{350.1}$	13.57	17.01	15.29
$Cunn_{1.1}$	3.72	4.48	4.10
$Tallis_{CVg}$	2.53	3.68	3.10
$Tallis_{350.1}$	2.53	3.68	3.10
$Tallis_{1.1}$	2.53	3.68	3.10
PB_{1-DPG}	2.66	3.80	3.23
Sub	2.16	2.78	2.47

+ Low C_i values indicate that the selection indices estimate high and consistent genetic gains for traits GY and PE.

In an analysis of the results of all selection cycles (Table 1), different conditions of genetic variability and heritability were noted, which allowed an estimation of genetic gains by selection indices and thereafter the application of the new statistical value C_i. The C_i values for the different indices are listed in Table 4. For both composites, the lowest C_i values were observed for the methodology proposed by Subandi (*UEM-Co1* 2.16 and *UEM-Co2* 2.78). The second highest index was that of Mulamba and Mock, when using CVg as economic weight, with C_i values of 2.40 and 2.90 for *UEM-Co1* and *UEM-Co2*, respectively. These C_i values, derived from the mean of composites and cycles, suggest good results in the estimation of high and balanced genetic gains for both PE and GY by the indices of Subandi and of Mulamba and Mock (Table 4), suggesting that these would improve the selection responses, in magnitude as well as in the balance between the two traits, and are therefore recommended for popcorn breeding. Among the multiplicative and restrictive indices, the indices of Subandi (multiplicative) and Tallis (restricted) achieved good results and allowed favorable genetic gain estimates for GY and PE in popcorn.

The use of the statistical value C_i eliminates the subjectivity of choice of the best indices, because this statistical value, genetic gain estimated by an index for many traits, are easily summarized in a single C_i value, based on the improvement possibilities of the study population. This study is a non- exhaustive comparison of efficiency of selection indices in popcorn. In this context, other complementary studies are suggested, to address other crops and traits, testing the C_i value as a tool in new combinations of economic weights and selection indices that allow greater genetic progress.

CONCLUSIONS

The selection indices were effective to estimate the genetic gains of half-sib families, indicating that their application can lead to high and balanced genetic gains.

According to the statistical value C_i, the index of Subandi, followed by that of Mulamba and Mock, estimated high and better balanced genetic gains for grain yield and popping expansion.

REFERENCES

Amaral Júnior AT, Freitas Júnior SP, Rangel RM, Pena GF, Ribeiro RM, Morais RC and Schuelter AR (2010) Improvement of a popcorn population using selection indexes from a fourth cycle of recurrent selection program carried out in two different environments. **Genetics and Molecular Research 9**: 340-347.

Amaral Júnior AT, Goncalves LSA, Freitas Júnior SP, Candido LS, Vitorazzi C, Pena GF, Ribeiro RM, Silva TRC, Pereira MG, Scapim CA, Viana AP and

Carvalho GF (2013) UENF 14: a new popcorn cultivar. **Crop Breeding and Applied Biotechnology 13**: 218-218.

Broccoli AM and Burak R (2004) Effect of genotype x environment interactions in popcorn maize yield and grain quality. **Spanish Journal of Agricultural Research 2**: 85-91.

Carpentieri-Pípolo V, Takahashi HW, Endo RM, Petek MR and Seifert AL (2002) Correlações entre caracteres quantitativos em milho-pipoca. **Horticultura Brasileira 20**: 551-554.

Coimbra RR, Miranda GV, Viana JMS, Cruz CD, Murakami DM, Souza LV and Fidelis RR (2002) Estimation of genetic parameters and prediction of gains for DFT1-Ribeirão popcorn population. **Crop Breeding and Applied Biotechnology 2**: 33-38.

Cruz CD (2013) GENES - A software package for analysis in experimental statistics and quantitative genetics. **Acta Scientiarum 35**: 271-276.

Cunningham EP, Moen RA and Gjedrem T (1970) Restriction of selection indexes. **Biometrics 26**: 67-74.

Daros M, Amaral Júnior AT, Pereira MG, Santos FS, Gabriel APC, Scapim CA, Freitas Júnior SP and Silvério L (2004a) Recurrent selection in inbred popcorn families. **Scientia Agricola 61**: 609-614.

Daros M, Amaral Júnior AT, Pereira MG, Santos FS, Scapim CA, Freitas Júnior SP, Daher RF and Ávila MR (2004b) Correlações entre caracteres agronômicos em dois ciclos de seleção recorrente em milho-pipoca. **Ciência Rural 34**: 1389-1394.

Faria VR, Viana JMS, Sobreira FM and Silva AC (2008) Seleção recorrente recíproca na obtenção de híbridos interpopulacionais de milho-pipoca. **Pesquisa Agropecuária Brasileira 43**: 1749-1755.

Freitas ILJ, Amaral Júnior AT, Viana AP, Pena GF, Cabral PDC, Vitorazzi C and Silva TRC (2013) Ganho genético avaliado com índices de seleção via REML/Blup no milho-pipoca UENF 14. **Pesquisa Agropecuária Brasileira 48**: 1464-1471.

Freitas ILJ, Amaral Júnior AT, Freitas Júnior SP, Cabral PDS, Ribeiro RM and Gonçalves LSA (2014) Genetic gains in the UENF-14 popcorn population with recurrent selection. **Genetics and Molecular Research 13**: 518-527.

Freitas Júnior SP, Amaral Júnior AT, Rangel RM and Viana AP (2009) Predição de ganhos genéticos na população de milho-pipoca UNB-2U sob seleção recorrente utilizando-se diferentes índices de seleção. **Revista Semina 30**: 803-814.

Granate MJ, Cruz CD and Pacheco CAP (2002) Predição de ganho genético com diferentes índices de seleção no milho-pipoca CMS-43. **Pesquisa Agropecuária Brasileira 37**: 101-108.

Hazel LN (1943) The genetic basis for constructing selection indexes. **Genetics 28**: 476-490.

Mulamba NN and Mock JJ (1978) Improvement of yield potential of the Eto Blanco maize (Zea mays L.) population by breeding for plant traits. **Egypt Journal of Genetics and Cytology 7**: 40-51.

Pacheco CAP, Gama EEG, Guimarães PEO, Santos MX and Ferreira AS (1998) Estimativas de parâmetros genéticos nas populações CMS-42 e CMS-43 de milho pipoca. **Pesquisa Agropecuária Brasileira 33**: 1995-2001.

Paterniani E (1967) Selection among and within half-sib families in a Brazilian population of maize (Zea mays L.). **Crop Science 7**: 212-216.

Pesek J and Baker RJ (1969) Desired improvement in relation to selected indices. **Canadian Journal of Plant Science 49**: 803-804.

Rangel RM, Amaral Júnior AT, Gonçalves LSA, Freitas Júnior SP and Candido LS (2011) Análise biométrica de ganhos por seleção em população de milho-pipoca de quinto ciclo de seleção recorrente. **Revista Ciência Agronômica 42**: 473-481.

Ribeiro RM, Amaral Júnior AT, Gonçalves LSA, Candido LS, Silva TR and Pena GF (2012) Genetic progress in the UNB-2U population of popcorn under recurrent selection in Rio de Janeiro. **Genetics and Molecular Research 11**: 1417-1423.

Ribeiro RM, Amaral Júnior AT, Pena GF, Vivas M, Kurosawa RN and Gonçalves LSA (2016) Effect of recurrent selection on the variability of the UENF-14 popcorn population. **Crop Breeding and Applied Biotechnology 16**: 123-131.

Santos FS, Amaral Júnior AT, Freitas Júnior SP, Rangel RM and Pereira MG (2007) Predição de ganhos genéticos por índices de seleção na população de milho-pipoca UNB-2U sob seleção recorrente. **Bragantia 66**: 389-396.

Santos FS, Amaral Júnior AT, Freitas Júnior SP, Rangel RM, Scapim CA and Mora F (2008) Genetic gain prediction of the third recurrent selection cycle in a popcorn population. **Acta Scientiarum Agronomy 30**: 651-655.

Sawazaki E, Castro JL, Gallo PB, Paterniani MEAGZ, Silva RM and Luder RR (2003) Potencial de híbridos temperados de milho pipoca em cruzamentos com o testador semitropical IAC 12. **Revista Brasileira de Milho e Sorgo 2**: 61-70.

Silva ES, Silva PSL, Nunes GHS and Silva KMB (2001) Estimação de parâmetros genéticos no composto de milho ESAM-1. **Caatinga 12**: 43-52.

Smith HF (1936) A discriminant function for plant selection. **Annals of Eugenics 7**: 240-250.

Subandi W, Compton A and Emeig LT (1973) Comparison of the efficiencies of selection indices for three traits in two variety crosses of corn. **Crop Science 13**: 184-186.

Tallis GM (1962) A selection index for optimum genotype. **Biometrics 22**: 120-122.

Vilarinho AA, Viana JMS, Santos JF and Câmara TMM (2003) Eficiência da seleção de progênies S_1 e S_2 de milho-pipoca, visando à produção de linhagens. **Bragantia 62**: 9-17.

Williams JS (1962) The evaluation of a selection index. **Biometrics 18**: 375-393.

Zinsly JR and Machado JA (1987) Milho-pipoca. In Paterniani E and Viégas GP (eds) **Melhoramento e produção do milho**. Editora Fundação Cargill, Campinas, p. 413-421.

9

Passiflora cristalina and *Passiflora miniata*: meiotic characterization of two wild species for use in breeding

Telma Nair Santana Pereira[1*], Ingrid Gaspar da Costa Geronimo[1], Ana Aparecida Bandini Rossi[2] and Messias Gonzaga Pereira[1]

Abstract: *Passiflora cristalina and Passiflora miniata are two new wild species found in the southern Amazon region. This study aimed to analyze the meiotic behavior of the two species, by meiotic analysis, meiotic index and pollen grain viability, using routine methodologies of the laboratory. By the meiotic analysis, the two species were diploid with 18 chromosomes, and nine bivalents were observed in diakinesis. Laggard chromosomes and fiber spindle problems were the abnormalities observed in both species. The recombination indices were 21.6 and 18.8 for P. cristalina and P. miniata, respectively. The most common abnormal post-meiotic products were triads. The meiotic index and the pollen grain viability for P. cristalina were 90.6% and 98.9%, respectively, and 91.6% and 82.2% for P. miniata, respectively. Based on the results, both species are fertile. Thus, gene transference to sour passion fruit by interspecific hybridization is possible in breeding programs.*

Key words: *Chromosome, meiotic abnormalities, recombination index, meiotic index.*

*Corresponding author:
E-mail: telmasp2012@gmail.com

[1] Universidade Estadual Norte Fluminense Darcy Ribeiro, Laboratório de Melhoramento Genético Vegetal, Avenida Alberto Lamego, 2000, Horto, 28.013-602, Campos dos Goytacazes, RJ, Brazil
[2] Universidade do Estado de Mato Grosso, Departamento de Ciências Biológicas, Fundação Universidade do Estado de Mato Grosso, Campus Universitário de Alta Floresta, Residencial Flaboyant, 78.580-000, Alta Floresta, MT, Brazil,

INTRODUCTION

The family Passifloraceae is composed of 18 genera, and the genus *Passiflora* is the richest in number of species within the family, comprising between 521 and 537 species (Feuillet and MacDougal 2004, Vanderplank 2007). Species of this genus are distributed in four subgenera: *Astrophea, Deidamioides, Decaloba* and *Passiflora* (Feuillet and MacDougal 2004). New species and mutants belonging to this genus are still being described (MacDougal 2001, Lira Júnior et al. 2014), and about 90% of them are native to the Americas (Lopes 1991). Ferreira (1994) reported the existence of more than 200 Brazilian native species, which places the country in a privileged position in relation to the genetic resources of passion fruit trees.

Two new species were described in 2011 and 2006, *Passiflora cristalina* and *Passiflora miniata*. The former belongs to the Super Section Diasthephana of the subgenus *Passiflora,* and it was named after being found in the Cristalino State Park, in the northeast of the state of Mato Grosso. The species presents red flowers, which are held erect before and during anthesis, becoming pendulous as the ovary develops (Figure 1A) (Vanderplank and Zappi 2011).

Passiflora miniata was described by Vanderplank (2006) as belonging to the subgenus *Passiflora*, super section Coccinea. It is originated and distributed in the Amazon region of Peru, Brazil, Colombia, and the Guianas (Lim 2012). The species presents red flower, and according to the author, it has three series of purple corona filaments, with small green and cream fruits (Figure 1B).

Wild species have attracted the attention of breeders due to their genetic potential, since they have genes that confer resistance to diseases or pests, besides agronomic traits of interest. Despite this, hybridization with cultivated species is not always possible, leading to underutilization (Hajjar and Hodgkin 2007). According to Nass et al. (2001), in spite of the potential as a gene repository, wild species related to the cultivars are used as last resource due to the lack of basic and useful information for breeding. McCouch et al. (2013) stated that wild species and landraces of different plant species are conserved in more than 1,700 germplasm banks; however, little genetic information on this type of germplasm is found in the literature, which means that breeders make either none, or very little use of this type of germplasm in breeding programs.

Plant breeding seeks to combine desirable alleles that are found in different genotypes in a single elite variety. The success of a breeding program depends on the ability of the breeder to transfer these desired alleles to a hybrid by constructing desired combinations of alleles in the chromosomes, and by projecting the right combination of chromosomes (Wijnker and Jong 2008). This can only be achieved by the occurrence of crossovers during meiosis. Thus, meiotic recombination plays a key role in the breeding success, since crossover occurs only during meiosis (Wijnker and Jong 2008). Therefore, meiosis is essential for breeding, since it becomes one of the sources of genetic variability, besides explaining reproductive phenomena and mechanisms of heredity (Caetano et al. 2003). In addition, basic studies of a species, such as chromosome number, ploidy level, and genome size, among others, are important for the most different fields of knowledge (Singh 2002).

The objectives of this work were to evaluate the meiotic behavior of two wild species, *Passiflora cristalina* and *Passiflora miniata*, with emphasis on meiotic analysis and on possible meiotic irregularities; and to estimate the recombination index, meiotic index and pollen viability.

Figure 1. Flowers of the species *Passiflora cristalina* (A) and *Passiflora miniata* (B).

MATERIAL AND METHODS

Flower buds were collected in five accessions of the two wild species, *Passiflora cristalina* and *Passiflora miniata*, native to the Brazilian Amazon, and naturally found in the municipality of Alta Floresta, extreme north of the state of Mato Grosso (lat 9º 53' 02" S, long 56º 14' 38" W, alt 320 m asl). The climate of the region is rainy tropical with temperatures ranging from 20 ºC to 38 ºC, with average annual temperature of 26 ºC (Köppen 1948).

For meiosis, flower buds at different stages of development were fixed in a 3:1 alcohol: acetic acid solution, and kept at 4 ºC. After 24h, the fixative solution was replaced with a 70% ethanol solution, and buds were kept in a freezer until the time of slides preparation. For slide preparation, anthers were macerated in 2% acetic carmine and, after removal of the debris and the coverslip, slides were observed using an optical microscope (Olympus BX 60). The different phases of meiosis were analyzed, and irregularities were counted.

Chiasmata were counted in 50 cells in diakinesis to estimate the recombination index (RI). The following expression was used to estimate the RI:

$RI = n(1 + X)$

in which n is the haploid number of chromosomes of the species, and X is the mean number of chiasmata per bivalent (Darlington 1958).

The meiotic index (MI) was estimated based on five slides/buds, and each slide was prepared with four anthers, which were macerated and stained with 2% acetic carmine for the visualization and counting of post-meiotic products. Tetrads were considered as a normal post-meiotic product, and monads, dyads, triads and polyads were considered as abnormal. Based on these data, the meiotic index was calculated according to Love (1951):

$$MI = \frac{Total\ normal\ post-meiotic\ products}{Total\ normal\ and\ abnormal\ post-meiotic\ products} \times 100$$

For the estimate of pollen viability, two anthers were macerated in Alexander's triple solution, composed of Orange G, acid fuchsin, and malachite green. In this way, pollen grains were classified as viable or unviable. The viable pollen grains were red/purple, while the unviable ones were green (Alexander 1969). Five slides were prepared per species, accounting for 200 pollen grains/slide, totaling 1000 pollen grains per species. Data were transformed into viability percentage (%).

RESULTS AND DISCUSSION

Meiotic analysis

Both species, *P. cristalina* (Figures 2A-D, and 2F) and *P. miniata* (Figures 3A-F), presented meiosis within the normality pattern. Nine pairs of bivalents were observed in prophase I cells of both species, which confirms that the species are

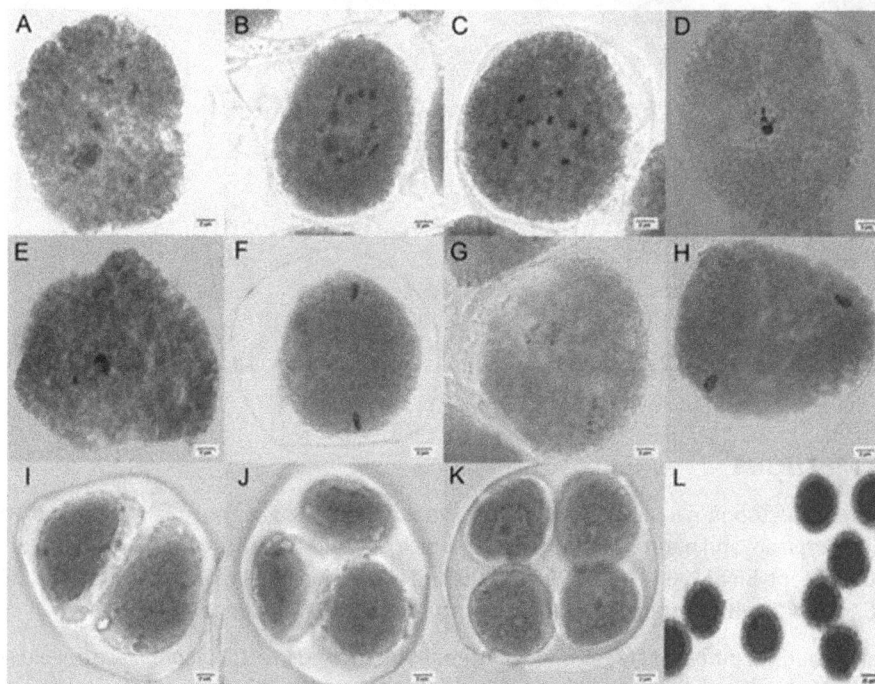

Figure 2. Meiosis of *P. cristalina* (A-C) Cells in diakinesis where nine bivalents are observed; (D) Metaphase I; (E) Metaphase I with a pair of laggard chromosomes; (F) Metaphase II; (G-H) cells in anaphase II (G) and anaphase II (H) with problems in the spindle fibers: (I-K) Post-meiotic products, dyad, triad and tetrad, respectively; (L) Viable pollen grains stained with Alexander's triple solution. Bar = 2 μm (A-K) and 20 μm (L).

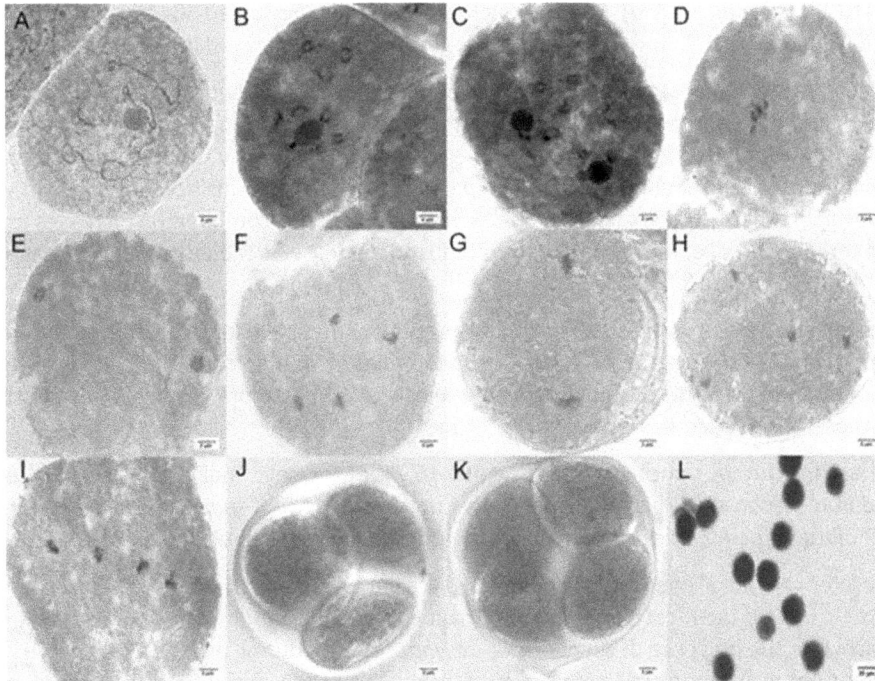

Figure 3. Meiosis of *P. miniata* (A) Pachytene where nine bivalents are observed; (B-C) diakinesis where bivalents and more than one nucleolus per cell are observed; (D) Metaphase I with a pair of laggard chromosomes, outside the equatorial plate; (E) Telophase I; (F-I) Cells in anaphase II (F), metaphase II (G), and in telophase II (H-I) with spindle fiber problems; (J-K) Post-meiotic products, triad and tetrad, respectively; (L) Viable pollen grains stained with Alexander's triple solution. Bar = 2 µm (A-K) and 20 µm (L).

diploid and have 2n=18 chromosomes. According to Hansen et al. (2006), the species of the subgenus *Passiflora* have the most representative basic chromosome number (x=9 chromosomes). Although passifloras can be classified into four karyological groups, represented by n=6, n=9, n=10 and n=12 (Melo and Guerra 2003), most species present 2n=2x=18 chromosomes and are diploid. However, polyploids and aneuploidshave been reported in the family (Melo et al. 2001).

More than one pair of chromosomes associated with the nucleolus were observed in both species, indicating the presence of at least two pairs of chromosomes containing the nucleolar organizing regions (NORs) (Figures 2A and 3A-C). Cells with more than one nucleolus were also observed (Figures 2B and 3C). Barbosa and Vieira (1997) studied sour passion fruit (*Passiflora edulis*), and observed that two pairs of chromosomes were associated with the nucleolus in 33% of the cells. According to the authors, the studied species has two NORs located in the secondary constrictions of chromosomes 8 and 9. NORs are repeated DNA sequences that encode ribosomal RNA 18S, 5.8S and 26S (18S-26S), and are located in secondary constrictions. NORs are usually associated with secondary constrictions; however, not all secondary constrictions are NORs (Battistin et al. 1999). During telophase and interphase, NORs are responsible for the nucleoli formation (Besendorfer et al. 2002).

Despite their importance, studies on chiasmata frequency are rare in Passifloras (Souza et al. 2008). The number of chiasmata is an important trait for meiotic stability, since they prevent the early migration of the chromosomes, and ensure that the bivalents are oriented to opposite poles (Wijnker and Jong 2008). The chiasmatas are originated from the crossover, and are the regions that keep the homologous chromosomes together during Prophase I (Mézard 2006). In the present work, rod and ring bivalents were observed. Rod bivalents (Figure 2B) present only one chiasm in one of the chromosome arms, while ring bivalents (Figure 3C) present chiasms in the two chromosome arms (Senda et al. 2005).

For *P. cristalina,* the estimated recombination index (RI) was of 18.8, and the chiasma mean per bivalent was of 1.08 chiasmata; and for *P. miniata*, the RI was 21.6, with 1.4 chiasmata per bivalent. Forni-Martins (1996) also observed low chiasma frequency in species of Erythrina (<1.5 per bivalent). The present results corroborate those reported by Souza and Pereira (2011), who analyzed 14 wild *Passiflora* species of the n = 9 chromosome group, and obtained RI ranging from 17.6 (*P. alata*) to 23.9 (*P. malacolhyla*). Considering that the authors evaluated 50 cells in diakinesis, chiasma mean per bivalent ranged from 0.96 to 1.7, and these values are close to those reported in this study.

The maximum recombination that can be obtained in species is determined by two factors: the number of chromosomes of the species, and the number and positions of crossovers in the homologous pair. Crossovers preferentially occur in certain areas, known as recombination hotspot. Conversely, areas with very low occurrence of crossovers are known as recombination cold spots (Wijnker and Jong 2008). In passifloras, these regions (hotspot and cold spot) have not yet been defined; however, in other species, such as maize, wheat and barley, recombination increases as it distances from the centromere. In tomato and rice, the recombination tends to decrease as the crossover approaches the telomeres (Mézard 2006).

Meiosis of both species was considered as normal. However, some irregularities were recorded, such as lagging chromosomes in metaphase I (Figures 2E and 3D), and anomalies in spindle fibers, which were the most frequent (Figures 2G-H-I). Souza and Pereira (2011) analyzed the meiosis of 14 passiflora species, and reported more meiotic irregularities than those observed in this study.

In *P. cristalina*, laggard chromosomes were observed in 16.7% of the cells analyzed in metaphase I; however, they were not observed in metaphase II. This abnormality was not observed in *P. miniata*. On the other hand, they have been observed in *P. alata, P. foetida, P. cincinnata* and *P. amethystina* in both in meiosis I and meiosis II (Kiihl et al. 2010).

Laggard chromosomes are not aligned on the equatorial plate during metaphase I or anaphase I. According to Pagliarini (2000), laggard chromosome is the most frequent abnormality during nuclear division, and a possible cause is the early segregation of chiasma, or the presence of *asynaptic* or *desynaptic* genes during prophase I. Laggard chromosomes are usually lost during nuclear division, resulting in micronuclei (Risso-Pascotto et al. 2003). These micronuclei are usually observed in post-meiotic products, and may lead to the formation of unbalanced gametes, egg abortion, and the consequent formation of non-viable gametes (Battistin et al. 2006).

The most frequent abnormality observed for both species was the lack of orientation of the spindle fibers, estimated in 17.8% for *P. cristalina,* and in 35.9% for *P. miniata*. Cells with this anomaly presented the chromosomal groups aligned in the T-shaped equatorial plate (Figure 2H, Figure 3G-H), which characterizes the transversal spindle (Souza et al. 2003, Shamina 2005). Some genes affect spindle formation during meiosis I and meiosis II (Shamina et al. 1999, Shamina 2005) and may cause problems in karyokinesis and cytokinesis, consequently generating abnormal post-meiotic products, such as dyads, triads and polyads. In passifloras, convergent spindle is directly related to the formation of triads, which are formed due to asynchrony during meiosis II (Souza and Pereira 2011).

The abnormalities observed in this study did not hinder the formation of post-meiotic products of the species (Figures 2I-K and 3J-K), since *P. cristalina* and *P. miniata* presented meiotic index of 90.6 and 91.6%, respectively. Plants with meiotic index between 90 and 100% are considered as cytologically stable (Love 1951), presenting no problems in generating fertile progeny. Triads were observed in 9.3% of the *P. cristalina* cells, and in 8.4% of the *P. miniata* cells, probably due to the anomalies of the spindle fibers described above. Only one dyad was observed in *P. cristalina* (Figura 2I), and none was observed in *P. miniata*. The determination of the meiotic index assists in the verification of the meiotic regularity. Thus, the higher the value of the meiotic index, the more regular is the meiotic behavior of the species (Love 1951).

Plants fertility depends on the meiotic regularity, which may occur during the formation of pollen grains. The normal course of meiosis guarantees the viability of the gamete. However, this event is controlled by several genes, and thus mutations may occur (Pagliarini 2000). The pollen viability estimated in this study was high, since it varied, on average, from 98.9% to 82.2% in *P. cristalina* and *P. miniata*, respectively. Figures 2L and 3L show the viable pollen grains. These values are in agreement with those reported by Souza et al. (2004), who observed pollen viability greater than 90% for most of the species studied, except for *P. pentagona*, which presented 78.22% pollen viability.

Based on the results obtained in this work, the two new *P. miniata* and *P. cristalina*, both belonging to the genus and sub-genus Passiflora, present n=x=9 chromosomes, and therefore are considered as two new diploid species. Meiosis of both species was generally normal, despite the irregularities observed. Nevertheless, for being wild, they are probably still under domestication process. Both species had good meiotic index and high pollen viability, and thus they can be used in future sour passion fruit (*P. edulis* Sims) breeding programs by means of interspecific hybridization.

ACKNOWLEDGEMENTS

The authors thank FAPERJ and CNPq for the financial support.

REFERENCES

Alexander MP (1969) Differential staining of aborted and nonaborted pollen. **Stain Technology 44**: 117-122.

Barbosa LV and Vieira MLC (1997) Meiotic behavior of passion fruit somatic hybrids, *Passiflora edulis* f. *flavicarpa* Degener + *P. amethystine*Mikan. **Euphytica 98**: 121-127.

Battistin A, Biondo E and Coelho LGM (1999) Chromosomal characterization of three native and one cultivated species of *Lathyrus* L. in southern Brazil. **Genetics and Molecular Biology 22**: 557-563.

Battistin A, Conterato IF, Pereira GM, Pereira BL and Silva MF (2006) Biologia floral, microsporogênese e número cromossômico em cinco espécies de plantas utilizadas na medicina popular no Rio Grande do Sul. **Revista Brasileira de Plantas Medicinais 8**: 56-62.

Besendorfer V, Samardzija M, Zoldos V, Solic ME and Papes D (2002) Chromosomal organization of ribosomal genes and NOR-associated heterochromatin, and NOR activity in some populations of *Allium commutatum* Guss. (Alliaceae). **Botanical Journal of the Linnean Society 139**: 99-108.

Caetano CM, d'Eeckenbrujgge GC, Olaya CA, Jimenez DR and Veja J (2003) Spindle absence in *Vasconcelleacundina marcensis* (Caricaceae). **The Nucleus 46**: 86-89.

Darlington CD (1958) **Evolution of genetic systems**. Oliver and Boyd, Endiburg, 256p.

Ferreira FR (1994) Germoplasma de *Passiflora* no Brasil. In São José AR (ed) **Maracujá: produção e mercado**. UESB, Vitória da Conquista, p. 24 -26.

Feuillet C and MacDougal JM (2004) A new infrageneric classification of *Passiflora*. **Passiflora13**: 34-38.

Forni-Martins ER (1996) Recombination indices in species of *Erythrina* L. (Leguminosaea, Papilionoideae). **Botanical Journal of Linnean Society 122**: 163-170.

Hajjar R and Hodgkin T (2007) The use of wild relative in crop improvement: a survey of developments over the last 20 years. **Euphytica 156**: 1-13.

Hansen AK, Lawrence G, Simpson BB, Downie SR, Stephen R, Cervi AC and Jansen RK (2006) Phylogenetic relationships and chromosome number evolution in *Passiflora*. **Systematic Botany 31**: 138-150.

Kiihl PRP, Barragan MF, Santos SP, Godoy SM, Alonso-Pereira AR, Stenzl NMC and Risso-Pascotto C (2010) Abnormal behavior of spindle during microsporogenesis of *Passiflora* (Passifloraceae). Arquivos de Ciências da Saúde da UNIPAR 14: 237-243.

Köppen W (1948) Climatologia: conunestudio de los climas de latierra. Fondo de Cultura Econômica, México, 479p.

Lim TK (2012) *Passiflora miniata*. **Edible medicinal and non-medicinal plants**. Springer, London, 1035p.

Lira Júnior JS, Flores PS and Bruckner CH (2014) UFV-M7: mutant yellow passionfruit genotype with photoperiod insensitivity for flowering. **Crop Breeding and Applied Biotechnology 14**: 128-131.

Lopes SC (1991) Citogenética do maracujá. In São José AR (ed) **A cultura do maracujá no Brasil**. Funepe, Jaboticabal, p. 201-209.

Love RM (1951) Varietal differences in meiotic chromosomes behavior of Brazilian wheats. **Agronomy Journal 43**: 72-76.

MacDougal JM (2001) Two new species of Passionflower (*Passiflora*, Passifloraceae) from southwestern Mexico. **Novon 11**: 69-75.

McCouch S (2013) Feeding the future. **Nature 499**: 23-24.

Melo NF and Guerra M (2003) Variability of the 5S and 45S rDNA sites in Passifloraceae L. species with distinct base chromosome numbers. **Annals of Botanical 92**: 309-316.

Melo NF, Cervi AC and Guerra M (2001) Karyology and cytotaxonomy of the genus *Passiflora* L. (Passifloraceae). **Plant Systematics and Evolution 226**: 69-84.

Mézard C (2006) Meiotic recombination hotspots in plants. **Biochemical Society 34**: 531-534.

Nass LL, Valois ACC, Melo IS and Valadares-Inglis MC (2001) Armazenamento de sementes de maracujá amarelo. **Revista Brasileira de Sementes 13**: 77-80.

Pagliarini MS (2000) Meiotic behavior of economically important plant species: the relationship between fertility and male sterility. **Genetics and Molecular Biology 23**: 997-1002.

Risso-Pascotto C, Pagliarini MS and Valle CB (2003) Mutation in the spindle checkpoint arresting meiosis II in *Brachiaria ruziziensis*. **Genome 46**: 724-728.

Senda T, Hiraoka Y and Tominagaa T (2005) Cytological affinities and infertilies between *Loliumtemulentum* and *L. persicum* (Poaceae) accessions. **Hereditas 142**: 45-50.

Shamina NV (2005) A catalogue of abnormalities in the division spindles of higher plants. **Cell Biology International 29**: 384 - 391.

Shamina NV, Dorogava N, Goncharov N, Orlova A and Trunova S (1999) Abnormalities of spindle and cytokine behavior leading to the formation of meiotic restitution nuclei in intergeneric cereal hybrids. **Cell Biology International 23**: 863-870.

Singh RJ (2002) **Plant cytogenetics**. CRC Press, Florida, 391p.

Souza MM and Pereira TNS (2011) Meiotic behavior in wild and domesticated species of *Passiflora*. **Revista Brasileira de Botânica 34**: 63-72.

Souza MM, Pereira TNS and Vieira MLC (2008) Cytogenetic studies in some species of *Passiflora*L. *(Passifloraceae)*: a review emphasizing Brazilian species. **Brazilian Archives of Biology and Technology 51**: 247-258.

Souza MM, Pereira TNS, Viana AP and Silva LC (2004) Pollen viability and fertility in wild and cultivated Passiflora species *(Passifloraceae)*. **Beiträgezur Biologie der Pflanzen 73**: 359-376.

Souza MM, Pereira TNS, Viana AP, Pereira MG, Bernacci LC, Sudré CP and Silva LC (2003) Meiotic irregularities and pollen viability in *Passiflora edmundoi* Sacco (Passifloraceae). **Caryologia 56**: 161-169.

Vanderplank RJR (2006) *Passiflora miniata*. Passifloraceae. **Curtis's Botanical Magazine 23**: 223-230.

Vanderplank RJR (2007) There are... lies, damned lies and statistics. A statistical look at the genus *Passiflora*. **Passiflora 17**: 14-15.

Vanderplank RJR and Zappi D (2011) *Passiflora cristalina,* a striking new species of *Passiflora* (Passifloraceae) from Mato Grosso, Brazil. **Curtis's Botanical Magazine 66**: 149-153.

Wijnker E and Jong H (2008) Managing meiotic recombination in plant breeding. **Trends in Plant Science 13**: 640-646.

10

MS INTA 416: A new Argentinean wheat cultivar carrying Fhb1 and Lr47 resistance genes

Carlos T. Bainotti[1], Silvina Lewis[2], Pablo Campos[3], Enrique Alberione[1], Nicolás Salines[1], Dionisio Gomez[1], Jorge Fraschina[1], José Salines[1], María B. Formica[1], Guillermo Donaire[1], Leonardo S. Vanzetti[1-4], Lucio Lombardo[1], María M. Nisi[1], Martha B. Cuniberti[1], Leticia Mir[1], María B. Conde[1] and Marcelo Helguera[1*]

Abstract: *MS INTA 416 is a hard red winter wheat selected for high yield potential and good bread-making quality, combined with moderate resistance to Fusarium-head-blight and high resistance to leaf-rust, due mainly to presence of resistance genes Fhb1 and Lr47. MS INTA 416 is adapted to main production areas of Central-Argentina.*

Key words: *Backcross, marker-assisted selection, yield, bread-making quality, disease control.*

*Corresponding author:
E-mail: helguera.marcelo@inta.gob.ar

[1] INTA EEA Marcos Juárez, Ruta 12, km 3, (2580) Marcos Juárez, Córdoba, Argentina
[2] Instituto de Recursos Biológicos, INTA, (1686) Hurlingham, Buenos Aires, Argentina
[3] INTA EEA Bordenave, Zona Rural, (8187) Bordenave, Buenos Aires, Argentina
[4] CONICET, Av. Rivadavia 1917 (C1033AAJ) CABA, Argentina ,

INTRODUCTION

The diseases Fusarium Head Blight (FHB), caused by *Fusarium graminearum* and leaf rust, caused by *Puccinia triticina,* are widespread and devastating for bread wheat production in the Southern cone of South America (Germán et al. 2007, Bainotti et al. 2013). Damage caused by FHB includes reductions in yield and seed quality, and toxin contamination by deoxynivalenol, threatening human health. Every year, according to the climatic conditions and the area in which susceptible cultivars are planted, leaf rust causes widespread epidemics if no chemical control is applied. Host resistance is considered an efficient and eco-friendly way to manage both diseases; however, progress in breeding for FHB resistance has been limited by the complex inheritance of the partial resistance currently available in wheat. Resistance to the spread of the disease within a spike (Type II resistance) is considered a stable form of FHB resistance, and one of the worldwide best known and most reliable sources of Type II FHB resistance is *Fhb1* from Sumai 3, a major QTL mapped on chromosome 3BS (Anderson et al. 2001, Buerstmayr et al. 2009), still not widely used in Argentina. In the case of leaf rust, several races are generally present in the *P. triticina* populations in the Southern cone of Latin America, probably due to a high acreage where wheat cultivars susceptible or moderately susceptible to leaf rust were sown (Germán and Kolmer 2014). Host resistance is governed by seedling and adult-plant resistance genes, among which the seedling resistance gene *Lr47* is particularly effective (Vanzetti et al. 2011).

MS INTA 416 is a hard red winter wheat (*Triticum aestivum* L.), developed and released by the INTA EEA Marcos Juárez, in 2016. MS INTA 416, previously designated JN12009, was selected for its high yield potential and good bread-making quality, aside from moderate resistance to Fusarium head blight and high resistance to leaf rust, conferred mainly by the resistance genes *Fhb1* and *Lr47* introgressed by artificial crosses and selected by marker-assisted selection (MAS). MS INTA 416 is adapted to rainfed and irrigated production areas in the sub-humid and humid plains of the provinces Córdoba, Santa Fé, Buenos Aires, and La Pampa, Argentina.

BREEDING METHODS

MS INTA 416 (JN12009) was selected from a population derived from two backcrosses, using the breeding line R4004 as recurrent parent and Sumai 3 as *Fhb1* donor. Breeding line R4004 was obtained from a population derived from six backcrosses using ProINTA Oasis (pedigree OASIS/TORIM-73) as recurrent parent and PI 603918 (pedigree Pavon 76 *8//T7AS-7S#1S-7S#1S/*ph1b*) as donor of the leaf rust resistance gene *Lr47*. ProINTA Oasis is a hard red spring wheat developed by INTA EEA Saenz Peña, released in 1989. Since the agronomic performance of this cultivar was very good and yields were high, it was readily adopted in the main wheat-producing areas of Argentina until 1997, when it became highly susceptible to leaf rust by the breakdown of resistance gene *Lr26* (Antonelli 2003). The development of line PI 603918 including a *Triticum speltoides* interstitial translocation carrying *Lr47* was previously described (Bainotti et al. 2009). *Lr47* is effective against field leaf rust infection in Argentina, according to information obtained from the commercial cultivar BIOINTA2004 released in Argentina in 2009, which carries this gene (Bainotti et al. 2009, Vanzetti et al. 2011, Campos 2013, Campos and Lopez 2015). For the development of R4004 in 1996, in greenhouse facilities of the Instituto de Recursos Biológicos (IRB) INTA, Hurlingham, PI 603918 was crossed with PROINTA Oasis and then backcrossed with the same cultivar for six generations. RFLP marker *Xabc465* (Dubcovsky et al. 1998) was used to select *Lr47* heterozygous plants from BC_1 to BC_3, and PCR markers (Helguera et al. 2000) from BC_4 to BC_6 (Figure 1). Then, at least three *Lr47*-heterozygous plants were self-pollinated producing $BC_6 F_2$ seeds. In 2000, $BC_6 F_2$ seeds were planted at INTA EEA Marcos Juárez for the selection of *Lr47* homozygous plants, using PCR markers as before. About 30 selected $BC_6 F_3$ head rows were planted in June 2001 in single 1-m rows, and evaluated in a non-replicated leaf rust screening nursery.

In June 2002, 18 $BC_4 F_5$ lines were advanced to a non-replicated observation plot trial in Marcos Juárez (plots of six 3-m rows) and in 2003, the same lines were advanced to a Multilocation Trial (MLT) at Pergamino, Corral de Bustos and Marcos Juárez. At this point, R4004 was selected for further backcrossing with Sumai 3 on the basis of grain yield, leaf rust resistance, uniformity, and general agronomic appearance. For the development of MS INTA 416, in 2004, by IRB INTA, R4004 was crossed with cultivar Sumai 3 (Funo/Taiwan Xiaomai), the *Fhb1* donor, kindly provided by Dr Jim Anderson (Dpt. of Agronomy and Plant Genetics, University of Minnesota), and backcrossed with R4004 for three generations. In heterozygous plants, locus *Fhb1* was traced with SSRs *Xgwm533* and *Xgwm493* (Figure 2). Then, at least three *Fhb1*-heterozygous plants were self-pollinated, producing $BC_3 F_2$ seeds. During 2007, the $BC_3 F_3$ plants were planted in Marcos Juárez for selection of homozygous *Fhb1* plants using SSRs *Xgwm533* and *Xgwm493,* as before. About 30 selected BC3 F4 head rows were planted in Marcos Juárez in June 2008 and 2009, in single 1-m rows under 15x2x1.5-m greenhouses, covered by nylon mesh 35, and the agronomic appearance of the plants was evaluated.

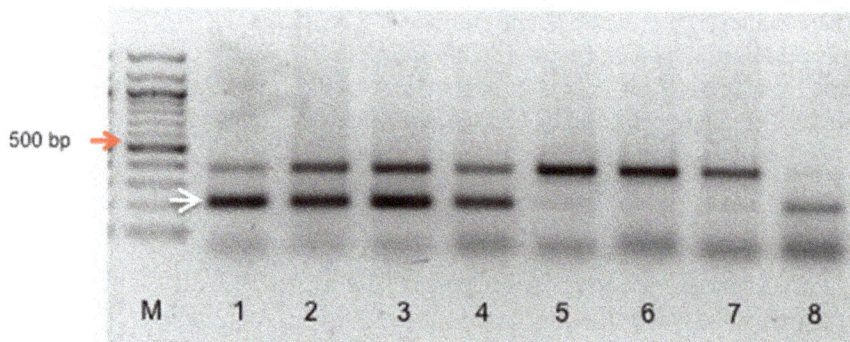

Figure 1. 1% agarose gel electrophoresis showing PCR amplification with *Lr47*-specific primers using genomic DNA from BC_6 plants. The *Lr47* DNA fragment is indicated by a white arrow. The 500bp fragment of the DNA marker (lane M) is indicated with a red arrow. Lanes 1-4: *Lr47*-heterozygous plants, 5-6: homozygous negative plants, 7: ProINTA Oasis, 8: Pavon S3 (*Lr47* donor).

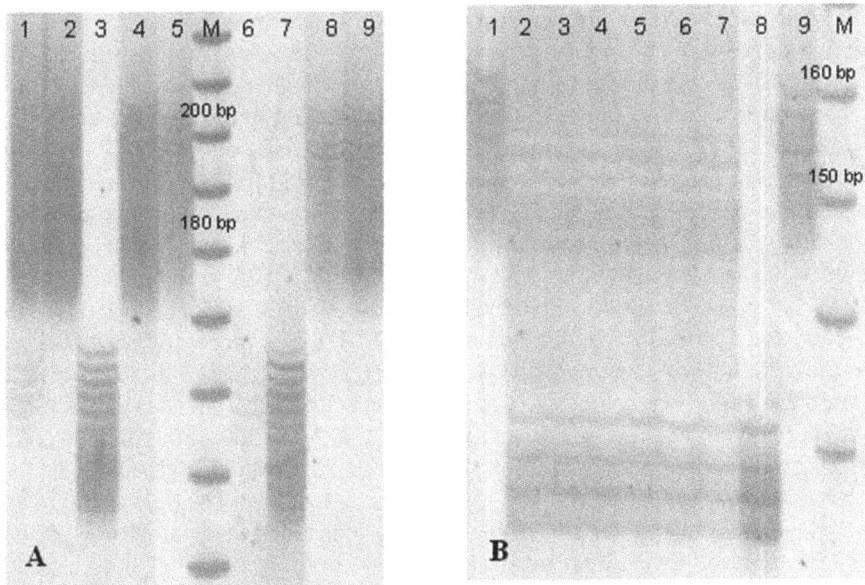

Figure 2. PCR profiles of *wms493* (A) and *wms533* (B) microsatellite markers flanking *Fhb1,* obtained from BC_3F_2 plants. For (A), in lane M, (size standard DNA), 200bp and 180bp fragments are indicated. Lane 1: *Fhb1* heterozygous plants; lanes 2, 4, 8, 9: *Fhb1* homozygous plants; lane 3: R4004; lane 5: Sumai 3 (*Fhb1* donor); lanes 6-7: negative homozygous plants. For (B), in lane M, 160bp and 150bp fragments are indicated. Lane1: *Fhb1* homozygous plant; lanes 2-7: *Fhb1* heterozygous plants; lane 8: R4004; lane 9 Sumai 3 (*Fhb1* donor).

In 2010, 28 BC_3 F_6 lines were planted at the same location in non-replicated observation plot trials (six 5-m rows) and, in 2011, 27 lines were advanced to the Preliminary Yield Trials (PYT) in Marcos Juárez. The PYT were arranged in a 10×9 alpha lattice design with two replications and the above plot size. In 2012, based on its yield potential, JN12009 was advanced to the Regional Yield Trials (RYT) in the provinces Buenos Aires (five locations), Córdoba (two locations), Entre Ríos (one location) and Chaco (one location), for three years (2012, 2013 and 2014), under rainfed conditions. A 6x8 alpha lattice design was used for the RYT at all locations, with three replications per trial (plots with seven 5-m rows). Seeding rates were standardized based on the seed size (300 seeds per m²). The RYT trials at Marcos Juárez were also used to measure: (1) days to heading (days from emergence until 50% of the spikes emerged from the boot), (2) plant height (measured at maturity as the mean stem length from the soil to the tip of the spike, excluding the awns), (3) leaf rust (*Puccinia triticina*) and stem rust (*Puccinia graminis tritici*) severities based on the modified Cobb Scale (Peterson et al. 1948), (4) head blight (*Fusarium graminearum spp*), tan spot (*Drechslera tritici* spp) and bacterial stripe (*Xanthomonas translucens* p.v. *undulosa*) incidence and severity, both on 0–9 scales (Stubbs et al. 1986).

MS INTA 416 in the seedling stage was also evaluated for resistance to stem rust and leaf rust at the Cereal Disease Laboratory in INTA Bordenave, in 2013 and 2014. Local leaf rust races MDP 10-20, MFP 10, MKT 10-20 were inoculated on seedlings as described by Long and Kolmer (1989). Leaf rust severity was evaluated on a 0 to 4 scale, as proposed by Stakman et al. (1962). The infection types identified by the symbols 0, 1, 2, or combinations were considered low infection types, indicating resistance, while 3 and 4 were considered high infection types, indicating susceptibility. Seedlings were also inoculated with the local stem rust races QHFTC, QRFTF, QRFTC, and evaluated according to Stakman et al. (1962). The molecular basis of vernalization and photoperiod response were determined using *Vrn-A1, Vrn-B1, Vrn-D1, Ppd-B1,* and *Ppd-D1* allele-specific PCR markers, as previously described (Yan et al. 2004, Fu et al. 2005, Beales et al. 2007, Díaz et al. 2012). To assess the bread-making quality, grain harvested in RYT trials in 2012, 2013 and 2014, in Marcos Juárez (unreplicated samples) was used. Samples were analyzed by standard AACC methods at the Quality Laboratory of INTA Marcos Juárez, for milling, volume weight, protein content, Chopin Alveograph, and bread baking as previously described (Bainotti et al. 2009). High-molecular-weight subunits of glutenin (HMWGs) in the composition of MS INTA 416 were determined by SDS-PAGE as before (Lawrence and Shepherd 1980).

AGRONOMIC AND BOTANICAL DESCRIPTION

Juvenile plants of MS INTA 416 (JN12009) (growth stages 22 to 29, according to Zadoks et al. (1974)) are semi-erect, with a recurved flag leaf at the beginning of inflorescence emergence (GS 52). At maturity (GS 90), the spikes are semi-short (81-100 mm), white yellow and dense, with inclined position. The glumes are white, long (9mm), have medium width (3.5 mm) and a straight shoulder shape. The kernel is red, vitreous, and ovate; the germ is medium-sized; the brush is medium-sized and has no collar. According to data from the RYT trials between 2012 and 2014 in Marcos Juárez, MS INTA 416 had a mean plant height of 91.6 cm and developed within 114.6 days from emergence to heading (HD). Comparisons of these values with other frequently grown varieties in Argentina are presented in Table 1. As expected, the HD values observed in our study were significantly longer than those in cooler regions of Brazil (Marchioro et al. 2007, Franco et al. 2014, Franco et al. 2015, Marchioro et al. 2016). MS INTA 416 is uniform for plant type, without obvious phenotypic variants, and remained stable over five generations of evaluation (2011-2015). Molecular data obtained from *Vrn-1* and *Ppd-1* adaptation genes defined MS INTA 416 as a winter (carrying the triple combination of "winter" -recessive- alleles within *Vrn-1* homoeologs) - insensitive (carrying at least one "insensitive" allele within *Ppd-1* homoeologs, with low photoperiod response - in this case *Ppd-D1*) wheat. Local cultivars with the same combination of *Vrn-1* /*Ppd-1* adaptation genes (winter insensitive) are Baguette 21, BIOINTA 2004, Buck Ranquel, PROINTA Puntal, SRM Nogal, and Themix, among others (Vanzetti et al. 2013, Gomez et al. 2014). Phenological data (heading time) obtained from a subset of local winter-insensitive wheats sown on six dates between April 29 (mid-autumn) and August 11 (mid-winter), at approximately fortnightly intervals between sowings, indicated two clear groups: BIOINTA 3003, BIOINTA 3005 and Baguette 31, with high vernalization requirements (did not flower when planted after June 26), and, SRM Nogal and MS INTA 416 with milder vernalization requirements, as they did not flower when planted on August 11, the last tested sowing date (Table 2). Variation in the duration of cold requirements to complete vernalization has been described previously for *Vrn-1* copy number variation (Díaz et al. 2012). Related with the photoperiod response, the relatively low frequency of winter-sensitive (1/11) compared to winter-insensitive (10/11) commercial wheats released in Argentina (Vanzetti et al. 2013) suggests a better adaptation of the second than the first group. Typical winter-sensitive wheats are normally grown in environments with a longer growing season than is being explored by most of the wheat cultivars sown in the wheat belt of Argentina (Gomez et al. 2014).

Table 1. Performance[1] of MS INTA 416 and other hard wheat cultivars in Marcos Juárez, Argentina, 2012-2014

Cultivars	HD	PH	LR	FHB	LS
MS INTA 416	114.6	91.6	0	0,5.1	7.2
Baguette Premium 11	119	80	60S	1.1	7.2
Baguette 17	116.3	85	70S	4.2	8.2
Klein Yarará	120	96.6	10MR	7.3	7.3
	TW	GP	LV	W	GW
MS INTA 416	74.73	137	712	264	27.9
Baguette Premium 11	72.27	128	658	264	25.9
Baguette 17	68.83	125	612	257	27.2
Klein Yarará	74.73	134	708	336	28.2

[1] HD = days from emergence to heading; PH = plant height without awns in cm; LR = leaf rust incidence and severity based on modified Cobb scale (Peterson et al. 1948); FHB = Fusarium head blight incidence and severity based on the scale of Stubbs et al. (1986); LS = leaf spot diseases incidence and severity based on the scale of Stubbs et al. (1986); TW = test weight in kg hl-1; GP = grain protein content in g kg-1; LV = loaf volume in cm³; W = general gluten strength by the Chopin Alveograph and GW = 1000-grain weight in grams.

Table 2. Heading date of Argentinian winter wheat cultivars considering eight successive sowing dates (in bold). Data obtained at the experimental station of Marcos Juárez, 2014

Cultivars	Apr 27	May 9	May 26	Jun 10	26 Jun	Jul 16	Jul 28	Aug 11
BIOINTA 3003	Sep 28	Sep 29	Oct 6	Oct 12	Oct 26	dnf	dnf	dnf
BIOINTA 3005	Sep 23	Sep 25	Oct 4	Oct 14	Oct 25	dnf	dnf	dnf
BAGUETTE 31	Oct 4	Nov 7	Oct 10	Oct 18	Nov 3	dnf	dnf	dnf
SRM NOGAL	Sep 13	Sep 15	Sep 24	Oct 3	Oct 13	Oct 24	Nov 5	dnf
MS INTA 416	Sep 9	Sep 10	Sep 20	Oct 1	Oct 13	Oct 26	Oct 31	dnf

dnf: did not flower

Table 3. Grain yield (kg ha^{-1}) of MS INTA 416 and controls at 18 trial locations (RYT), in 2012, 2013 and 2014. In bold, grain yields of MS INTA 416 ranked highest or second highest

Cultivars	Marcos Juárez	Paraná	Pergamino	Balcarce	Tres Arroyos	Bordenave
MS INTA 416 (2012)	**3272**	2742	**3222**	4576	4396	5517
Baguette Premium 11	2774	3079	3663	4911	4901	5719
Baguette 17	2954	2978	2814	5573	5124	5615
Klein Yarará	2636	2967	2693	4895	4593	5258
Mean	2826	2467	2561	4703	4811	5688
LSD $_{0.05}$ (kg ha^{-1})	401	367	357	639	586	541
CV (%)	9.38	10.9	9.09	8.38	9.81	7.83
MS INTA 416 (2013)	**4295**	2981	4916	**8961**	4638	**1985**
Baguette Premium 11	4261	3557	5576	9718	4655	1110
Baguette 17	4294	3461	4446	8528	4680	1622
Klein Yarará	4126	4320	5386	8355	4575	1452
Mean	4432	3254	5024	8397	4736	1972
LSD $_{0.05}$ (kg ha^{-1})	570	330	691	802	688	530
CV (%)	9.37	7.52	8.49	5.9	8.96	14.3
MS INTA 416 (2014)	**5008**	**3800**	**4905**	**4580**	5695	**5659**
Baguette Premium 11	3223	2589	4798	4331	4543	4307
Baguette 17	3308	3166	4863	4832	4466	4370
Klein Yarará	3669	3486	5315	4520	6321	6048
Mean	4091	3219	4757	4721	5610	5199
LSD $_{0.05}$ (kg ha^{-1})	368	369	595	580	522	709
CV (%)	5.50	7.07	8.41	7.00	6.90	9.14

Yield performance

MS INTA 416 was tested at 18 trial locations of the provinces Buenos Aires, Córdoba, Santa Fé and Entre Ríos (RYT trials, in 2012, 2013 and 2014). In this analysis, the yield of MS INTA 416 was ranked highest or second highest in relation to the test control varieties Baguette Premium 11, Baguette 17 and Klein Yarará in 11 trials, including in Marcos Juárez in 2012, 2013, 2014; Pergamino in 2012, 2014; Balcarce in 2013, 2014, Bordenave in 2013, 2014, Paraná in 2014 and Tres Arroyos in 2014 (Table 3). In view of this good performance, cultivar MS INTA 416 was indicated for cultivation in rainfed and irrigated production areas in the sub-humid and humid plains of the provinces Córdoba, Santa Fé, Buenos Aires, and La Pampa, Argentina.

Disease resistance

Evaluations of dominant leaf rust races on seedlings in Argentina showed that MS INTA 416 is highly resistant, with an infection type (IT) of 0 (Table 4). The IT indicating high resistance observed in MS INTA 416 can be explained by the presence of *Lr47,* from R4004, confirmed by a gene-specific PCR marker (Helguera et al. 2000). In line with this hypothesis, in advanced breeding lines carrying *Lr47* J12012, J13013 and JN12015, IT was 0. Our data agree with previous studies describing no virulence of the major *P. triticina* populations in relation to *Lr47* in Argentina, México and the Southern cone (Huerta-Espino et al. 2011, Vanzetti et al. 2011, Campos 2013, Campos and Lopez 2015). Evaluations of seedlings infected with dominant stem rust races in Argentina also showed that MS INTA 416 is highly resistant, with ITs between 0 and 1+ (Table 4), based on an unknown source of genetic resistance. As expected, field observations for disease resistance showed that MS INTA 416 was highly resistant to the prevalent leaf rust races in Marcos Juárez in 2012, 2013 and 2014. A similar situation was observed for stem rust in Marcos Juárez in 2014. Leaf spot diseases (tan spot and Septoria leaf blotch) were detected in Marcos Juárez, in 2012, and MS INTA 416 showed moderate resistance (7/2), similar to Baguette Premium 11 (7/2), Baguette 17 (8/2) and Klein Yarará (7/3). Bacterial stripe severity indices were intermediate to high (9/6), classifying the cultivar as moderately susceptible. In 2012, FHB was observed in Marcos Juárez and MS INTA 416 showed moderate resistance (0.5/1). Fusarium head blight was also evaluated in Marcos Juárez, in 2014, in a nursery with natural infection, where MS INTA proved moderately resistant (2/2), similar to Baguette Premium 11 (2/2) and Klein Yarará (4/2) (data not shown). The good performance of MS INTA 416 in response to FHB can be explained by the

Table 4. Seedling stage infection types[1] of wheat cultivars and MS INTA 416 after inoculation with dominant leaf rust and stem rust races in Argentina in 2014

Cultivars	Leaf rust strains			Stem rust strains		
	MDP 10-20	MFP 10	MKT 10-20	QHFTC	QRFTF	QRFTC
MS INTA 416	0	0	0	0	0;	1+
SY 041	3-	3	0	4	3	4
SY 015	2=	1	3	4	4	4
Buck Bellaco	;1	0;	1	0	;1	1
Klein Serpiente	1+	22+	3	2=	0	2=
ACA 307	0;	3+	4	4	4	4
Cambium	3	1++	4	1+	0	11+

[1] '0'= no visible uredia; ';'= hypersensitive flecks; '1'= small uredia with necrosis; '2'= small to moderate-sized uredia with green islands, surrounded by necrosis or chlorosis; '3'= moderate uredia size, with or without chlorosis; '4'= large uredia without chlorosis; 'N'= necrosis. Symbols '-' and '+' indicate smaller or larger uredinia. The most common infection is listed first. For example, 11+; indicates infection types 1 and 1+.

presence of *Fhb1* confirmed by the flanking markers *Xgwm533* and *Xgwm493*. Marker-assisted selection of *Fhb1* has been used successfully used in the development of germplasm adapted to specific environments (Bainotti et al. 2013, Bernardo et al. 2014, Anderson et al. 2015). MS INTA 416 is the first wheat cultivar released in Argentina carrying the genetic resistance sources *Fhb1* and *Lr47* against FHB and leaf rust, respectively.

Milling and baking quality

Since 1998, bread wheat cultivars from Argentina are being classified based on their commercial and industrial quality performance as Quality Group (QG)1, cultivars with extra strong gluten suitable for blending; QG 2, cultivars adapted to traditional baking (fermentation time longer than 8 hours); and QG 3, cultivars suitable for direct baking methods (fermentation time less than 8 hours) (Cuniberti and Ottamendi 2004). MS INTA 416 belongs to QG 2, with W values of 264, similarly to Baguette Premium 11 (QG 2 check), lower than Klein Yarará (W=336, QG 1 check) but higher than Baguette 17 (W= 257, QG 3 check). The mean test weight of MS INTA 416 and Klein Yarará was 74.73 kg hL^{-1}, higher than that of Baguette Premium 11 and Baguette 17 (72.27 and 68.83 kg hL^{-1}, respectively). The mean grain protein content (137 g kg^{-1}) was higher than of the check cultivars Klein Yarará, Baguette Premium 11 and Baguette 17 (134, 128 and 125 g kg^{-1}, respectively). Additional quality parameters for MS INTA 416 and check cultivars are listed in Table 1.

MS INTA 416 contains the subunits 1, 7+9 and 5+10 in *Glu-A1*, *Glu-B1* and *Glu-D1* loci respectively, where the subunits 1 and 5+10 were associated with good and 7+9 with intermediate bread-making quality (Payne 1987). The possibility of 1BL/1RS rye translocation affecting wheat quality (Dhaliwal et al. 1987) was not tested and can therefore not be excluded.

BASIC SEED PRODUCTION

The Argentinian National Institute of Agricultural Technology - INTA (Rivadavia 1439, Ciudad Autónoma de Buenos Aires, Argentina) has licensed the seed company LDC Semillas SA (Olga Cossettini 240, Ciudad Autónoma de Buenos Aires, Argentina) to multiply and sell protected wheat cultivars developed by the INTA Wheat Breeding Program for 10 years, as of June 14, 2014. Cultivar MS INTA 416 will be released on the market in 2018.

REFERENCES

Anderson JA, Stack RW, Liu S, Waldron BL, Fjeld AD, Coyne C, Moreno-Sevilla B, Fetch J, Mitchell Song QJ, Cregan PB and Frohberg RC (2001) DNA markers for Fusarium head blight resistance QTLs in two wheat populations. **Theoretical and Applied Genetics 102**: 1164-1168.

Anderson JA, Wiersma JJ, Linkert GL, Reynolds S, Kolmer JA, Jin Y, Dill-Macky R and Hareland GA (2015) Registration of "Rollag" spring wheat. **Journal of Plant Registrations 9**: 201-207.

Antonelli EF (2003) **La roya anaranjada (*Puccinia triticina* Erikss) sobre** **la efimera resistencia observada en la última década en cultivares comerciales de trigo de amplia difusión en la Argentina.** Grafos SRL, Necochea, 22p.

Bainotti C, Alberione E, Lewis S, Cativelli M, Nisi M, Lombardo L, Vanzetti L and Helguera M (2013) Genetic resistance to Fusarium head blight in wheat (*Triticum aestivum* L.). Current status in Argentina. In Alconada Magliano TM and Chulze SN (eds) **Fusarium head blight in Latin America.** Springer Netherlands, Dordrecht, p. 231-240.

Bainotti C, Fraschina J, Salines JH, Nisi JE, Dubcovsky J, Lewis SM, Bullrich L, Vanzetti L, Cuniberti M, Campos P, Formica MB, Masiero B, Alberione

E and Helguera M (2009) Registration of "BIOINTA 2004" Wheat. **Journal of Plant Registrations 3**: 165.

Beales J, Turner A, Griffiths S, Snape JW and Laurie DA (2007) A pseudo-response regulator is misexpressed in the photoperiod insensitive Ppd-D1a mutant of wheat (*Triticum aestivum* L.). **Theoretical and Applied Genetics 115**: 721-33.

Bernardo A, Bai G, Yu J, Kolb F, Bockus W, and Dong Y (2014) Registration of Near-Isogenic Winter Wheat Germplasm Contrasting in for Fusarium Head Blight Resistance. **Journal of Plant Registrations 8**: 106-108.

Buerstmayr H, Ban T and Anderson JA (2009) QTL mapping and marker-assisted selection for Fusarium head blight resistance in wheat: a review. **Plant Breeding 128**: 1-26.

Campos P (2013) Physiological specialization of *Puccinia triticina* on wheat in Argentina in 2011. **Borlaug Global Rust Initiative Technical Workshop**. New Dehli, p 107.

Campos P and Lopez J (2015) Physiological specialization of *Puccinia triticina* on wheat in Argentina in 2013. **Borlaug global rust initiative technical workshop.** Available at <http://www.globalrust.org/content/physiological-specialization-puccinia-triticina-wheat-argentina-2013>. Accessed in May 2016.

Cuniberti M and Ottamendi M (2004) Creating class distinction. **World Grain.** Available at <http://www.world–grain.com>. Accessed in May 2016.

Dhaliwal AS, Mares DJ and Marshall DR (1987) Effect of 1B-1R chromosome translocation on milling and quality characteristics of bread wheats. **Cereal Chemistry 64**: 72-76.

Díaz A, Zikhali M, Turner AS, Isaac P and Laurie DA (2012) Copy number variation affecting the photoperiod-B1 and vernalization-A1 genes is associated with altered flowering time in wheat (*Triticum aestivum*). **PloS one 7**: e33234.

Dubcovsky J, Lukaszewski AJ, Echaide M, Antonelli EF and Porter DR (1998) Molecular characterization of two *Triticum speltoides* interstitial translocations carrying leaf rust and greenbug resistance genes. **Crop science 38**: 1655-1660.

Franco FA, Marchioro VS, Schuster I, Dalla Nora T, Alcantara de Lima FJ, Evangelista A, Polo M, and do Prado CM (2014) CD 122 - Bread wheat, suitable for cultivation across southern Brazil. **Crop Breeding and Applied Biotechnology 14**: 136-138.

Franco FA, Marchioro VS, Schuster I, Dalla Nora T, Polo M, Alcântara de Lima FJ, Evangelista A and dos Santos DA (2015) CD 1550: bread wheat cultivar with high gluten strength for the cooler regions of Brazil. **Crop Breeding and Applied Biotechnology 15**: 48-50.

Fu D, Szucs P, Yan L, Helguera M, Skinner JS, Von Zitzewitz J, Hayes PM and Dubcovsky J (2005) Large deletions within the first intron in VRN-1 are associated with spring growth habit in barley and wheat. **Molecular genetics and genomics 273**: 54-65.

Germán S, Barcellos A, Chaves M, Kohli M, Campos P and de Viedma L (2007) The situation of common wheat rusts in the Southern Cone of America and perspectives for control. **Australian Journal of Agricultural Research 58**: 620-630.

Germán SE and Kolmer JA (2014) Leaf rust resistance in selected late maturity, common wheat cultivars from Uruguay. **Euphytica 195**: 57-67.

Gomez D, Vanzetti L, Helguera M, Lombardo L, Fraschina J and Miralles DJ (2014) Effect of Vrn-1, Ppd-1 genes and earliness *per se* on heading time in Argentinean bread wheat cultivars. **Field Crops Research 158**: 73-81.

Helguera M, Khan IA and Dubcovsky J (2000) Development of PCR markers for the wheat leaf rust resistance gene *Lr47*. **Theoretical and Applied Genetics 100**: 1137-1143.

Huerta-Espino J, Singh RP, Germán S, McCallum BD, Park RF, Chen WQ, Bhardwaj SC and Goyeau H (2011) Global status of wheat leaf rust caused by *Puccinia triticina*. **Euphytica 179**: 143-160.

Lawrence GJ and Shepherd KW (1980) Variation in glutenin protein subunits of wheat. **Australian Journal of Biological Sciences 33**: 221-234.

Long DL and Kolmer JA (1989) A north American system of nomenclature for *Puccinia recondita* f. sp. *tritici*. **Phytopathology 79**: 525.

Marchioro VS, Franco AF, Dalla Nora T, Schuster I, Oliveira EF and Alves Sobrinho A (2007) CD 114: Wheat cultivar for colder regions. **Crop Breeding and Applied Biotechnology 7**: 100-102.

Marchioro VS, Franco AF, Schuster I, Montecelli TDN, Polo M, Lima FJA, Evangelista A and Santos DA (2016) CD 1104 - Extra strong wheat with high yield potential. **Crop Breeding and Applied Biotechnology 16**: 246-249.

Peterson RF, Campbell AB and Hannah AE (1948) A diagrammatic scale for estimating rust intensity on leaves and stems of cereals. **Canadian Journal of Research 26**: 496-500.

Stakman EC, Stewart DM and Loegering WQ (1962) **Identification of physiologic races of *Puccinia graminis* var. *tritici*.** USDA-ARS, E617 (Revised 1962), 54p.

Stubbs R, Prescott JM, Saari EE, Dubin HJ (1986) **Cereal disease methodology manual.** CIMMYT, Mexico, 51p.

Vanzetti LS, Campos P, Demichelis M, Lombardo LA, Aurelia PR, Vaschetto LM, Bainotti CT and Helguera M (2011) Identification of leaf rust resistance genes in selected Argentinean bread wheat cultivars by gene postulation and molecular markers. **Electronic Journal of Biotechnology 14:** doi: 10.2225/vol14-issue3-fulltext-14

Vanzetti LS, Yerkovich N, Chialvo E, Lombardo L, Vaschetto L and Helguera M (2013) Genetic structure of Argentinean hexaploid wheat germplasm. **Genetics and Molecular Biology 36**: 391-399.

Yan L, Helguera M, Kato K, Fukuyama S, Sherman J and Dubcovsky J (2004) Allelic variation at the VRN-1 promoter region in polyploid wheat. **Theoretical and Applied Genetics 109**: 1677-1686.

Zadoks JC, Chang TT and Konzak CF (1974) A decimal code for the growth stages of cereals. **Weed Research 14**: 415-421.

11

IAC OL 5 - New high oleic runner peanut cultivar

Ignácio José de Godoy[1*], João Francisco dos Santos[1], Marcos Doniseti Michelotto[2], Andrea Rocha Almeida de Moraes[1], Denizart Bolonhezi[2], Rogério Soares de Freitas[1], Cassia Regina Limonta de Carvalho[1] , Everton Luis Finoto[2] and Antonio Lúcio Melo Martins[2]

Abstract: *IAC OL 5 is a new peanut cultivar recommended to growers of peanut regions of the state of São Paulo as another option for planting during the intervals of sugarcane renewal. Its main traits are its runner growing habit, its moderate resistance to virus and foliar diseases, and the high oleic trait.*

Key words: *Peanut cultivar, Arachis hypogaea L., runner growing habit, high oleic trait.*

*Corresponding author:
E-mail: ijgodoy@iac.sp.gov.br

[1] Instituto Agronômico (IAC)/Apta, Av. Barão de Itapura, 1481, CP 28, 13.020-902, Campinas, SP, Brazil
[2] Apta, Departamento de Descentralização do Desenvolvimento (DDD), Av. Barão de Itapura, 1481, CP 28, 13.020-902, Campinas, SP, Brazil

INTRODUCTION

Peanuts are the third most important annual crop, in terms of planting area, of the state of São Paulo, comprising over one hundred thousand hectares (IEA 2016). Most of this area is located in sugarcane cropping regions, where peanuts are planted as a rotation crop.

Currently, peanut production in these areas is highly technified, and uses cultivars of runner growing habit, whose cycle duration varies from 130 to 150 days. Planting peanuts in plots for sugarcane renewal requires that the cycle of the cultivar be no longer than 130 days.

The peanut cultivars that have the high oleic trait (over 70% of oleic acid in the oil fraction of the seed) are demanded by the peanut industry. Oleic acid content in cultivars that do not have the "high oleic" trait ranges from 40 to 50%. High oleic acid content extends the product's shelf life (Mozingo et al. 2004).

The first high oleic cultivars released in Brazil, IAC 503 and IAC 505, have cycle duration longer than 130 days, from planting to harvesting. This imposes a limitation to their use, due to the requirements in rotation with sugarcane (Godoy et al. 2009). IAC has recently released two other high oleic cultivars, IAC OL3 and IAC OL4 (Godoy et al. 2014), whose cycles do not exceed 130 days, allowing better adjustment to the sugarcane areas. Besides the shorter cycle, these cultivars present high yielding performance at the presence of efficient chemical control of foliar diseases.

IAC OL5 has been released to meet the demand for high oleic runner cultivars, with cycle duration adjusted to the sugarcane rotation system, moderate resistance to virus, and tolerance to foliar diseases.

BREEDING METHOD

IAC OL5 was obtained from a cross between the breeding line IAC 23 A – 65/3 and the accession 2562 of the IAC germplasm collection. IAC 23 A – 65/3 is a component line of the cultivar IAC Caiapó, considered moderately resistant to foliar diseases. The accession 2562 was the source of the "high oleic" trait. The F_2 population was planted in the field and subjected to individual plant selection for yield and pod/seed. From each selected plant, seeds samples ($F_{2:3}$ progenies) were taken to the laboratory and analyzed by gas chromatography for fatty acid content. Samples showing 70-80% oleic acid (meaning homozygous recessives for the "high oleic" trait) were selected.

The selected high oleic progenies were subjected to individual plant selection from the F_3 to F_5 generation for plant yield and pod/seed physical quality. In the F_6 generation, individual plant selection was performed based on the degree of seed maturity upon digging the plants at 100 days after planting (early harvesting). In the following 3 years, the best lines of the selected plants were evaluated for yield and seed maturity under anticipated harvesting (digging the plants before 130 days). One of these lines was indicated as the new cultivar IAC OL5.

AGRONOMIC PERFORMANCE

Cultivar IAC OL5 (tested as line 825) was selected for yield performance out of a group of 15 high oleic lines in 6 experiments carried out in 2 growing seasons (2011/12 and 2012/13), in 3 locations of the state of São Paulo: Ribeirão Preto (lat 21° 2' S, long 47° 87' W, alt 546 m asl), Pindorama (lat 21° 22' S, long 48° 91' W, alt 527 m asl), and Votuporanga (lat 20° 46' S, long 50° 06' W, alt 510 m asl) (Table 1). Plots were harvested between 125 and 130 days after planting, and yield was compared with 3 control cutivars under these conditions, in which the cycle should not exceed 130 days. The new cultivar presented mean of 5.416 kg ha⁻¹ of unshelled peanuts, and outyielded the control cultivars IAC 503 and IAC 505, whose cycles are longer than 130 days, and the not high oleic IAC 886, whose cycle is of approximately 130 days.

The percentage of mature kernels in the new cultivar was evaluated in the 6 experiments, and the data was compared with the 3 control cultivars (Table 2). On average, cv. IAC OL5 showed maturity close to 80%, indicating that its cycle, from planting to harvesting, is adjusted to the sugarcane rotation, since this situation requires cycles no longer than

Table 1. Yield of peanut lines and cultivars in 6 experiments and 3 locations in the state of São Paulo

Genotype	Experiments[1]						Mean
	Pindorama	Ribeirão Preto	Votuporanga	Pindorama	Ribeirão Preto	Votuporanga	
	2011/12			2012/13			
L. 802	4,808	3,508 a	5,328 a	5,534 a	4,661 b	6,351 a	5,032
L. 807	4,890	3,146 b	5,683 a	4,383 b	4,512 b	5,008 b	4,604
L. 814	4,872	4,097 a	5,039 a	5,446 a	4,946 b	4,784 b	4,864
L. 815	4,864	3,802 a	5,230 a	5,494 a	5,240 a	4,907 b	4,923
L. 817	5,101	3,049 b	5,289 a	4,855 b	4,401 b	5,024 b	4,620
L. 818	4,998	3,408 a	5,741 a	5,169 a	5,044 b	5,290 b	4,942
L. 821	4,858	3,933 a	6,004 a	6,010 a	4,702 b	5,444 a	5,158
L. 823	4,846	3,719 a	5,697 a	6,040 a	4,857 b	6,228 a	5,231
L. 824	4,763	3,476 a	5,886 a	5,218 a	4,474 b	5,589 a	4,901
L. 829	4,395	3,107 b	5,694 a	5,133 a	4,702 b	5,222 b	4,694
L. 845	4,194	2,922 b	3,424 b	5,552 a	4,607 b	4,718 b	4,236
L. 846	5,209	2,570 b	5,430 a	4,490 b	5,816 a	4,708 b	4,704
L. 847	5,188	3,483 a	5,295 a	5,716 a	5,738 a	4,200 b	4,937
L. 849	4,828	2,806 b	5,776 a	4,504 b	4,141 b	4,978 b	4,505
IAC OL5	5,339	4,156 a	6,119 a	5,264 a	6,101 a	5,518 a	5,417
IAC 503	4,966	3,028 b	4,292 b	3,730 b	5,280 a	4,845 b	4,357
IAC 505	4,527	3,676 a	5,487 a	4,627 b	4,425 b	5,448 a	4,698
IAC 886	4,445	3,256 b	5,512 a	5,012 a	5,133 a	5,220 b	4,763
Mean	4,838	3,392	5,385	5,121	4,932	5,194	
CV (%)	8.7	18.3	10	13.0	12.3	12.8	

[1] Means followed by the same letter in the column do not significantly differ from each other according to the Tukey's test at 5% probability.

130 days. Cultivars IAC 505 and IAC 886 presented maturity values of 74.9 and 74.1%, respectively, and the long cycle cv. IAC 503 presented maturity value of 66.9%.

The growing season of 2015/16 was characterized by an exceptional amount of rain and a high pressure of foliar diseases for the peanut crop, especially late leafspot (*Cercosporidium personatum*). Under these conditions, yield of cv. IAC OL5 was compared with the recently released high oleic 130 day-cycle cultivars IAC OL3 and IAC OL4, which are susceptible to the disease, in a three-location experiment (Table 3). The new cultivar had yield ranging from 3.279 to 6.528 kg ha⁻¹ of unshelled peanuts. IAC OL3 yielded 3.132 to 5.790, and IAC OL4 yielded 2.510 to 5.877 kg ha⁻¹. In Ribeirão Preto, the most affected location, IAC OL5 outyielded IAC OL3 by 16%, and IAC OL4, by 45%. These data indicate that IAC OL5 has some tolerance to the disease, when compared with the other two cultivars.

Over the past 3 years, peanut crops in the state of São Paulo have been infected, although in moderate intensity, by viruses of the tospovirus group, TSWV (Tomato Spotted Wilt Virus), and GRSV (Groundnut Ringspot Virus). TSWV is widely spread in the peanut region in the USA, where the virus epidemics are high. The most efficient method of control is varietal resistance. In 2015 and 2016, IAC OL5 was tested for resistance to TSWV in field conditions under high severity of the disease, in Tifton, GA, USA (Table 4). As compared with Tifguard, a highly resistant American cultivar, and Sun Oleic 97, an American standard for susceptibility to the disease, IAC OL5 showed moderate resistance level. Other cultivars commercially known in Brazil, such as Granoleico, IAC OL3, and IAC 886 were classified as susceptible.

Table 2. Percentage of mature kernels in 4 peanut cultivars in experiments harvested between 125 and 130 days after planting

Genotype	Experiments[1]						Mean
	Pindorama	Ribeirão Preto	Votuporanga	Pindorama	Ribeirão Preto	Votuporanga	
	2011/12			2012/13			
IAC OL5	81.9 a	76.6 a	69.1 a	80.9 a	85.7 a	85.4 a	79.9
IAC 503	75.7 c	62.5 c	39.5 b	66.1 b	76.9 a	80.5 a	66.9
IAC 505	79.1 ab	73.9 ab	60.2 a	78.6 ab	77.5 a	80.1 a	74.9
IAC 886	77.2 bc	68.8 bc	58.4 ab	76.4 ab	82.8 a	80.8 a	74.1
Mean	78.5	70.4	56.8	75.5	80.7	81.7	73.9
CV (%)	2.4	14.2	16.5	10.8	11.5	6.5	

[1] Means followed by the same letter in the column do not significantly differ from each other according to the Tukey's test at 5% probability.

Table 3. Yield of cv. IAC OL5 compared with cvs. IAC OL3 and IAC OL4 in 3 locations, in 2015/16

Genotype	Locations				Index
	Pindorama	Ribeirão Preto	Votuporanga	Mean	
IAC OL5	6,528	3,279	4,162	4,656	111
IAC OL3	5,790	3,132	4,340	4,421	103
IAC OL4	5,877	2,510	4,513	4,300	100
Mean	6,025	2,973	4,338	4,445	
CV (%)	11.0	17.4	11.0	13.1	

Table 4. Reactions of peanut cultivars to the virus TSWV (Tifton, GA, USA, 2015/2016)

Cultivar	Disease Index[1]	Resistance Score[2]
Granoleico	20.1	S
IAC OL3	17.7	S
Sun Oleic 97	17.3	S
IAC 886	13.9	MS
IAC OL5	10.3	MRMS
Tifguard	1.0	R

[1] Disease index of cv. Tifguard (Index=1,0), standard genotype for high resistance to TSWV
[2] Score scale based on the indices of cv. Tifguard: 1-4=R (Resistant); 4.1-8.0=MR (moderately resistant); 8.1-12.0=MR-MS (Moderately Resistant to Moderately Susceptible); 12.1-16.0=MS (Moderately Susceptible); 16.1-20.1=S (Susceptible).

TECHNOLOGICAL TRAITS

Kernels of IAC OL5 are round-shaped, commercially classified as "runner" type for their size (mean weight of 100 kernels = 60 to 70 g), and have tan testa color. Their oil content is of 48-49%, and they present the high oleic trait (70-80%).

Seed production

IAC OL5 was registered in Registro Nacional de Cultivares (Ministry of Agriculture, Brazil) in 2016; IAC is the creator and maintainer of the cultivar, and produces the genetic (breeder's) seeds.

ACKNOWLEDGEMENTS

The authors thank the following peanut companies for the financial support to the IAC breeding program responsible for the development of the cultivar: Balsamo Indústria de Alimentos, Brumau Exportadora de Óleo e Grãos, CAP Agroindustrial Exportação, Cooperativa Camap, Cooperativa Copercana, Cooperativa Coplana, Manduca Produção de Sementes, Mars Brasil Indústria de Alimentos, MIAC Máquinas Agrícolas, Santa Helena Indústria de Alimentos, and Sementes Esperança.

REFERENCES

Godoy IJ, Carvalho CL, Martins ALM, Bolonhezi D, Freitas RS, Kasai FS, Ticelli M, Santos JF, Oliveira EJ and Morais LK (2009) IAC 503 e IAC 505: cultivares de amendoim com a característica "alto oleico". In Anais do 5° Congresso Brasileiro de Melhoramento de Plantas. SBMP, Guaraparí (CD-ROM).

Godoy IJ, Santos JF, Carvalho CRL, Michelotto MD, Bolonhezi D, Freitas RS, Ticelli M, Finoto EL and Martins ALM (2014) IAC OL3 e IAC OL4: new Brazilian peanut cultivars with the high oleic trait. Crop Breeding and Applied Biotechnology 14: 200-203.

IEA - Instituto de Economia Agrícola (2016) Estatísticas da produção paulista/ amendoim. Available at < www.iea.sp.gov.br >. Accessed on November 11, 2016.

Mozingo RW, O'Keefe SF, Sanders TH and Hendrix KW (2004) Improving shelf life of roasted and salted inshell peanuts using high oleic fatty acid chemistry. Peanut Science 31: 40-45.

12

BRS 331 – Early cycle double-haploid wheat cultivar

Pedro Luiz Scheeren[1], Eduardo Caierão[1]*, Márcio Só e Silva[1], Luiz Eichelberger[1], Alfredo do Nascimento Júnior[1], Eliana Maria Guarienti[1], Martha Zavariz de Miranda[1], Ricardo Lima de Castro[1], Leila Costamilan[1], Flávio Martins Santana[1], João Leodato Nunes Maciel[1], Maria Imaculada Pontes Moreira Lima[1], João Leonardo Fernandes Pires[1], Douglas Lau[1], Paulo Roberto Valle da Silva Pereira[1] and Gilberto Rocca da Cunha[1]

Abstract: *The wheat cultivar 'BRS 331' was developed by Embrapa. It results from an interspecific cross between wheat and maize by double-hapolid method. 'BRS 331' shows solid stem in the base of the plant, short leaves and super-early cycle to maturity. It is classified as bread wheat in all of the regions that is recommended in the States of Rio Grande do Sul, Paraná and Santa Catarina, Brazil.*

Key words: *Triticum aestivum, crop breeding.*

*Corresponding author:
E-mail: eduardo.caierao @embrapa.br

[1] Embrapa Trigo, Rodovia BR 285, km 294, CP 451, 99.001-970, Passo Fundo, RS, Brazil

INTRODUCTION

Wheat (*Triticum aestivum* L.) is an autogamous species of worldwide adaptation and highly relevant for the Brazilian agriculture. Currently, Brazil consumes around 11 million tons of wheat per year, higher than the national cereal production, which reached 5.3 million tons in 2014 (CONAB 2015).

The purpose of the breeding program of Embrapa Wheat is to provide producers with cultivars that are competitive at the agronomic level and have an industrial quality appropriate for the different segments of the milling industry. Since 1974, more than 70 new cultivars were released for cultivation (Só e Silva et al. 2010, Caierão et al. 2014, Scheeren et al. 2014). Cultivar BRS 331 was developed in the framework of a partnership between Embrapa Wheat and the Fundação Pró-Sementes de Apoio à Pesquisa (Pro-Seed Research Support Foundation), which supervised the testing, marketing and distribution.

The objective of this study was to describe the yield performance, main agronomic traits and end-use profile of industrial suitability of the Embrapa Wheat cultivar BRS 331.

BREEDING METHOD

'BRS 331' was derived from the cross F68675 made in the winter of 2000, in a greenhouse of Embrapa Wheat, in Passo Fundo, RS. The parent lines were PF 990602, originated from a genotype developed in Pelotas, by Embrapa Clima Temperado, and WT 98109, a line developed by Embrapa Soja, in Londrina, PR.

In 2001, the F_1 generation was grown in a greenhouse, in Passo Fundo, RS. In 2002, it was included in a plant collection called DHM (Double-Haploid line of maize) highlights. One ear was emasculated and pollination performed with maize pollen. In the laboratory, the resulting embryo was transferred to an appropriate growth medium, where a seedling was grown that originated $F1_{DH}$ (DH 16914). Three spikes of the F_1 generation were sown in the winter of 2003, of which one gave rise to PF 015733-C with characteristics of medium-sized with early cycle. In the winter of 2004, PF 015733-C was sown on an experimental field of Embrapa Wheat and included in the collection of new double-haploid wheat lines. In each generation, after threshing of the selected plots, the seeds were visually selected, and PF 015733-C had a superior performance in terms of grain filling. It was selected as outstanding and separated for testing.

PERFORMANCE CHARACTERISTICS

In 2005, line PF 015733-C was evaluated in the 18th preliminary test set of Embrapa Wheat lines. In view of the continuously superior performance in the Preliminary Test Network (EPR) in 2006, the line was advanced for testing for the Cultivation and Use Value (VCU) in 2007. The VCU was determined in the tests of 2007, 2008 and 2010. All trials were arranged in a randomized block design with three replications. Each experimental unit, consisting of one genotype, was sown in five 5-m-long rows, spaced 0.2 m apart, with a total evaluated area of 5 m². All cultural practices were applied according to the technical instructions of the commission for wheat and triticale research in Brazil (Comissão Brasileira de Pesquisa de Trigo e Triticale 2006, 2008 and 2009). Prior to sowing, the seeds of the tests were treated with triadimenol + imidacloprid. The trials were carried out in the states of Rio Grande do Sul, Santa Catarina and southern Paraná, in the regions of wheat adaptation 1 and 2 in different sites and sowing dates.

In 2007, 2008 and 2010, cultivar BRS 331 was compared with the controls BRS 208 and BRS Guamirim (in terms of relative percentage for the variable grain yield). The relative percentage of the consolidated performance of 'BRS 331' for the variable grain yield was 98% (2007), 97% (2008) and 103% (2010) of the mean of the two controls each year. In the period of participation in the VCU tests, the relative percentage of the consolidated performance was 100% (Table 1). The highest mean grain yield of the cultivar was 4,950 kg ha⁻¹ in 2010, while the general mean was 4,181 kg ha⁻¹.

'BRS 331' is a medium-sized cultivar with a super-short cycle (on average 125 days from emergence to maturity in Passo Fundo), moderately resistant to lodging and frost in the vegetative stage; susceptible to shattering and moderately resistant to pre-harvest sprouting; and susceptible to blight (Table 2). In relation to the main biotic stresses, it is resistant to *Soilborne Wheat Mosaic Virus* (SBWMV), moderately resistant to powdery mildew (*Blumeria graminis*), spot blotch (*Bipolaris sorokiniana*) and scab (*Fusarium graminearum*). The reaction to tan spot and BYDV (*Barley Yellow Dwarf Virus*) was characterized as moderately susceptible. To leaf rust (*Puccinia triticina*), the reaction was classified as susceptible to the races B48, B53, B55, B57, and B59 in the seedling stage, in tests under controlled environmental conditions (growth chamber). The "field reaction" of the cultivar to leaf rust was classified as moderately susceptible.

Preliminarily, 'BRS 331' (line PF 015733-C) was classified as bread wheat (W ≥ 220 x 10⁻⁴ J or stability ≥ 10 min; falling number ≥ 220 s), based on Instruction No. 38 of 30/11/2010 (Table 3). For the samples analyzed between 2006 and 2010, the Grain Quality Laboratory of Embrapa Wheat determined a mean gluten strength (W) of 223 x 10⁻⁴ J and Elasticity Index (EI) of 50.1% in 37 samples from VCU test locations. For the Adaptation Region (AR) 1 the mean W value was 222 x 10⁻⁴ J

Table 1. Grain yield means (kg ha⁻¹) of BRS 331 and the controls BRS 208 and BRS Guamirim, in 2007, 2008 and 2010

Genotype	2007		2008		2010		Mean	
	(kg ha⁻¹)	(%¹)	(kg ha⁻¹)	(%¹)	(kg ha⁻¹)	(%¹)	(kg ha⁻¹)	(%¹)
Locations	8		9		10		27	
BRS 331	3.286	98	4.306	97	4.950	103	4.181	100
BRS Guamirim	3.443	103	4.654	105	4.640	97	4.246	102
BRS 208	3.262	97	4.229	95	4.963	103	4.151	98
TM²	3.352	100	4.442	100	4.801	100	4.198	100

¹ % = mean percentage of the two best controls.
² TM = Mean of controls BRS Guamirim and BRS 208.
Locations of evaluation in 2007: Passo Fundo/RS (2 sowing times), São Borja/RS, Três de Maio/RS, Vacaria/RS, Chapecó/SC, Campos Novos/SC and Guarapuava/PR. Locations of evaluation in 2008: Passo Fundo/RS (2 sowing times), São Borja/RS (2 sowing times), Três de Maio/RS (2 sowing times), Vacaria/RS, Chapecó/SC and Campos Novos/SC. Locations of evaluation in 2009: Passo Fundo/RS (2 sowing times), São Borja/RS (2 sowing times), Três de Maio/RS (2 sowing times), Vacaria/RS, Victor Graeff/RS, Abelardo Luz/SC and Canoinhas/SC.

Table 2. Description of the main agronomic characteristics of wheat cultivar BRS 331 compared with the controls BRS Guamirim and BRS 208

Agronomic traits	BRS 331	BRS Guamirim	BRS 208
Plant height	Medium	Low	Medium/High
Cycle to heading	Super-early	Super-early	Medium
Cycle to maturity	Super-early	Super-early	Medium
Reaction to pre-harvest sprouting	S	MS	S
Reaction to lodging	R	MR	S/MS
Reaction to common bacterial blight	S	MR	R
Reaction to powdery mildew	MR	S	MR
Reaction to spot blotch	MR	MR	MS
Reaction to tan spot	MS	MR	MS
Reaction to leaf rust	S[1] and MS[2]	MR/MS	MR
Reaction to scab	MR	MR	MS
Reaction to BYDV	MS	MS	MR
Reaction to SBWMV	R	S	MS

R = Resistant; MR = Moderately resistant; MS = Moderately susceptible; S = Susceptible
[1] Races B48, B53, B55, B57 and B59 (in seedling stage).
[2] in adult plants.

Table 3. Industrial quality of BRS 331 in the wheat adaptation regions 1 and 2

Characteristics	Mean of region 1[1]	Mean of region 2[2]	General Mean
Samples	20	17	37
Mean W	222	226	223
Mean L*	92	91.5	91.71
Mean b*	11.2	10.8	11.01
Tenacity (P)	89	93	91
Extensibility (L)	71	65	68
Mean P/L	1.3	1.5	1.4
Mean EI	49.6	50.7	50.1

Samples = number of samples of each region; W = Gluten strength (x 10^{-4}Joules); L* = luminosity *or* flour whiteness (Minolta) – "0" = black and "100" = white; b* = color b (Minolta) – "+" = yellow and "-" = blue; P = Tenacity or maximum resistance to extension; L= Extensibility (mm); P/L = Tenacity/extensibility ratio; EI = Elasticity index in percentage.
[1] Representative locations of Region 1 - Passo Fundo/RS, Vacaria/RS, Victor Graeff/RS, Campos Novos/SC, Canoinhas/SC and Guarapuava/PR. [2] Representative locations of Region 2 - São Borja/RS, Três de Maio/RS, Chapecó/SC and Abelardo Luz/SC

in 20 samples from the states of Rio Grande do Sul, Santa Catarina and Paraná, preliminarily classified as bread wheat. For AR 2, the mean W was 226 x 10^{-4}J in 17 samples from Rio Grande do Sul and Southern Santa Catarina, classified in the same way as for AR 1. The other relevant traits for the industrial quality profile of the cultivar are presented in Table 3.

Cultivar BRS 331 has red grain with good resistance to "mottled grains". The grains are extra hard (mean hardness of 91, in SKCS equipment). Classified as bread wheat, it is suggested that this wheat is used for baking in general, fresh pasta, for cookies such as "crackers", domestic use (homemade cakes, cookies and bread), and for blends.

This cultivar represents a significant advance in terms of earliness and plant architecture, mainly about stem thickness, the short and erect leaves and the ear size.

Wheat cultivar 'BRS 331' was suitable for cultivation in AR 1 and 2 of the States of Rio Grande do Sul and Santa Catarina, as well as in AR 1 of Paraná.

BASE SEED PRODUCTION

BRS 331 is registered and protected by the Ministry of Agriculture, Livestock and Supply (MAPA) under numbers 28233 and 20120129, respectively. Embrapa Wheat is responsible for the genetic seed of the cultivar, the Embrapa Produtos e Mercado (SPM) for the Basic seed and the founders of the Fundação de Apoio a Pesquisa Pró-Sementes for the certified seed, in partnership with Embrapa.

REFERENCES

Caierao E, Scheeren PL, Só e Silva M, Castro RL (2014) History of wheat cultivars released by Embrapa in forty years of research. **Crop Breeding and Applied Biotechnology 14**: 216-223.

Comissão Brasileira de Pesquisa de Trigo e Triticale (2006) **Indicações técnicas para trigo e triticale – safra 2007**. Embrapa Trigo, Passo Fundo, 114p.

Comissão Brasileira de Pesquisa de Trigo e Triticale (2008) **Indicações técnicas para trigo e triticale – safra 2008**. Embrapa Trigo, Londrina, 147p.

Comissão Brasileira de Pesquisa de Trigo e Triticale (2009) **Indicações técnicas para trigo e triticale – safra 2010**. Embrapa Trigo, Veranópolis, 170p.

CONAB (2015) **Série histórica de área semeada de trigo.** Available at <http://www.conab.gov.br/conabweb>. Accessed in April 2015.

Scheeren PL, Caetano VR, Caierão E, Só e Silva M, Nascimento Jr A, Eichelberger L, Miranda MZ and Brammer SP (2014) BRS 328 – Double haploid bread wheat cultivar. **Crop Breeding and Applied Biotechnology 14**: 65-67.

Só e Silva M, Caierão E, Scheeren PL, Eichelberger, L, Nascimento Jr A and Miranda MZ (2010) BRS 327 – a new bread wheat cultivar. **Crop Breeding and Applied Biotechnology 10**: 370-373.

13

Morphological and physiological characteristics in vitro anthurium plantlets exposed to silicon

Gabrielen de Maria Gomes Dias[1*], Joyce Dória Rodrigues Soares[2], Suelen Francisca Ribeiro[3], Adalvan Daniel Martins[2], Moacir Pasqual[2] and Eduardo Alves[4]

Abstract: *The objective was to evaluate morphological and physiological differences in anthurium plants* in vitro, *with the use of silicon added to the culture medium. Nodal segments were inoculated in Pierik with different sodium silicate concentrations (0.0, 0.5, 1.0 or 2.0 mg L^{-1}). After 100 days in a growth room, phytotechnical characteristics were evaluated, physiological concentrations of silicon, photosynthetic pigments and microanalysis X-ray. A higher yield of chlorophyll a and b, was observed in plants supplemented with 2.0 mg L^{-1} of sodium silicate. Anthurium plants showed better growth development with an increase in the number of leaves, dry weight and length of the aereal part in plants supplemented with 0.5 and 2.0 mg L^{-1} sodium silicate. It was also observed an increase in the number of roots of plants supplemented with 1.0 mg L^{-1} of sodium silicate. The use of sodium silicate in culture medium improves the quality of plantlets with an increased absorption of nutrients in the plant Anthurium.*

Key words: *Anthurium andraeanum, growth, development, sodium silicate.*

*Corresponding author:
E-mail: gabriellen@gmail.com

[1] Universidade da Integração da Lusofonia Afro-Brasileira (UNILAB), Instituto de Desenvolvimento Rural, Avenida da Abolição, 3, 62.790-000, Redenção, CE, Brazil
[2] Universidade Federal de Lavras (UFLA), Departamento de Agricultura, CP 3037, 37.200-000, Lavras, MG, Brazil
[3] UFLA, Departamento de Biologia
[4] UFLA, Departamento de Fitopatologia

INTRODUCTION

Anthurium is currently considered to be a product of great business potential in domestic and foreign floriculture markets. *Anthurium* is prized not only for its natural beauty and the color structure of its flowers, but also for its high durability topost-harvest (Junqueira and Peetz 2012).

Several studies have been conducted using *Anthurium* tissue culture to evaluate the potential of obtaining plantlets using this method. In general, the most important factor has been the cultivar specificity, resulting in the need to develop micropropagation protocols unique to each genotype (Gantait and Mandal 2010).

Puchoa (2005) observed that, in general, researchers working with tissue culture have found a wide variation in the responses obtained *in vitro* when using different genotypes of anthurium. Consequently, adjustments in the culture medium are essential for obtaining vigorous plants of commercial quality.

The mineral elements that stimulate the growth of plants and that are essential for only a few plant species or under specific conditions are considered beneficial elements (Furlani 2004). These elements are important for the normal growth and development of some plants, but their absence is not considered a limiting factor. One of the elements considered to be beneficial is silicon (Korndörfer 2006).

Although silicon is not considered to be an essential element, the benefits of using this element in agriculture are being increasingly recognized by researchers worldwide. For example, the use of silicates in agriculture, in addition to direct effects on productivity, can a significant contribute in the enhancement and, consequently, reduce the use of pesticides (Reis et al. 2008). The structural role of silicon (Si) in the cell wall can increase the concentrations of hemicellulose and lignin, increasing the rigidity of the cell (Camargo et al. 2007).

Studies on the effects of silicon applied to the *in vitro* cultivation of floriculture segments are still quite limited. The most recent studies examining the role of Si have been performed in flowers with post-harvest and production, such as calla (Almeida et al. 2009), rose (Locarno et al. 2011), gerbera (Guerrero et al. 2012), chrysanthemum (Carvalho-Zanao et al. 2012, Sivanesan et al. 2013) and ornamental sunflower (Carvalho et al. 2009, Oliveira et al. 2013). The only work examining the role of Si in ornamental plants grown *in vitro* has been performed in orchids (Soares et al. 2011, Soares et al. 2012). Despite the wide variety of ornamental plants grown commercially, relatively few species have been evaluated for their potential for silicon absorption (Frantz et al. 2010).

Considering the benefits observed for other species, the effect of silicon on the development of plants grown *in vitro* needs to be assessed, particularly for species with commercial importance, such as anthurium. The present study aimed to evaluate the use of silicon on the growth of *Anthurium andraeanum* cv. Rubi plants *in vitro* and to evaluate the morphological and physiological characteristics of the leaves when using concentrations of sodium silicate.

MATERIAL AND METHODS

Nodal segments of *Anthurium andraeanum* cv. Rubi plantlets established *in vitro* were inoculated in Pierik solution (Pierik 1976) supplemented with 30 g L^{-1} sucrose and solidified with 1.8 g L^{-1} Phytagel™. Sodium silicate (Na_2SiO_3) was added to the culture medium at the following concentrations: 0.0, 0.5, 1.0 and 2.0 mg L^{-1}. The pH of the culture medium was adjusted to 5.8, and the culture medium was then autoclaved at 121 °C and 1.2 atm for 20 minutes.

Subsequently, in a laminar flow hood, nodal segments were inoculated in 400 mL flasks containing 50 mL of culture medium and the different sodium silicate concentrations. The vials were maintained in a growth chamber with a photoperiod of 16 hours at 25 ± 2 °C temperature, with a light intensity of 52.5 µmol m^{-2} s^{-1}. After 100 days, we evaluated the parameters described below:

Phytotechnical features - evaluated all experimental plantlets by observing the number of leaves, the number of roots, the length of the aereal part (cm), the fresh and dry weights of the shoots (g), the root length (cm) and the fresh and dry weights of the roots (g). The dry weight of the plant material was evaluated after drying in an oven at 60 °C temperature for 72 hours or until a constant weight was obtained.

Chlorophyll contents - The contents of chlorophyll *a*, *b* and the total chlorophyll contents were quantified according to the method described by Arnon (1949). For this purpose, leaves were collected from 10 plants per treatment, and 0.1 g of leaf tissue was macerated in liquid nitrogen and placed in 80% acetone. The material was then centrifuged at 8,000x g for 15 minutes, the supernatant was collected and diluted to 25 ml, and the absorbance was measured in a spectrophotometer at 663 nm and 645 nm. The pigment content was calculated as described by Lichtenthaler (1987).

Silicon concentration - leaves were collected from 12 plants, dried in a forced-circulation oven at 60 °C temperature to a constant weight and ground separately. The determination of silicon by the molybdenum-blue colorimetric method was performed according to the methodology proposed by Gallo and Furlani (1978).

X-ray microanalysis - samples from the middle third of the second sheet of three plants were fixed in Karnovsky solution (Karnovsky 1965), dehydrated in increasing concentrations of acetone (30%, 50 %, 70 %, 90% and 100%) and then subjected to critical point drying, using liquid CO_2 as a transition medium (Robards 1978). The samples were subsequently coated with carbon and analyzed using a Baltec CED 020 LEO-EVO 40 scanning electron microscope, following the protocol described by Alves and Perina (2012).

Experimental design and statistical analyses - The experimental design was completely randomized, with four treatments consisting of 15 vials that each contained 2 explants. The data were subjected to analyses of variance and to regression analyses, with $p < 0.05$, using the software SISVAR (Ferreira 2011).

RESULTS AND DISCUSSION

The concentrations of sodium silicate applied to the Pierik culture medium affected anthurium seedling growth *in vitro*. As demonstrated in Figure 1, there was a significant effect of the sodium silicate concentration on the number of leaves, the length of the aereal part, the number of roots and the dry weight of the shoots. Increasing concentrations of sodium silicate appeared to behave quadratically, with the peak of the function indicating the increased production of sheets up to the estimated concentration of 0.7 mg L^{-1} and further increases in the concentration of sodium silicate resulting in a marked decrease in the number of leaves. Similar results were also observed by Braga et al. (2009) working with strawberry *in vitro*. However, unlike the results of the present study, Pasqual et al. (2011), working with orchids *in vitro*, found a greater number of sheets with the use of calcium silicate. Moreover, Asmar et al. (2011), working with *in vitro* banana, and Oliveira et al. (2013), working with ornamental sunflower, found no differences in the number of leaves when applying sodium silicate and calcium.

With increasing concentrations of sodium silicate, a linear increase in the length of the aerial part of anthurium was observed, with plants treated with the concentration of 2.0 mg L^{-1} achieving the maximum length of 1.82 cm (Figure 1). For Asmar et al. (2011), the use of sodium silicate provided the largest shoot growth for *in vitro* banana. Pasqual et al. (2011) obtained higher length of the aereal part for *in vitro* orchids using 2.0 mg L^{-1} calcium silicate.

The dry mass of the aerial part of the plant demonstrated a quadratic increase according to increase in the concentration of sodium silicate. The maximum yield was estimated at the concentration of 1.4 mg L^{-1} sodium silicate (Figure 1). Similarly, Almeida et al. (2009) and Sivanesan et al. (2013) reported that different concentrations of silicon promoted increased dry matter in the aerial parts of the chrysanthemum and chrysanthemum plants. Asmar et al. (2011), working with *in vitro* banana, reported higher shoot dries mass when using sodium silicate in the culture medium.

A larger number of roots were observed in plants supplemented with 1.0 mg L^{-1} sodium silicate, with a quadratic response observed for this concentration, followed by a sharp decline in root numbers for plants treated with higher concentrations of sodium silicate (Figure 1). Similar results were observed by Soares et al. (2008) when they worked with sodium silicate in the acclimatization of orchids.

The decrease in the number of roots observed in plants cultivated in higher concentrations of sodium silicate may stem from the fact that any nutrient taken in excess can cause a nutritional imbalance in the plant, including silicon. In

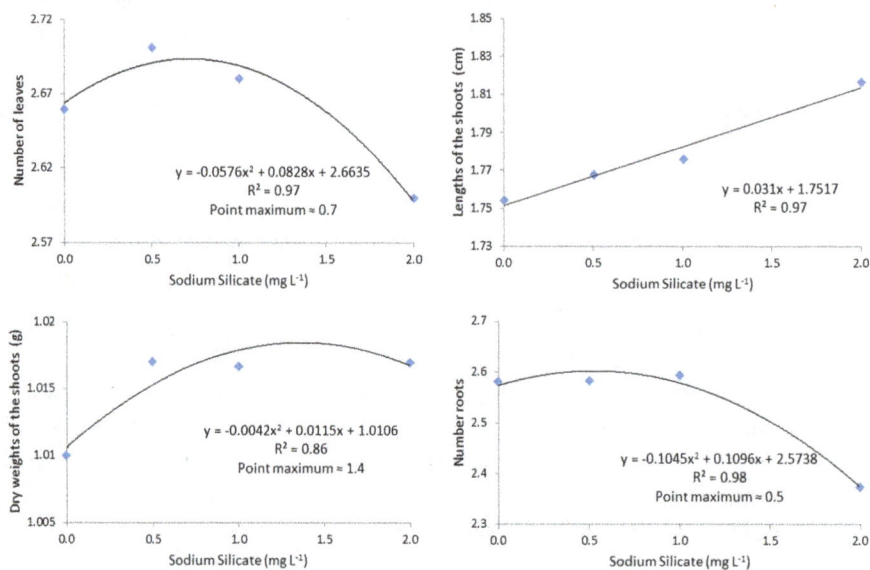

Figure 1. Phytotechnical features in *Anthurium andraeanum* cv. Rubi plantlets after 100 days in culture with different concentrations of sodium silicate.

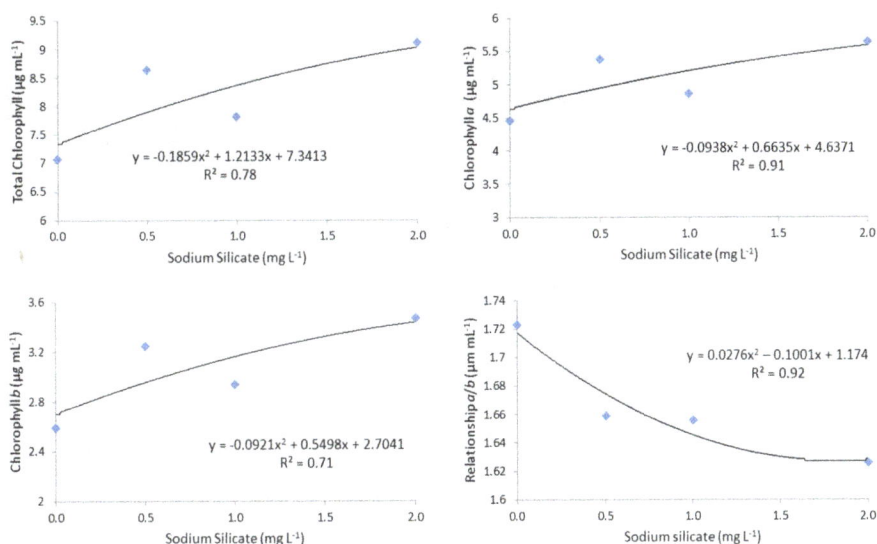

Figure 2. Chlorophyll *a* and *b* contents, total chlorophyll and *a/b* ratio in the leaves of *Anthurium andraeanum* cv. Rubi plantlets grown for 100 days *in vitro* in a culture medium containing sodium silicate.

studies of mineral nutrition, it is necessary to consider the nutrients as a whole because, in the process of absorption, one nutrient can exert influence over other nutrients, given the possible interactions that may occur (Malavolta et al. 1997).

It can be observed in Figure 2 that there is an increasing trend in the levels of chlorophyll *a* and *b* and total chlorophyll levels with increasing concentrations of sodium silicate, and the best result was obtained with 2.0 mg L^{-1} sodium silicate. The highest levels of chlorophyll *b* were observed in the control, which was culture medium without the addition of silicon, followed by decreasing levels with increasing concentrations of sodium silicate.

The results of the present study demonstrated the effectiveness of using silicon on the chlorophyll content of anthurium, similar to the data reported by Gong et al. (2005), Sousa et al. (2010), Locarno et al. (2011) and Sivanesan et al. (2013). These results are likely due to silicon being associated with the maintenance of photosynthesis, the chlorophyll distribution and the protection and preservation of the structural and functional deterioration of cell membranes (Agarie et al. 1998). Increased chlorophyll content with silicon supplementation was also observed by Al-Aghabary et al. (2005) in tomato plants.

The silicon content in the leaves of anthurium was significantly influenced by the concentrations of sodium silicate used. As shown in Figure 3, there was a proportional increase in the concentration silicon in the leaves of anthurium plantlets at higher concentrations of sodium silicate. The highest concentration was estimated to occur at 1.5 mg L^{-1} sodium silicate, with a quadratic response to increasing concentrations.

In general, plants are considered to be silicon accumulating plants if they have foliar silicon levels greater than 1% and are considered to be non-accumulating plants if they have foliar silicon levels lower than 0.5 % (Ma et al. 2001). Although anthurium are monocotyledonous plants, they are classified as non-accumulative for silicon (0.23 to 0.40%) (Figure 3). Similar results were also observed by Almeida et al. (2009) when working with silicon in the mineral nutrition of calla. Thus, anthurium plantlets have no

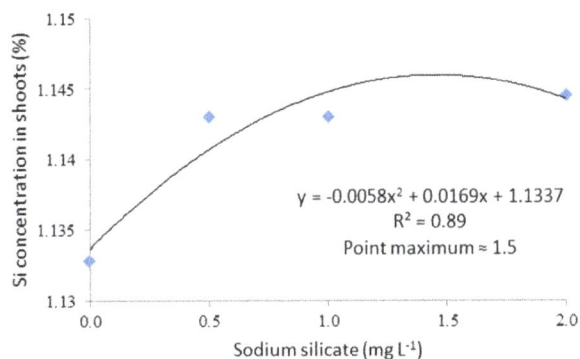

Figure 3. Silicon concentrations in shoots of *Anthurium andraeanum* cv. Rubi cultured *in vitro* for 100 days.

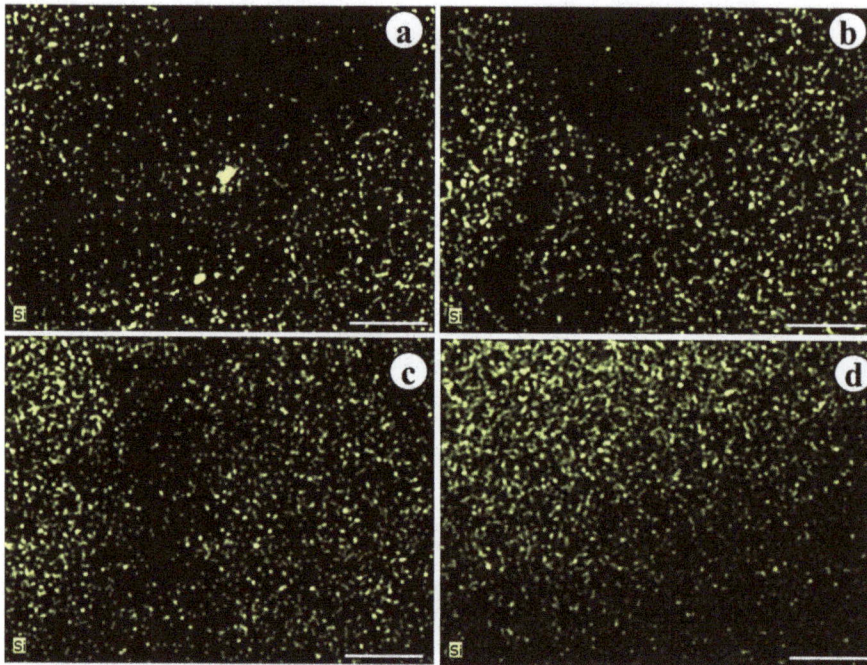

Figure 4. Mapping of Si on the abaxial epidermis of leaves from *Anthurium andraeanum* cv. Rubi plantlets cultured *in vitro* for 100 days, demonstrating the presence and the evolution of the distribution of silicon with increasing concentrations of sodium silicate in the culture medium. a) Control, b) 0.5 mg L^{-1}, c) 1.0 mg L^{-1}, d) 2.0 mg L^{-1}. Bar = 20 μm.

ability to accumulate silicon in shoots at high concentrations but can benefit from this element even in small quantities.

The presence of silicon in the control can be explained by the fact that silicon is an abundant element in nature that is present everywhere, even on water and its amount varies according to the region (Luz et al. 2006, Larezzini and Bonotto 2014). Si identified in the treatments without sodium silicate likely originated from the water used during the experiment that, although deionized, contained some neutral forms of Si that could not be completely eliminated. However, it is noteworthy that Si was absorbed in insignificant amounts in the control when compared to treatments receiving sodium silicate.

In the X-ray microanalysis (MAX), Si was found on the abaxial side of leaves treated with sodium silicate. Detection of Si was observed in all treatments, including the control. The data confirm the chemical analysis of the leaves, which also identified the presence of silicon in all treatments.

When mapping was performed for silicon (Si), there was an evolutionary distribution related to the concentration of sodium silicate applied to the culture medium (Figures 4B, C and D). The presence of Si in the control was also detected during mapping of the abaxial side of the leaf (Figure 4A).

The polymerization of silicon on the lower surface of the sheet, termed the silicification process, is common in grasses (Lux et al. 2002) and may occur as observed in the dicotyledonous coffee (Pozza et al. 2004). Previously, there have been no reports in the literature regarding the deposition of Si in anthurium. The MAX demonstrated that Si is positioned primarily in the epidermis because, when observed in cross-section, the Si levels in the sheet appeared much lower than in the control.

CONCLUSIONS

Levels of photosynthetic pigments were higher when plants of *Anthurium andraeanum* cv. Rubi were grown in culture medium supplemented with 2.0 mg L^{-1} of sodium silicate.

The use of sodium silicate in culture medium improves the quality of plantlets with an increased absorption of nutrients in the plant *Anthurium andraeanum* cv. Rubi.

ACKNOWLEDGEMENTS

We thank the Coordination of Improvement of Higher Education Personnel (CAPES) for funding, and the National Council for Scientific and Technological Development (CNPq) for granting the scholarship.

REFERENCES

Agarie S, Agata W and Kaufman PB (1998) Involvement of silicone in the senescence of Rice leaves. **Plant Production Science 1**: 104-105.

Al-Aghabary K, Zhujun Z and Qinhua S (2005) Influence of silicon supply on chlorophyll content, chlorophyll fluorescence, and antioxidative enzyme activities in tomato plants under salt stress. **Journal of Plant Nutrition 27**: 2101-2115.

Almeida EFA, Paiva PDO, Carvalho JG, Oliveira NP, Fonseca J and Carneiro DNM (2009) Efeito do silício no desenvolvimento e na nutrição mineral de copo-de-leite. **Revista Brasileira de Horticultura Ornamental 15**: 103-113.

Alves E and Perina FJ (2012) Apostila do curso introdutório à microscopia eletrônica de varredura e microanálise de raios-X. UFLA/FAEPE, Lavras, 63p.

Arnon DI (1949) Copper enzymes in isolates choroplasts. Polyphenoloxidade in *Beta vulgaris*. **Plant Physiology 24**: 1-15.

Asmar AS, Pasqual M, Rodrigues FA, Araujo AG, Pio LAS and Silva SO (2011) Fontes de silício no desenvolvimento de plântulas de bananeira 'Maçã' micropropagadas. **Ciência Rural 41**: 1127-1131.

Braga FT, Nunes CF, Favero AC, Pasqual M, Carvalho JG and Castro EM (2009) Anatomical characteristics of the strawberry seedlings micropropagated using different sources of silicon. **Pesquisa Agropecuária Brasileira 44**: 128-132.

Camargo MS, Korndörfer GH and Pereira HS (2007) Solubilidade do silício em solos: influência do calcário e ácido silícico aplicados. **Bragantia 66**: 637-647.

Carvalho PC, Júnior LAZ, Grossi JAS and Barbosa GB (2009) Silício melhora produção e qualidade do girassol ornamental em vaso. **Ciência Rural 39**: 2394-2399.

Carvalho-Zanao MP, Júnior LAZ, Barbosa JG, Grossi JAS and Ávila VT (2012) Yield and shelf life of chrysanthemum in response to the silicon application. **Horticultura Brasileira 30**: 403-408.

Ferreira DF (2011) SISVAR: a computer statistical analysis system. **Ciência & Agrotecnologia 35**: 1039-1042.

Frantz JM, Locke JC, Sturtz DS and Leisner S (2010) Silicon in ornamental crops: detection, delivery and function. In Rodrigues FA (ed) **Silício na agricultura**. UFV, DPF, Viçosa, p. 111-134.

Furlani AMC (2004) Nutrição mineral. In Kerbauy GB (ed) **Fisiologia vegetal**. Guanabara Koogan, Rio de Janeiro, p. 40-75.

Gallo JR and Furlani PR (1978) Determinação de silício em material vegetal

pelo método colorimétrico do azul de molibdênio. **Bragantia 37**: 5-11.

Gantait S and Mandal N (2010) Tissue culture of *Anthurium andraeanum*: a significant review and future prospective. **International Journal of Botany 6**: 207-219.

Gong HJ, Zhu XY, Chen KM, Wang SM and Zhang CC (2005) Silicon alleviates oxidative damage of wheat plants in pots under drought. **Plant Science 169**: 313-321.

Guerrero AC, Fernandes DM and Ludwig F (2012) Acúmulo de nutrientes em gérbera de vaso em função de fontes e doses de potássio. **Horticultura Brasileira 30**: 201-208.

Junqueira AH and Peetz MS (2012) Comercialização de antúrios no Brasil: aspectos relevantes dos mercados interno e externo. In Castro ACR, Terao D, Carvalho ACPP and Loges V (eds) **Antúrio**. Embrapa, Brasília, p. 163.

Karnovsky MJ (1965) A formaldehyde-glutaraldehyde fixative of high osmolality for use in eletron microscopy. **Journal of Cellular Biology 27**: 137-138.

Korndörfer GH (2006) Elementos benéficos. In Fernandes MS (ed) **Nutrição mineral de plantas**. SBCS, Viçosa, p.355-374.

Larezzini FT and Bonotto DM (2014) O silício em águas subterrâneas do Brasil. **Ciência e Natura 36**: 159-168.

Lichtenthaler HK (1987) Chlorophylls and carotenoids: pigment photosynthetic biomembranes. **Methods in Enzymology 148**: 362-385.

Locarno M, Fochi CG and Paiva PDO (2011) Influência da adubação silicatada no teor de clorofila em folhas de roseira. **Ciência e Agrotecnologia 35**: 287-290.

Lux A, Luxova M, Hattori T, Inanaga S and Sugimoto Y (2002) Silicification in sorghum (*Sorghum bicolor*) cultivars with different drought tolerance. **Physiologic Plantarum 115**: 87-92.

Luz JMQ, Guimarães STMR and Korndörfer GH (2006) Produção hidropônica de alface em solução nutritiva com e sem silício. **Horticultura Brasileira 24**: 295-300.

Ma JF, Miyake Y and Takahashi E (2001) Silicon as a beneficial element for crop plant. In Datnoff LE, Korndörfer GH and Snyder G (eds) **Silicon in agriculture**. Elsevier Science, New York, p. 17-39.

Malavolta E, Vitti GC and Oliveira AS (1997) **Avaliação do estado nutricional das plantas. Princípios e aplicações**. 2nd edn, POTAFOS, Piracicaba, 319p.

Oliveira JTL, Campos VBC, Chaves LHGC and Guedes Filho DH (2013)

Crescimento de cultivares de girassol ornamental influenciado por doses de silício no solo. **Revista Brasileira de Engenharia Agrícola e Ambiental 17**: 123-128.

Pasqual M, Soares JDR, Rodrigues FA, Araujo AG and Santos RR (2011) Influência da qualidade de luz e silício no crescimento *in vitro* de orquídeas nativas e híbridas. **Horticultura Brasileira 29**: 324-329.

Pierik RLM (1976) *Anthurium andraeanum* Lindl. Plantles produced from callus tissues cultivated *in vitro*. **Physiologia Platarum 37**: 80-82.

Pozza AAA, Alves E, Pozza EA, Carvalho JG, Montanari M, Guimarães PTG and Santos DM (2004) Efeito do silício no controle da cercosporiose em três variedades de cafeeiro. **Fitopatologia Brasileira 29**: 185-188.

Puchoa D (2005) *In vitro* mutation breeding of anthurium by gamma radiation. **International Journal of Agriculture & Biology 7**: 11- 20.

Reis THP, Figueiredo FC, Guimarães PTG, Botrel PP and Rodrigues CR (2008) Efeito da associação silício líquido solúvel com fungicida no controle fitossanitário do cafeeiro. **Coffee Science 3**: 76-80.

Robards AW (1978) An introduction to techniques for scanning electron microscopy of plant cells. In Hall JL (ed) **Electron microscopy and cytochemistry of plant cells**. Elsevier, New York, p. 343-444.

Sivanesan I, Son MS, Song JY and Jeong BR (2013) Silicon supply through the subirrigation system affects growth of three chrysanthemum cultivars. **Horticulture, Environment, and Biotechnology 54**: 14-19.

Soares JDR, Pasqual M, Rodrigues FA, Villa F and Carvalho JG (2008) Adubação com silício via foliar na aclimatização de um híbrido de orquídea. **Ciência & Agrotecnologia 32**: 626-629.

Soares JDR, Pasqual M, Rodrigues FA, Villa F and Araujo AG (2011) Fontes de silício na micropropagação de orquídea do grupo *Cattleya*. **Acta Scientiarum Agronomy 33**: 503-507.

Soares JDR, Pasqual M, Araujo AG, Castro EM, Pereira FJ and Braga FT (2012) Leaf anatomy of orchids micropropagated with different silicon concentrations. **Acta Scientiarum Agronomy 34**: 413-421.

Sousa JV, Rodrigues CR, Luz JMQ, Carvalho PC, Rodrigues TM and Brito CH (2010) Silicato de potássio via foliar no milho: fotossíntese, crescimento e produtividade. **Bioscience Journal 26**: 502-513.

14

Adjusting the Scott-Knott cluster analyses for unbalanced designs

Thiago Vincenzi Conrado[1*], Daniel Furtado Ferreira[1], Carlos Alberto Scapim[3] and Wilson Roberto Maluf[2]

Abstract: *The Scott-Knott cluster analysis is an alternative approach to mean comparisons with high power and no subset overlapping. It is well suited for the statistical challenges in agronomy associated with testing new cultivars, crop treatments, or methods. The original Scott-Knott test was developed to be used under balanced designs; therefore, the loss of a single plot can significantly increase the rate of type I error. In order to avoid type I error inflation from missing plots, we propose an adjustment that maintains power similar to the original test while adding error protection. The proposed adjustment was validated from more than 40 million simulated experiments following the Monte Carlo method. The results indicate a minimal loss of power with a satisfactory type I error control, while keeping the features of the original procedure. A user-friendly SAS macro is provided for this analysis.*

Key words: *Type I error rate, unequal number of observations, Monte Carlo simulations, means clustering procedures, SAS macro.*

*Corresponding author:
E-mail: tconrado@hotmail.com

[1] Universidade Federal de Lavras (UFLA), Departamento de Biologia, Campus, CP 3037, 37.200-000, Lavras, MG, Brazil
2 UFLA, Departamento de Agricultura
3 Universidade Estadual de Maringá (UEM), Agronomia, 87.080-000, Maringá, PR, Brazil

INTRODUCTION

A common problem in plant breeding is comparison of new genetic combinations. In order to detect significant difference among treatments, several Multiple Comparison Procedures (MCP) were developed: LSD (Fisher 1935), Tukey (1949), SNK (Student 1908, Newman 1939, Keuls 1952), Scheffé (1953), and Duncan (1955). Nonetheless, all these procedures can result in groups overlapping, where one treatment ends up belonging to two or more groups simultaneously (Calinski and Corsten 1985). This behavior usually prevents a clear division of the whole set into two or more groups of treatments and also leads to a more complex simultaneous analysis of multiple variables due to the presence of overlapping subsets. Thus, selection for advancement of new genetic combinations to the next step in the plant breeding program requires extra effort to overcome this statistical issue.

Cluster analysis is a promising solution to avoid subset overlapping from widely-used MCP (O'Neill and Wetherill 1971, Plackett 1971). One example of an intuitive and satisfactory approach, avoiding subset overlapping, is the use of cluster analysis over the Mahalanobis generalized distance (Rao 1952). Additionally, clustering techniques can be applied to taxonomy purposes since they have high affinity to Hotelling's Principal Component Analysis and Fisher's Discriminant Analysis (Hotelling 1933, Fisher 1936, Edwards and Cavalli-Sforza 1965).

In 1974, Alastair J. Scott and Martin Knott publicized their idea of using the Maximum Likelihood (ML) ratio test to evaluate the significance of partitions from cluster analysis of sample treatment means in designs with an equal number of observations per treatment (Scott and Knott 1974). The first review of methods for Scott-Knott means separation suggesting their use in agronomics was provided several years afterward (Chew 1976). The Scott-Knott approach is an alternative to the MCP in a situation in which two or more internally homogenous subsets of sample treatment means are expected. It uses a univariate form of the divisive clustering procedure (Edwards and Cavalli-Sforza 1965) with a likelihood ratio test for determining when to stop the clustering process to create non-overlapping, distinct, and exclusive subsets of sample treatment means. The process orders the treatment means to minimize the number of possible treatment mean partitions to be pondered (Fisher 1958) and then maximizes the sum of squares between clusters to determine the best partitioning. Despite a significant increase in the calculation volume for every additional treatment even after the ordering of treatment means, it is still feasible, even manually, if the number of partitions remains lower than 12 (Scott and Knott 1974). Indeed, the computations are more onerous than an MCP (Carmer and Walker 1985). Nevertheless, it should not be a problem for any modern computer (Gates and Bilbro 1978).

Some procedures with the same idea of partitioning means into non-overlapping groups were published after Scott-Knott (1974). These procedures presented variations in regard to the decision-making process and the clustering logic, ranging from agglomerative to divisive, hierarchical to non-hierarchical, but all of them ensure groups with no overlapping (Jolliffe 1975, Cox and Spjotvoll 1982, Calinski and Corsten 1985, Bozdogan 1986, Bautista et al. 1997, Di Renzo et al. 2002, Ciampi et al. 2008).

Many researchers prefer cluster analysis in order to facilitate interpretation and presentation of results since it results in non-overlapping, distinct, mutually exclusive groupings of the observed treatment means (Gates and Bilbro 1978, Carmer and Lin 1983, Calinski and Corsten 1985, Carmer and Walker 1985). This advantage is very clear when it is necessary to evaluate more than one variable simultaneously because the test easily allows for a positive selection of primary traits and a negative selection for any traits remaining to be evaluated. It can be effortlessly performed over the clustered data with multiple variables by initially applying filters to keep only higher performance clusters for the most important trait (i.e. yield) and then by removing some clusters of lower performance in the variables of secondary importance (i.e. plant height, biomass, etc.). This procedure should result in a highly reduced subset of treatments that present higher performance for the top priority trait, with a desirable level for the secondary traits.

An early evaluation of the Scott-Knott test with agglomerative procedures under scenarios where there is more than one true group of treatment means, or partial true null hypothesis (p-H_0), exposed the lack of an appropriate experimentwise type I error control. The result of simulations suggested that the test should be used only when the experiment has been performed with great precision, and it may be unsuitable for experiments where use of MCP would be considered inappropriate, such as those whose design and purpose suggest meaningful, orthogonal, linear contrasts with a single degree of freedom among the treatment means. However, the Scott-Knott test exhibited a higher ability to correctly reject the null hypothesis (power) and detect small differences between treatments than even the LSD test (Willavise et al. 1980).

Moreover, the Scott-Knott test has the highest rate of correct decisions and aptitude for improving performance as the number of treatments increases, in comparison with the SNK, Duncan, t-student, and Tukey tests (Silva et al. 1999, Borges and Ferreira 2003). The test exhibits higher than nominal type I error rate when evaluated in simulated scenarios in which the null hypothesis (H_0) is false for some treatments (p-H_0), although for scenarios where the null hypothesis is true for all treatments, the empirical type I error rate is under nominal levels even for the experimentwise type I error rate (Di Renzo et al. 2002, Borges and Ferreira 2003).

The Scott-Knott test also provides higher robustness compared to the MCP tests for mean separation in non-Gaussian distributions (Borges and Ferreira 2003). Despite the lack of control of type I error, the test demonstrates much higher Power than any MCP, although these two features, high robustness and power, are very common to most cluster analyses (Bautista et al. 1997, Silva et al. 1999, Di Renzo et al. 2002, Borges and Ferreira 2003). The Scott-Knott test displays similar type I and type II error in comparison to Bautista et al. (1997) and Di Renzo et al. (2002). However, its performance is superior to that of Jollieffe (1975) (Di Renzo et al. 2002).

Group homogeneity can be improved by changing the clustering approach from divisive to non-grouped treatment clustering (Bhering et al. 2008). It usually reduces the number of significantly different clusters - slightly increasing the number of treatments grouped in each one of the different clusters. In spite of this drawback, this consequence can be useful in plant breeding scenarios in which positive selection followed by retesting is applied, since it can shift a small number of treatments from an inferior cluster to a superior one.

Since most plant breeding designs are unbalanced, the objective of this research is to adjust and validate the Scott-Knott test in order to allow its use in experiments under partially balanced incomplete block designs or balanced designs with missing plots, since the non-adjusted procedure is only applicable to balanced designs. This paper proposes a novel solution for use of the Scott-Knott test under unbalanced designs followed by its validation. In order to ease its use, a user-friendly macro for the SAS/STAT® software is also provided.

MATERIAL AND METHODS

Description of the proposed adjustment procedure

The original Scott-Knott (1974) test begins by ranking all the k treatment means to be grouped and then by calculating B_0 from the k treatments partitioned in two smaller subsets. The B_0 value is calculated for every $k-1$ possible partition, and the partition with the highest value of B_0 is tested using λ as two distinct subsets of treatment means. The test uses the circumference constant π (=3.14159…) and related adjusts to approximate the λ distribution to the χ^2 distribution. If the chi-square test with $\left(\frac{k}{\pi-2}\right)$ degrees of freedom rejects the null hypothesis, the process repeats; each one of these distinct subsets is, in turn, further subdivided until each of the final clusters is shown to be homogeneous by a likelihood ratio test on λ.

$$\lambda = \frac{\pi}{2(\pi-2)}\left.B_0\right/\hat{\sigma}_0^2 \tag{i}$$

The statistic λ (i) depends on B_0, which is the maximum value from the sum of squares of all the possible partitions of k treatments into two groups, and on $\hat{\sigma}_0^2$, which is the maximum likelihood estimator of σ^2 for treatments under the null hypothesis.

Equation (ii) shows how vs^2 is used where s^2 represents an unbiased estimator of σ^2 associated with v degrees of freedom, y_i is the treatment mean i, and $\bar{y}$ is the mean of all k treatments. The variable n is the number of replications, or the total number of blocks according to the experimental design.

$$\hat{\sigma}_0^2 = \left[\sum_{i=1}^{k}(y_i-\bar{y})^2 + vs^2\right]\bigg/(k+v)\ ;\ s^2 = \frac{MSE}{n} \tag{ii}$$

Since the full Means Square Error (MSE) model is a good measure of variance, it is used as a satisfactory term for estimation of s^2.

Equation (iii) shows the relation between the unbiased estimator s^2 and the Standard Error of the Mean SE_y, where RMSE is Root Mean Square Error. It is valid only under an equal number of observations for every treatment ($n_1 = n_2 = … = n_k$). Additionally, under a balanced experimental design, SE_y has the very same value for every treatment and leads to equation (iv), the base of the proposed adjustment, where the mean of the sum of the squares of SE_y estimates s^2.

$$s^2 = \frac{MSE}{n} = \left(\sqrt{\frac{MSE}{n}}\right)^2, \text{and } SE_{\bar{y}} = \frac{RMSE}{\sqrt{n}} = \sqrt{\frac{MSE}{n}}, \text{thus } s^2 = \left(SE_{\bar{y}}\right)^2 \tag{iii}$$

$$s^2 = \left(SE_{\bar{y}}\right)^2, \text{hence } s^2 = \frac{1}{k}\sum_{i=1}^{k}\left(SE_{\bar{y}_i}\right)^2 \tag{iv}$$

Moreover, equation (iv) used in a balanced experimental design can be modified and expressed as equation (v), where a different number of observations for every treatment is also permitted. After the modification, the corrected unbiased estimator of s_c^2 can change according to the $SE_{\bar{y}_i}$ of treatments in the partitioned set. Thus, in order to accommodate subsets of treatments with unequal and equal numbers of observations, s_c^2 should be calculated for every null hypothesis before testing the statistic λ against a χ^2 distribution with the associated υ degrees of freedom. Hence, for every clustering step, s_c^2 can change to adapt to the number of observation of each treatment in the current clustering process.

$$s_c^2 = \frac{1}{k}\sum_{i=1}^{k}\left(SE_{\bar{y}_i}\right)^2 \tag{v}$$

Along with correction of s_c^2, the raw treatment mean y_i should be replaced by $\hat{y}_i$, which is the treatment mean adjusted to the effect of the unequal number of replications/blocks. The following changes in the original procedure are minimal and are disclosed in equations (vi). The notation λ_c should be used to identify λ statistics while using the correction even though the testing process against the χ^2 distribution remains the same as the original procedure.

$$\lambda_c = \frac{\pi}{2(\pi-2)}\,B_0\Big/\hat{\sigma}_{0c}^2\,, where\;\hat{\sigma}_{0c}^2 = \left[\sum_{i=1}^{k}(\hat{y}_i - \bar{y})^2 + vs_c^2\right]\Big/(k+v)\,; \tag{vi}$$

As expected, the correction increases the $\hat{\sigma}_{0c}^2$ value as the number of observations per treatment decreases - lowering the final λ_c value. This leads to a lower probability of rejecting the null hypothesis, which protects the test from the type I error. The unbalanced treatment adjustment maintains the same features and results as the original method in balanced treatment scenarios. Indeed, s_c^2 only changes for clusters in an unbalanced condition (*i.e.*, missing plots). When clustering the same experiment, after partitioning all treatment means with missing plots, the remaining clusters should have the same s_c^2 value. It is important to keep in mind that since the process follows a hierarchical clustering sequence, the very same subset of treatment means with an unequal number of observations can be partitioned multiple times before composing the final specific cluster. Indeed, the calculation of s_c^2 for every candidate partition that challenges the χ^2 distribution makes the adjustment hard to be calculated manually, but it provides satisfactory protection to the original Scott-Knott test without a significant reduction in power.

Validation of the proposed adjustment procedure

The s_c^2 deduction can indicate how the correction affects the Scott-Knott test; nevertheless, it is necessary to quantify and compare the power and type I error of the adjustment while using it. in order to validate the proposed adjustment, use of the Monte Carlo method (Metropolis and Ulan 1949) is a suitable option to simulate experiments with known parameters and then evaluate the results by comparing the original test against the adjusted solution for unbalanced designs (Carmer and Swanson 1971, Silva et al. 1999, Borges and Ferreira 2003). For that purpose, more than 40 million experiments were simulated for multiple unbalanced levels combined with several α values. The simulation scheme is composed of three main branches: complete H_0 ($\mu_1 = \mu_2 = \mu_3 = ... \mu_i$), partial H_0 ($\mu_1 = ... = \mu_{1/2} \neq \mu_{(1/2+1)} = ... \mu_i$), and complete alternative hypothesis H_1 ($\mu_1 \neq \mu_2 \neq \mu_3 \neq ... \mu_i$). The first branch was used only to quantify type I error and the third only to measure power, while the second branch measures type I error and power.

All three branches contained nine levels of α (0.01, 0.02, 0.05, 0.08, 0.10, 0.12, 0.15, 0.18, and 0.20). Within each α level, there were ten levels of missing data (0.00, 0.01, 0.02, 0.05, 0.08, 0.10, 0.12, 0.15, 0.18, and 0.20). Since the second and third branches were used to evaluate the test Power they also exhibited four (1, 2, 3, and 4) levels of δ (true difference between two treatment means). In order to improve the robustness of the study, 50,000 experiments were simulated for all 810 Monte Carlo simulation setups across all three branches, culminating in a total of 40.5 million simulated experiments.

Furthermore, every simulated experiment was composed of a random number of blocks (3 to 20) and a random number of treatments (4 to 100). Experiments with a number of observations lower than 50 were replaced to avoid a small number of degrees of freedom after data removal at random to reach the required missing level. The number of both blocks and treatments were from a uniform distribution. The effects of block and observation error were from a normal distribution with a mean of zero and a standard deviation of one. The differences between subsets were defined as the product of δ and $\sigma_{\bar{x}}$ (standard error of the mean). After each experiment was generated, some plot values were removed at random. As the simulation removed plots randomly with no restriction, the minimum number of plots was set to one per treatment to avoid treatments with no plots.

Instead of measuring type I error per comparison, it was measured per experiment, a situation in which rejection of a single incorrect null hypothesis in an experiment scores as experimentwise type I error. This approach is more severe and general because it does not consider the number of treatments in the experiment (*i.e.*, a higher number of treatments promotes an even higher number of contrasts, and it implies a higher probability of type I error). However, this approach should be able to make a better distinction between the original and adjusted procedures. Converging results were expected for both procedures (original and adjusted) under balanced designs. Thus, contrast can be observed only between balanced and unbalanced designs.

All 40.5 million experiments were simulated in SAS/IML® and analyzed with SAS System for Windows 9.3 (SAS Institute 2011). The data were evaluated using the Generalized Linear Models Procedure (Proc GLM). Output of the adjusted means was grouped by a compiled macro. A recursive SAS local host multithread approach with isolated workplaces was used to speed up the simulation run time.

Stability of the process and the ability to suspend it was ensured by the use of macros capable of error handling, also oriented to processing batches of 5,000 experiments and logging all the processing responses.

Regarding the accuracy of the estimated type I error rates using Monte Carlo simulations, the exact binomial test was applied, contrasting the nominal significance level against the obtained empirical rate (Leemis and Trivedi 1996). In scenarios in which the exact binomial test rejected the null hypothesis ($p < 0.01$), the performance of the Scott-Knott test should be considered *conservative* when the empirical rate is lower than the nominal rate and should be considered *liberal* if higher. In scenarios in which the exact binomial test did not reject the null hypothesis, the tests were classified as *precise*. The F-value was obtained using equation (vii), where y represents the number of experiments with at least one type I error, α is the nominal significance level, and N is the number of simulated experiments (50,000). The *p-value* was found using $\upsilon_1 = 2(N - y)$ and $\upsilon_2 = 2(y + 1)$ degrees of freedom (Santos et al. 2001).

$$F = \left(\frac{y + 1}{N - y}\right)\left(\frac{1 - \alpha}{\alpha}\right) \qquad \text{(vii)}$$

RESULTS AND DISCUSSION

Table 1 summarizes the results of 4.5 million simulated experiments. These experiments were simulated under the complete H_0 hypothesis (no real difference among treatments). For experiments with a balanced design (no missing plots), as the nominal α level increased, the empirical experimentwise type I error became higher. This persisted under experiments with missing plots using the proposed Scott-Knott adjustment, but reduced when the level of imbalance increased. It can be observed that the empirical values obtained using the Monte Carlo method for a balanced design (0%

Table 1. Empirical experimentwise type I error under no true difference between treatments

Nominal Alpha	Unbalance level (%)									
	0	1	2	5	8	10	12	15	18	20
1	0.932	0.926	0.834[†]	0.820[†]	0.896	0.760[†]	0.776[†]	0.760[†]	0.778[†]	0.672[†]
2	1.910	1.920	1.768	1.758	1.746[†]	1.728[†]	1.724[†]	1.736[†]	1.554[†]	1.692[†]
5	4.854	4.762	4.918	4.914	4.804	4.524[†]	4.358[†]	4.558[†]	4.318[†]	4.316[†]
8	8.046	8.168	7.832	7.760	7.686[†]	7.596[†]	7.556[†]	7.356[†]	7.190[†]	7.106[†]
10	10.184	10.334	10.284	9.830	9.936	9.546[†]	9.634[†]	9.498[†]	9.500[†]	9.514[†]
12	12.436[†]	12.374	12.166	12.024	12.018	11.728	11.814	11.576[†]	11.192[†]	11.234[†]
15	15.366	15.728[†]	15.430[†]	15.248	15.052	15.058	15.062	14.602	14.580[†]	14.060[†]
18	18.686[†]	18.910[†]	18.394	18.446[†]	18.284	18.120	18.200	17.658	17.750	17.382[†]
20	20.982[†]	20.900[†]	20.614[†]	20.508[†]	20.370	20.444	19.878	19.706	19.800	19.840

[†] represents scenarios where the exact binomial test rejected the null hypothesis

of missing plots), in which the adjusted and non-adjusted procedures exhibit the same results, are below the nominal α level for values smaller than 0.05, but according to the exact binomial test, the difference is not significant. In contrast, the empirical value is significantly higher than the nominal value for some α levels higher than 0.10, which means that the original procedure should be considered liberal at these levels since it does not properly control the type I error even under the complete H_0 hypothesis. The intermittent classification for the alpha levels of 0.12 and 0.15 as a trend for empirical rates to surpass nominal rates as the nominal alpha level increases could be caused by approximation to the χ^2 distribution used by Scott-Knott (1974), but this thesis should be evaluated in further studies and does not belong to the scope of this study.

Moreover, in half of the simulated combinations, the experimentwise type I error was evaluated as significantly different from the nominal value by the exact binomial test. As expected, the adjustment led to a more conservative approach as the level of missing plots increased. This result suggested that in order to use the proposed adjustment, the user must take into account the level of imbalance (either from the planned design or from random loss of plots) before selecting the nominal α level.

In contrast, the adjusted and non-adjusted (original) Scott-Knott test exhibited a higher empirical experimentwise type I error rate than the nominal rate under p-H_0 (Table 2). It also showed a small increase in the experimentwise type I error rate when the level of missing plots became higher, but the magnitude of the experimentwise type I error rate reduced as the α level increased. This result validated the findings of Silva et al. (1999) and exposed the weakest point of the Scott-Knott test - the lack of control of experimentwise type I error under a p-H_0.

Additionally, lower values of δ culminated in smaller differences in the experimentwise type I error rate between the adjusted and non-adjusted results of the Scott-Knott procedure (Figure 1). This trend persisted upon increasing the nominal α. Increasing α or δ led to a reduction in the difference in Power among balanced and unbalanced experimental designs (Table 3). It is also important to keep in mind that a higher value of δ indicates larger differences among the treatment values. Hence, it is easier for both procedures to detect these differences and reject the null hypothesis for any level of imbalance. The adjusted and non-adjusted tests exhibited lower Power for δ ≤ 1. No significant differences in Power between the adjusted and non-adjusted procedures were noticed for δ > 1. Additionally, the adjusted Scott-Knott test maintained very high Power, even with a small α value under a complete H_1 (Figure 2).

However, as the level of imbalance got higher, there was

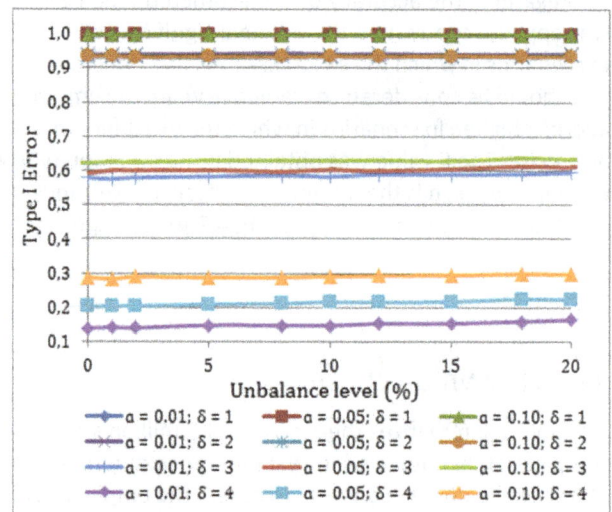

Figure 1. Empirical experimentwise error under the partial null hypothesis in the combination of three significance levels (α) by four levels of true difference between two treatment means (δ).

Table 2. Empirical experimentwise type I error under true difference between treatments of four standard errors of the mean ($4\sigma_{\bar{x}}$)

Nominal Alpha	Unbalance level (%)									
	0	1	2	5	8	10	12	15	18	20
1	13.842	14.136	13.962	14.748	14.722	14.740	15.280	15.398	15.986	16.482
2	15.124	15.560	15.740	15.870	16.474	16.504	16.860	17.132	17.374	17.894
5	20.218	20.280	20.456	20.830	21.100	21.558	21.532	21.692	22.472	22.246
8	25.406	25.408	25.136	25.244	25.944	25.798	25.974	26.114	26.246	26.952
10	28.676	28.178	28.818	28.674	28.706	29.046	29.222	29.440	29.756	29.684
12	31.684	31.522	31.628	31.670	31.874	31.722	32.100	31.938	32.492	32.448
15	36.538	36.356	36.696	36.192	36.470	36.186	36.368	36.632	36.688	36.554
18	40.600	40.778	40.530	40.698	40.960	40.770	40.602	40.984	41.238	41.174
20	43.680	43.630	43.438	43.448	43.514	43.486	43.530	43.846	43.260	43.600

Table 3. Power of Adjusted Scott-Knott in several unbalance levels under the partial null hypothesis (H_0) under four levels of true difference between two treatment means (δ)

δ	Unbalance level (%)									
	0	1	2	5	8	10	12	15	18	20
p=0.01										
1	32.525	32.652	32.233	31.884	31.453	30.982	31.306	30.748	30.236	29.735
2	84.938	84.993	85.082	85.029	85.062	85.107	85.015	85.014	85.067	85.027
3	96.582	96.574	96.566	96.537	96.516	96.513	96.491	96.459	96.452	96.433
4	99.519	99.513	99.515	99.484	99.469	99.477	99.454	99.438	99.415	99.397
p=0.05										
1	48.049	47.38	47.468	47.305	46.570	46.849	46.455	46.399	45.756	45.600
2	85.206	85.256	85.253	85.295	85.259	85.289	85.289	85.292	85.339	85.228
3	96.662	96.673	96.605	96.627	96.638	96.625	96.596	96.565	96.529	96.534
4	99.552	99.546	99.538	99.514	99.500	99.475	99.476	99.456	99.430	99.430
p=0.10										
1	53.764	53.616	53.668	53.400	53.436	52.877	53.281	52.663	52.678	52.265
2	85.406	85.338	85.369	85.365	85.362	85.396	85.386	85.453	85.439	85.464
3	96.792	96.786	96.794	96.757	96.710	96.718	96.694	96.690	96.659	96.653
4	99.560	99.559	99.542	99.542	99.532	99.509	99.502	99.479	99.459	99.459

a small loss of power when using the proposed adjustment. This performance was expected since missing information causes lower ability to reject the null hypothesis due to the additional protection required to control type I error. The small loss of power is a suitable indicator for adjustment efficiency, which is very important since the Scott-Knott test is recognized for its high power, with superior performance over the LSD and other widely used MCP (Willavise et al. 1980, Silva et al. 1999, Borges and Ferreira 2003). In spite of that, there is a trend of power reduction as the number of members per cluster decreases. This has already been pointed out and is similarities between hierarchical and non-hierarchical procedures (Tasaki et al. 1987), but it should not be assumed to be common to all clustering procedures since the clustering procedure of Bozdogan (1986) shows exactly the opposite response.

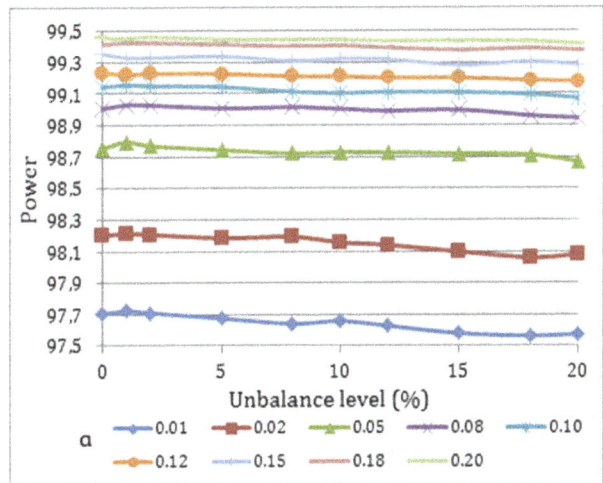

Figure 2. Empirical power under the complete H_1 hypothesis in nine significance levels (α) across ten unbalance levels.

Although the loss of power lowers the total number of clusters, it is a tolerable deficiency for scenarios where the entries that are wrongly clustered together should be retested in the next stage of research. Since the retesting routine is often used in plant breeding programs, this error is preferable to the possibility of the error of discarding an entry without a satisfactory level of confidence. Thus, as for the non-adjusted Scott-Knott procedure, it is necessary to understand the error tolerance of the experiment under evaluation before using the proposed adjustment.

It is noteworthy that even using the proposed adjustment, the most common cause of the type I error under p-H_0 for the Scott-Knott test is late compensation for incorrect partitioning in the previous step, as a consequence of divisive binary partitioning. This usually occurs in scenarios where the true number of clusters is different from powers of 2 or from the geometric sequences with common ratio 2 (data not shown). This unsatisfactory compensation is very noticeable when the true number of clusters is 3, which is a weakness common to various clustering procedures (Tasaki et al. 1987). If the gap between clusters is not clear enough, the maximum likelihood test may select a splitting point around the median by mistake. Then, in the next step, while it seeks for the point that maximizes the likelihood, it has a chance to correctly split the subset between the first and second clusters. A clear demonstration of this is an experiment with 9 treatments truly distributed in 3 clusters, for example ABC/DEF/GHI, in which the test incorrectly performs the first partitioning as ABCDE/FGHI and then it differentiates the first true cluster from the rest of the subset, resulting in (ABC/DE)/FGHI. In following, for the same reason, the test can correctly discriminate the third true cluster from treatment F, culminating in 4 clusters: (ABC/DE)/[F/GHI]. Although the first and third clusters are correct, the second cluster is improperly divided, increasing the type I error rate. This type of result is a consequence of adoption of a divisive hierarchical approach in order to allow comparison of the selected critical value which was obtained by empirical approximation, and afterwards, to declare the computed statistic significant or not (Carmer and Lin 1983). Some approaches avoiding hierarchical clustering have been published to avert this undesirable feature by simply allowing the creation of completely new clusters in every step of evaluation (Cox and Spjotvoll 1982, Calinsk and Corsten 1985, Bozdogan 1986). Despite that, the divisive hierarchical approach is still used for clustering (Di Rienzo et al. 2002, Valdano and Di Rienzo 2007).

Within plant breeding applications, the use of non-overlapping, mutually-exclusive subsets such as Scott-Knott creates a clear cutoff for the genotype advancement procedure, while results with multiple distinct subsets can help in financial management by assigning the right subset to an appropriate testing pipeline. Using the proposed adjustment procedure, this distinguishing feature is extended to experiments with missing data, which are very common in yield trials. For example, using cluster analysis on an unbalanced yield trial that results in 6 distinct subsets, the breeder would be able to submit solely the genotype subset partitioned in the highest category, "Group A", to be tested in the most accurate and expensive Pipeline I (the maximum number of locations in a randomized complete block design). Group B of genotypes could be placed in the intermediate Pipeline II (a smaller set of locations), and Group C and D could be tested in the lower cost Pipeline III (augmented blocks in the same locations as Pipeline II), while discarding the genotypes in Groups E and F (that have inferior performance compared to the commercial checks, clustered in Group C). After harvesting, the breeder can choose to retest only the superior genotypes from Pipeline III together with the new entries to be tested in Pipeline II or I.

A small drawback to the use of the proposed adjustment procedure is the increased complexity and volume of calculations in comparison to the non-adjusted procedure. Thus, in order to promote better dissemination of the proposed adjustment, a free compiled SAS GLM macro was developed and can be downloaded at http://www.tconrado.com/sas/sk.zip. The compressed file also contains an example to provide better understanding of the macro options and about how to use the software.

The proposed adjusted Scott-Knott procedure had performance similar to the original procedure under unbalanced experimental designs, with minimal loss of power, while maintaining satisfactory control of the experimentwise type I error and improved performance at $\alpha \geq 0.05$. This adjustment increases the spectrum for use of the test, providing the researcher with an alternative to the MCP, even under a significant loss of experimental data (missing plots), and it is readily available for use in SAS.

ACKNOWLEDGMENTS

We appreciate the helpful comments made by Mr. Gregory Reeves and the support from the Brazilian Government offered by the National Council for Scientific and Technological Development (CNPq) and the National Council for the Improvement of Higher Education (CAPES).

REFERENCES

Bautista MG, Smith DW and Steiner RL (1997) A cluster-based approach to means separation. **Journal of Agricultural, Biological and Environmental Statistics 2**: 179-197.

Bhering L, Cruz CD, Vasconcelos ES, Ferreira A and Resende MFR (2008) Alternative methodology for Scott-Knott test. **Crop Breeding and Applied Technology 8**: 9-16.

Borges LC and Ferreira DF (2003) Poder e taxas de erro tipo I dos testes Scott-Knott, Tukey e Student-Newman-Keuls sob distribuições normal e não normais dos resíduos. **Revista de Matemática e Estatística**

21: 67-83.

Bozdogan H (1986) Multi-sample cluster analysis as an alternative to multiple comparison procedures. **Bulletin of Informatics and Cybernetics 22**: 95-130.

Calinski T and Corsten LCA (1985) Clustering means in ANOVA by simultaneous testing. **Biometrics 41**: 39-48.

Carmer SG and Lin WT (1983) Type I error rates for divisive clustering methods for grouping means in analysis of variance **Communications in Statistics - Simulation and Computation 12**: 451-466.

Carmer SG and Swanson MR (1971) Detection of differences between means: a Monte Carlo study of five pairwise multiple comparison procedures. **Agronomy Journal 63**: 940-945.

Carmer SG and Walker WM (1985) Pairwise multiple comparisons of treatment means in agronomic research. **Journal of Agronomic Education 14**: 19-26.

Chew V (1976) Comparing treatment means: a compendium. **Hortscience 11**: 348-357.

Ciampi A, Lechevallier Y, Limas MC and Marcos AC (2008) Hierarchical clustering of subpopulations with a dissimilarity based on the likelihood ratio statistic: application to clustering massive data sets. **Pattern Analysis and Applications 11**: 199-220.

Cox DR and Spjotvoll E (1982) On partitioning means into groups. **Scandinavian Journal of Statistics 9**: 147-152.

Di Rienzo JA, Guzmán AW and Casanoves F (2002) A multiple-comparisons method based on the distribution of the root node distance of a binary tree. **Journal of Agricultural, Biological, and Environmental Statistics 7**: 129-142.

Duncan DB (1955) Multiple range and multiple F tests. **Biometrics 11**: 1-42.

Edwards AWF and Cavalli-Sforza LL (1965) A method for cluster analysis. **Biometrics 21**: 362-375.

Fisher RA (1935) **The design of experiments**. Oliver and Boyd, London, 252p.

Fisher RA (1936) The use of multiple measurements in taxonomic problem. **Annals of Eugenics 7**: 179-188.

Fisher RA (1958) On grouping for maximum homogeneity. **Journal of the American Statistical Association 55**: 789-98.

Gates CE and Bilbro JD (1978) Illustration of a cluster analysis method for mean separation. **Agronomy Journal 70**: 462-465.

Hotelling H (1933) Analysis of a complex of statistical variables into principal components. **Journal of Educational Psychology 24**: 417-441.

Jolliffe IT (1975) Cluster analysis as a multiple comparison method. **Applied Statistics, Proceedings of Conference at Dalhousie University 1**: 159-168.

Keuls M (1952) The use of the "studentized range" in connection with an analysis of variance. **Euphytica 1**: 112-122.

Leemis L and Trivedi KS (1996) A comparison of approximate interval estimators for the Bernoulli parameter. **The American Statistician Alexandria 50**: 63-68.

Metropolis N and Ulam S (1949) The Monte Carlo method. **Journal of the American Statistical Association 44**: 335-341.

Newman D (1939) The distribution of range in samples from a normal population expressed in terms of an independent estimate of standard deviation. **Biometrika 31**: 20-30.

O'Neill R and Wetherill GB (1971) The present state of multiple comparison methods. **Journal of the Royal Statistical Society 33**: 218-250.

Plackett RL (1971) The discussion on R O'Neill and G B Wetherill present state of multiple comparison methods. **Journal of the Royal Statistical Society 33**: 242-243.

Rao CR (1952) **Advanced statistical methods in biometric research**. John Wiley, New York, 390p.

Scheffé H (1953) A method for judging all contrasts in the analysis of variance. **Biometrika 40**: 87-110.

Santos C, Ferreira DF and Bueno-Filho JSS (2001) Novas alternativas de testes de agrupamento avaliadas por meio de simulação de Monte Carlo. **Ciência e Agrotecnologia 25**: 1382-1392.

SAS Institute (2011). SAS/IML 9.3 User's Guide. Sas Institute.

Scott AJ and Knott M (1974) A cluster analysis method for grouping means in the analysis of variance. **Biometrics 30**: 507-512.

Silva EC, Ferreira DF and Bearzotti E (1999) Avaliação do poder e taxas de erro tipo I do teste de Scott-Knott por meio do método de Monte Carlo. **Ciência e Agrotecnologia 23**: 687-696.

Student (1908) The probable error of a mean. **Biometrika 6**: 1-25.

Tasaki T, Yoden A and Goto M (1987) Graphical data analysis in comparative experimental studies. **Computational Statistics & Data Analysis 5**: 113-125.

Tukey JW (1949) Comparing individual means in the analysis of variance. **Biometrics 5**: 99-114.

Valdano SG and Di Rienzo J (2007) Discovering meaningful groups in hierarchical cluster analysis. An extension to the multivariate case of a multiple comparison method based on cluster analysis. **InterStat**: 1-28.

Willavise SA, Carmer SG and Walker WM (1980) Evaluation of cluster analysis for comparing treatment means. **Agronomy Journal 72**: 317-320.

15

Environmental stratification in cotton in the presence or absence of genotypes with high ecovalence

João Luís da Silva Filho[1*], Camilo de Lelis Morello[1], Nelson Dias Suassuna[1], Francisco José Correia Farias[1], Fernando Mendes Lamas[2], Murilo Barros Pedrosa[3], José Lopes Ribeiro[4] and Taís de Moraes Falleiro Suassuna[1]

Abstract: *Cottonseed yield data of 17 genotypes, of which two genotypes contributed to more than 30% of the total ecovalence, and grown in 23 locations, were used to compare four methods of disjoint environmental stratification: a) environmental index (I_e): favorable or unfavorable environments; b) stratification in partitions that maximize the sum of squares of the genotype x partition interaction [$(GP)_m$]; c) environmental scores of the second principal component of the GGE analysis (PC2-GGE); d) environmental scores of the first principal component of the AMMI analysis (PC1-AMMI). Scenarios were simulated (10,000 simulations per scenario) using combination of nine or 13 environments and 11 genotypes, either including or excluding those with the highest ecovalence values. In all scenarios, the greatest selection gains were obtained via PC2-GGE stratification, and the lowest selection gains were obtained via I_e. The ecovalences of the genotypes influenced the results obtained using the stratification methods.*

Key words: *Gossypium hirsutum, genotype x environment interaction, AMMI and GGE biplot.*

*Corresponding author:
E-mail: joao.silva-filho@embrapa.br

[1] Embrapa Algodão, CP 147, 58.428-095, Campina Grande, PB, Brazil
[2] Embrapa Agropecuária Oeste, CP 661, 79.804-970, Dourados, MS, Brazil
[3] Fundação Bahia, Rodovia BR 020/242, S/N, km 50,7, 47.850-000, Zona Rural, Luís Eduardo Magalhães, BA, Brazil
[4] Embrapa Meio-Norte, CP 01, 64.006-220, Teresina, PI, Brazil

INTRODUCTION

The development and release of cotton cultivars in Brazil follows growers' demands for competitive lint yield, and fulfill industrial textile requirements, in particular in the Cerrado environment, the largest cotton growing area in Brazil (Morello et al. 2010, Morello et al. 2012, Morello et al. 2015). Prior to cultivars being recommended, multiple trials (different locations and/or growing seasons) are carried out to evaluate genotypes in different growing conditions. However, in general, the performance of genotypes across environments may not be consistent, due to the genotype x environment interaction (GEI).

Several approaches are used to study the GEI, e.g. linear regression models (Eberhart and Russell 1966), non-linear models (Toler and Burrows 1998), and multivariate methods via principal component (PC) analysis, such as the additive main effects and multiplicative interaction model - AMMI (Gauch and Zobel 1996, Gauch 2013), and the genotype main effect plus genotype x environment interaction - GGE (Yan et al. 2000, Yan et al. 2011).

In regression models, the genotype performance is adjusted in function of the environmental index (I_e), categorized as unfavorable or favorable, according to their respective performance related to the overall means. It is assumed that selection gains are higher when cultivar recommendation assigns them to one of these categories. The models of Toler and Burrows (1998) and AMMI may be used simultaneously to study and estimate the effects of phenotypic stability (Ferreira et al. 2006).

On the other hand, for AMMI and GGE, PC analysis is carried out to capture patterns of the GEI and make inferences from graphic dispersion. The difference between models is the data matrix used. AMMI is a double centered PC analysis and GGE is environment-centered PC analysis (Gauch et al. 2008). Considerations and comparisons of both methods are found in the literature (Gauch 2006, Yan et al. 2007, Gauch et al. 2008, Yang et al. 2009, Balestre et al. 2009, Oliveira et al. 2010).

Although the mathematical and statistical properties of the models used in the GEI studies are well established, it is important to know how genotypes and environments affect the interaction (Crossa et al. 2015). The ecovalence is a relative contribution of each genotype for the GEI and a measure of the agronomic stability. The genotypes that follow the mean environment performance are considered more stable (Ramalho et al. 1993). The smaller the estimated ecovalance, the more stable is the genotype.

However, there is lack of information about the influence of genotype ecovalence on the efficiency of stratification methods. In a simulation study, Ferraudo and Perecin (2014) concluded that the AMMI and Eberhart and Russel models, and mixed models, in addition of detecting different aspects of GEI, detected the interactions only for the genotypes with considerable ranking exchanges across environments. Furthermore, although the methods based on multivariate analysis are widely used to discard environments, there is little information on whether such methods are influenced by the number of genotypes and of environments.

This study aimed at comparing four models of environmental stratification, simulating the presence or absence of high ecovalence genotypes, when the number of environments is higher or lower than the number of genotypes.

MATERIAL AND METHODS

Experimental data

Cottonseed yield data obtained from 17 cotton cultivars evaluated in 23 locations in the Brazilian Cerrado were used in this study (Silva Filho et al. 2008). The ecovalence of two cultivars, BRS Ipê and Delta Penta, hereinafter referred to as high ecovalence cultivars (HEC), contributed for more than 30% of the sum of squares of the GEI (Table 1). This information is relevant for the understanding of the methodology described below.

Stratification methods

Using a process of resampling of the original data, four methods of environmental stratification and a control with no stratification were compared. Disjoint classifications in two groups were used to facilitate the implementation of the routines and the categorization of environments in one of the groups. If the study aimed at grouping the environments into macro-regions, it is probable that grouping in more than two strata would be necessary. In each resampling (simulation), the following methods were applied:

i) control (no environmental stratification): equivalent to selecting the genotype with the best mean performance across environments;

ii) stratification by environmental index (I_e): locations with yield above the overall mean constituted a group (favorable environment), and locations below the overall mean constituted the other group (unfavorable environment);

iii) stratification by maximization of the sum of squares of genotype x partition - $(GP)_m$: the sum of squares of the GEI can be decomposed into $SSGEI = SSGEI_1(P_1) + SSGEI_2(P_2) + SSGP$, in which $SSGEI_1(P_1)$ is the sum of the squares of the GEI within partition 1; $SSGEI_2(P_2)$ is the sum of squares of the GEI within partition 2; and SSGP is the sum of squares of genotype x partition interaction. For each simulated sample, all possible partitions of environments in two groups were carried out, then the partition with the highest SSGP was chosen as $(GP)_m$. The number of possibilities for "n" environments is $C_{n,1} + C_{n,2} + ... + C_{n,n/2}$ taking the whole part to the decimal "n/2" values. This method was used only as a maximization reference of the interaction between the two groups, since it is impractical when a high number of environments is evaluated.

iv) stratification by environment PC2 scores of the GGE analysis (PC2-GGE): according to Yan and Hunt (2001), scores of the first PC of the GGE analysis are associated with the simple part of the GEI, while the scores of the second PC are associated with the crossover GEI. Thus, the environment PC2 scores were adopted as reference, considering the environments with negative score in a group, and those with positive scores in another group.

v) stratification by environment PC1 scores of the AMMI analysis (PC1-AMMI): Gauch et al. (2008) reported that PC1-AMMI and PC2-GGE are highly correlated. Thus, the environment PC1 scores were used considering the environments with negative score in a group, and those with positive scores in another group.

Scenarios in which the methods were compared

The highest number of possible combinations of n elements taken k at a time, $C_{n,k}$, occurs if k is equal to n/2 (or near the whole number). In this case, with 23 environments, k = 11 or 12; for 17 genotypes, k = 8 or 9. We assume that the higher the values of $C_{n,k}$, the higher are the chances of representing different situations of the GEI. Considering that one of the goals was to evaluate scenarios with higher or lower number of environments than the number of genotypes, nine or 13 environments and 11 genotypes, which included or not the HEC, were resampled. Therefore, the four methods and the control were compared in four scenarios:

i) number of environments lower than the number of genotypes, with the presence of HEC - 10,000 simulations by resampling of the original matrix (17 genotypes x 23 environments), considering nine environments and 11 genotypes. In this case, in each resampling, nine environments were randomly sampled among the 23; the two HEC (BRS Ipê and Delta Penta) were always present, and the other nine genotypes were randomly selected among the remaining 15.

ii) number of environments lower than the number of genotypes, without the presence of HEC - 10,000 simulations by resampling of the original matrix, considering nine environments and 11 genotypes. Similarly to the previous scenario, nine environments were randomly selected among the 23; the two HEC were excluded from the resampling process. The 11 genotypes were randomly sampled among the remaining 15.

iii) number of environments higher than the number of genotypes, with the presence of HEC - 10,000 simulations considering 11 genotypes and 13 environments randomly sampled among the 23 environments. The two HEC were always present, and the other nine genotypes were sampled among the remaining 15;

iv) number of environments higher than the number of genotypes, without the presence of HEC - 10,000 simulations considering 11 genotypes and 13 environments randomly selected among the 23 environments, and 11 genotypes selected among 15, excluding the two HEC.

Regarding the stratification method (iii), $(GP)_m$, in each one of the 10,000 simulations per scenario, all the 255 or 4095 stratification possibilities (nine or 13 environments, respectively) were included. Thus, for the latter situation, more than 40 million tests per scenario were computed.

Criteria for comparison between the stratification methods

At the end of the 10,000 simulations in each scenario, the following criteria were calculated for each method:

i) mean selection gains (SG): in each simulation, for each group, the genotype with the highest cottonseed yield was assumed to be the best genotype. Then, the weighted mean of the best genotypes was calculated with the number of environments as weight. The SG (as a percentage) was calculated by the difference between the weighted mean and the overall mean of the sample, divided by the overall mean of the sample. The mean values of these percentages in 10,000 simulations were then calculated for each stratification method. For the control method, this criterion corresponds to selecting the genotypes with the best mean performance in the sample.

ii) percentage of the sum of squares of the GEI captured by the sum of square of the genotype x partition interaction (GP): since GEI can be decomposed into $SSGEI = SSGEI_1(P_1) + SSGEI_2(P_2) + SSGP$, the ratio SSGP/SSGEI was calculated in each simulation, taking the mean value of the simulations for each stratification method. The values of this criterion for the method (iii), $(GP)_m$, are reference values, since, they are the mean of the maximum values for this statistic obtained in the simulations.

iii) percentage of wins (PW): number of simulations where the stratification method have the highest SG value, in the 10,000 simulations for each scenario;

iv) percentage of exclusive wins (PEW): number of simulations where the stratification method had exclusively the highest SG value, in the 10,000 simulations for each scenario;

v) percentage of common wins between the pairs of methods (PCW): number of simulations where each pair of stratification methods had the highest GS value, in 10,000 simulations for each scenario;

vi) Pearson's correlation between the GS and the GP: Pearson's correlation was calculated between SG and GP, for each method, in each scenario, considering the 10,000 simulations.

All analyses were carried out in the R environment, including the routines for each one of the stratification method. All routines were based on functions and commands previously available in R ("sample", "colMeans", "rowMeans", "SVD", among others).

RESULTS AND DISCUSSION

Table 1 shows the mean cottonseed yield and ecovalences estimates for 17 cotton cultivars evaluated in 23 locations. The two cultivars with the lowest yield, BRS Ipê and Delta Penta, contributed with more than 30% to the GEI, and according to Silva Filho et al. (2008), they are not well adapted to any of the 23 locations evaluated.

In such situations, one may question whether the presence of genotypes with high ecovalence can influence the stratification patterns given by AMMI or GGE analysis either on the magnitude of the GEI captured by the eigenvalues, or on the dispersion graphic.

Table 2 shows the mean values of the comparison criteria for the stratification methods in different scenarios simulated. As expected, the SG values in the absence of environmental stratification (control) were the lowest in all scenarios, since, at worst, the best genotype of a group should be the best genotype in the overall mean whatever the partitioning of the groups of environment may be. Thus, there is no PEW for the control method, and PW indicates the percentage of simulations in which all the methods simultaneously provided the same SG. The simultaneous agreement ranged from 4.5 to 7.8%, being higher, in absolute terms, in the presence of the HEC and, or, in more environments. According to the results, the more complex the GEI, the more similar to each other the methods tend to be, possibly due to greater difficulty in identifying patterns.

Table 1. Mean cottonseed yield (Y) and ecovalences estimates in percentage (W_i) for 17 cotton cultivars evaluated in 23 locations of the Brazilian Cerrado

Cultivars	Y (kg ha^{-1})	W_i (%)	Cultivars	Y (kg ha^{-1})	W_i (%)
BRS Buriti	4.185	5.03	Fibermax 977	3.701	6.20
FMT 701	4.120	4.25	BRS Cedro	3.695	2.12
CNPA GO 999	3.994	5.77	Fabrika	3.673	2.25
BRS Araçá	3.972	5.23	SL 506	3.642	4.69
BRS Aroeira	3.958	3.71	Fibermax 966	3.539	4.87
CNPACO 01-56818	3.893	3.11	Coodetec 406	3.462	2.42
Delta Opal	3.863	6.08	Delta Penta	3.339	17.6
CNPA CO 00-337	3.797	6.77	BRS Ipê	3.149	15.7
Coodetec 409	3.709	4.16	-	-	-

Adapted from Silva Filho et al. (2008).

Table 2. Mean percentage values of selection gains (SG), ratio of sum of squares of the genotype x partitions interaction and the sum of the squares of the total interaction (GP), percentage of wins (PW), percentage of exclusive wins (PEW), and Pearson's correlation (r) between SG and GP, obtained for two sample sizes, in the presence or absence of high ecovalence cultivars (HEC), for each of the stratification methods

Sample size	Stratification method	Presence of HEC					Absence of HEC				
		GP (%)	SG (%)	PW (%)	PEW (%)	r	GP (%)	SG (%)	PW (%)	PEW (%)	r
	$(GP)_m$	47.2	13.2	25.0	7.5	0.06	34.7	11.4	30.7	12.8	0.05
11 genotypes	PC1-AMMI	42.6	13.5	29.9	12.5	0.12	31.1	11.6	41.2	19.0	0.16
9 enviroments	PC2-GGE	23.8	14.1	57.8	46.7	0.12	23.7	11.7	49.0	34.0	0.27
	I_e	16.7	13.1	20.6	13.8	-0.03	13.6	10.6	15.1	10.1	-0.13
	*Control	-	12.2	6.4	-	-	-	9.7	4.5	-	-
	$(GP)_m$	42.8	12.5	23.7	9.4	0.03	28.4	10.6	28.8	18.4	0.06
11 genotypes	PC1-AMMI	38.3	12.6	26.1	11.8	0.13	25.1	10.7	33.9	18.9	0.11
13 environments	PC2-GGE	18.6	13.3	63.3	54.2	0.12	19.5	11.0	52.3	40.0	0.26
	I_e	13.3	12.3	17.6	9.7	-0.04	9.4	9.8	12.3	6.8	-0.18
	*Control	-	11.8	7.8	-		-	9.3	5.4	-	-

* No stratification

It should be noted that identical SG values do not necessarily imply that the methods identified identical partitions, since the same genotype could be the winner in the partitions obtained by each method, or different partitions with different genotypes selected could provide identical SG value. In all scenarios, the descending order of PW and SG was: PC2-GGE, PC1-AMMI, $(GP)_m$, and I_e. Therefore, discussion was organized by comparing a given stratification method with the one immediately superior in performance.

Comparing I_e and $(GP)_m$ is the same as comparing a strategy based on scale effect, I_e, and another that maximizes the variance of the interaction between groups, $(GP)_m$. In all scenarios, the GP values captured by the method $(GP)_m$ were almost three times the value of I_e, and PW of $(GP)_m$ were always higher than those of I_e (Table 2). In situations in which HEC were present, the PEW obtained in I_e were higher.

The I_e was the method that correlated less with the others, since PCW were very close to the PW of the control (Table 3). For instance, in the presence of HEC with 11 genotypes and nine environments, PCW between $(GP)_m$ and I_e was 6.5%, while the PW of the control was 6.4%. This may due to the fact that I_e does not use variance in the stratification process.

Despite retaining little of the GEI between partitions, the I_e method was the only one with negative Pearson's correlations between SG and GP, with higher values, in modulus, in the absence of HEC (Table 2).

Table 3. Percentage of common wins between the pairs of methods, in different simulated scenarios

Pairs of stratification methods	* Scenarios			
	Presence of HEC	Absence of HEC	Presence of HEC	Absence of HEC
	11 gen; 9 env	11 gen; 9 env	11 gen; 13 env	11 gen; 13 env
$(GP)_m$: PC1-AMMI	15.60	16.00	13.80	9.60
$(GP)_m$: PC2-GGE	9.21	8.89	8.61	6.81
$(GP)_m$: I_e	6.46	4.68	7.83	5.39
PC1-AMMI: PC2-GGE	9.10	13.10	8.50	11.40
PC1-AMMI: I_e	6.57	4.76	7.86	5.44
PC2-GGE: I_e	6.64	4.67	7.89	5.46

* gen: genotypes; env: environments; HEC: high ecovalence cultivars

The results of the present study do not support the widely diffused recommendation of cultivars based on defining favorable or unfavorable environments. However, it should be noted that the PEW obtained in the different scenarios (from 6.8 to 13.8%) shows that, in some situations, the more complex models were not efficient to evaluate SG.

Although $(GP)_m$ was slightly superior to I_e, showing advantage in incorporating variance information on environmental stratification, maximizing GP was not efficient when the objective was to increase SG, since Pearson's correlations between SG and GP were very low, ranging from 0.03 to 0.06 (Table 2). Even if this method was advantageous in relation to the others, the computational cost to try all possible environments partitions into two groups would make it impractical in situations with high numbers of environments; therefore, it should be interpreted as a reference in relation to the capture of the GP.

In all scenarios, the PC1-AMMI method had higher SG, PW, and PEW than $(GP)_m$. Results confirm the classic assumption conferred to the AMMI analysis, which is able to separate the variation of interaction into pattern and noise, and it is not appropriate to completely explore the GEI, at least for partitions disjoined into two groups.

PC1-AMMI and $(GP)_m$ had the greatest PCW between each other in three of the four scenarios (Table 3), always superior to 9.5%. Taking the ratio between PCW of the two methods and the PW of $(GP)_m$, a measure of how PC1-AMMI could be used in substitution of $(GP)_m$ was obtained. In this regard, the best scenario is that where the number of environments is lower than that of genotypes, and where HEC are present: 62.4% ($15.6 \div 25.0$); and the worst scenario, 33.3% ($9.6 \div 28.8$), is when the number of environments is superior to that of genotypes, and HEC are absent. Thus, it is useful to consider whether, and how, more complex AMMI models would be able to identify the partitions in which $(GP)_m$ was superior to PC1-AMMI.

On the other hand, our results did not corroborate some of the consolidated and well disseminated statements regarding the use of AMMI analysis. In the evaluation of an AMMI model, besides the number of PCs to be chosen, a primary consideration is how much of the variation is explained by the first PC capture (Lavoranti et al. 2007, Yang et al. 2009, Gauch 2013). There were higher values for GP when HEC was present, while higher PW and PEW were observed in the absence of HEC. Indeed, scenarios with higher captures of the GEI via GP were not those in which PC1-AMMI had better evaluations when compared with the others (Table 2). Pearson's correlations between SG and GP obtained for the PC1-AMMI method was low (≤ 0.16), but higher than those for $(GP)_m$.

The presence of HEC may have influenced the pattern and magnitude of the GEI captured in the first PC of the AMMI analysis. With GEI concentrated in a few genotypes, decomposition via SVD (singular values decomposition) seems to

Table 4. Selection gains (%) obtained by each stratification method in simulations that a given method was exclusive winner in the different scenarios simulated (two sample sizes, in the presence or absence of high ecovalence cultivars - HEC)

Sample size	Stratification method	Winning method in the presence of HEC				Winning method in the absence of HEC			
		$(GP)_m$	PC1-AMMI	PC2-GGE	I_e	$(GP)_m$	PC1-AMMI	PC2-GGE	I_e
	$(GP)_m$	15.3	12.2	12.8	12.6	12.5	10.7	11.1	10.3
11 gen	PC1-AMMI	14.3	13.8	12.9	12.8	11.4	12.0	11.2	10.4
9 env	PC2-GGE	14.5	12.5	14.6	13.0	11.2	10.5	12.7	10.5
	I_e	14.0	11.9	12.6	14.1	10.5	9.7	10.6	11.6
	*Control	13.6	11.0	11.8	11.9	9.8	8.8	9.8	9.3
	Mean	14.3	12.3	13.0	12.9	11.1	10.3	11.1	10.4
	$(GP)_m$	14.2	11.7	12.0	11.9	11.5	10.0	10.3	9.7
11 gen	PC1-AMMI	13.4	12.9	12.1	12.0	10.6	11.1	10.4	9.8
13 env	PC2-GGE	13.5	11.9	13.6	12.2	10.5	9.9	11.9	9.9
	I_e	13.3	11.4	11.9	13.1	9.6	9.1	9.8	10.8
	*Control	13.1	11.0	11.4	11.6	9.3	8.6	9.4	9.2
	Mean	13.5	11.8	12.2	12.1	10.3	9.8	10.4	9.9

* No stratification; gen: Genotypes; env: environments; HEC: high ecovalence cultivars

seek this pattern. In the study of Silva Filho et al. (2008), the two HEC presented, in modulus, the highest scores of the first PC. These considerations are supported by Ferraudo and Perecin (2014), who claim that the AMMI model is efficient to capture interaction when there is strong fluctuation in the ranking of genotypes along the environment, i.e., with high ecovalence.

In all scenarios, when comparing PC2-GGE and PC1-AMMI, the estimates of SG, PW and PEW were higher in PC2-GGE, while those of GP were higher in PC1-AMMI. An explanation for GP values is that, in SVD analyzes, the first PC always captures more variation than the second, although the data matrices used by PC1-AMMI and PC2-GGE are different. Including the genotype's effects in the GGE analysis may lead to a buffering effect to the estimates of GP, since there was little variation in the presence or absence of HEC for the same scenario of sample size (Table 2).

Gauch et al. (2008) stated that the PC1 of the AMMI analysis is highly correlated with the PC2 of the GGE analysis. However, the PCW values between the two methods were slightly lower than 10% in the presence of HEC, and slightly greater than this value in the absence of HEC (Table 3). The ratio between the PCW values of the two methods and PW values of the PC1-AMMI is an estimate on how close PC2-GGE is from PC1-AMMI. When including HEC, these values were 30.4% (9.1 ÷ 29.9), when 11 genotypes and nine environments were analyzed, and 32.6% (8.5 ÷ 26.1), when 11 genotypes and 13 environments were simulated. In the absence of HEC, these values were 31.8% (13.1 ÷ 41.2) and 33.6% (11.4 ÷ 33.9), respectively.

The combined use of AMMI and regression models to address different aspects of GEI has been reported (Ferreira et al. 2006, Ferraudo and Perecin 2014). This approach supports that stratification-PC2 via GGE as the most satisfactory method. Indeed, Yan et al. (2007) stated that the original model of the GGE analysis is also called regression model sites or locations (SREG - linear-bilinear regression model site), a multivariate regression.

However, results do not allow the exclusive recommendation of PC2-GGE. In the scenario with the highest PW and PEW for this method (with 11 genotypes, 13 environments and including HEC), there was 63.3% PW, and 54.2% PWE. Thus, in approximately 36% of the cases, there are potentially more efficient models regarding cultivars recommendation. Furthermore, PC1-AMMI and/or PC2-GGE identified the greatest SG of 74.8 to 80.9% ($PW_{PC1-AMMI}$ + $PW_{PC2-GGE}$ − $PCW_{PC1-AMMI:PC2-GGE}$). Thus, in at least 20% of the cases, the partitions that provided greater SG were not identified by these methods.

The PC2-GGE method presented its lower estimates of PW and PEW in scenarios where HEC were excluded from the simulations. In the data set used, HEC were among the low yielding cotton cultivars. The inclusion or exclusion of HEC may have changed the magnitude of GEI in the samples. However, these considerations may not be extended in cases where the HEC are among the high yielding, a matter of great practical concern. Unfortunately, no data set with such configuration was available. Thus, choosing the stratification methods goes beyond measuring, decomposing, and capturing the GEI in the partitions found by a particular method. It is important to know the nature of the GEI, how it is affected by genotypes and whether specific levels of treatments are decisive for the interaction, as mentioned by Crossa et al. (2015).

In order to verify if the stratification methods explore different aspects of GEI, we computed the SG, for each method, when a given method was exclusive winner (Table 4). When PC1-AMMI was exclusive winner, in general, smaller SG were obtained for all methods; on the other hand, when $(GP)_m$ was exclusive winner, higher SG were observed. Increasing the number of environments reduces the SG by any method. The SG in the presence and absence of HEC are not comparable, since the simulations were carried out on the data set with different means.

Specific levels of treatment, such as genotypes of high ecovalence, can be decisive in the study of the GEI, influencing the choice of the stratification method. The I_e method is the least recommended for a general use, and it presents less correlation with the other methods. While bearing in mind that there are situations in which other methods are potentially more advantageous, we believe that stratification by PC2-GGE should be recommended.

ACKNOWLEDGEMENTS

The authors thank to Paulo Jean Nogueira da Silva for converting routines developed in SAS/IML to R environment and to Dr. Marc Giband (CIRAD) for reviewing the manuscript. The authors would like to thank the anonymous reviewers for their constructive comments that greatly contributed to improving the manuscript.

REFERENCES

Balestre M, Pinho RG, Souza J C and Oliveira RL (2009) Genotypic stability and adaptability in tropical maize based on AMMI and GGE biplot analysis. **Genetics and Molecular Research 8**: 1311-1322.

Crossa J, Vargas M, Cossani CM, Alvarado G, Burgueño J, Mathews KL and Reynolds MP (2015) Evaluation and interpretation of interactions. **Agronomy Journal 107**: 736-747.

Eberhart SA and Russell WA (1966) Stability parameters for comparing varieties. **Crop Science 6**: 36-40.

Ferraudo GM and Perecin D (2014) Mixed model, AMMI and Eberhart-Russell comparison via simulation on genotype × environment interaction study in sugarcane. **Applied Mathematics 5**: 2107-2119.

Ferreira DF, Demétrio CGB, Manly BFJ, Machado AA and Vencovsky R (2006) Statistical models in agriculture: biometrical methods for evaluating phenotypic stability in plant breeding. **Cerne 12**: 373-388.

Gauch HG (2006) Statistical analysis of yield trials by AMMI and GGE. **Crop Science 46**: 1488-1500.

Gauch HG (2013) A simple protocol for AMMI analysis of yield trials. **Crop Science 53**: 1860-1869.

Gauch HG and Zobel RW (1996) AMMI analysis of yield trials. In Kang MS and Gauch HG (ed) **Genotype by environment interaction**. CRC Press, New York, p. 85-122.

Gauch HG, Piepho HP and Annicchiarico P (2008) Statistical analysis of yield trials by AMMI and GGE: further considerations. **Crop Science 48**: 866-889.

Lavoranti OJ, Dias CTS and Kraznowski WJ (2007) Phenotypic stability via AMMI model with Bootstrap re-sampling. **Boletim de Pesquisa Florestal 2**: 45-52.

Morello CL, Pedrosa MB, Suassuna ND, Lamas FM, Chitarra LG, Silva JL, Andrade FP, Barroso PAV, Ribeiro JL, Godinho VPC and Lanza MA (2012) BRS 336: a high-quality fiber upland cotton cultivar for Brazilian savanna and semi-arid conditions. **Crop Breeding and Applied Biotechnology 12**: 92-95.

Morello CL, Suassuna ND, Barroso PAV, Silva Filho JL, Ferreira ACB, Lamas FM, Pedrosa MB, Chitarra LG, Ribeiro JL, Godinho VPC and Lanza MA (2015) BRS 369RF and BRS 370RF: Glyphosate tolerant, high-yielding upland cotton cultivars for central Brazilian savanna. **Crop Breeding and Applied Biotechnology 15**: 290-294.

Morello CL, Suassuna ND, Farias FJC, Lamas FM, Pedrosa MB, Ribeiro JL, Godinho VPC and Freire EC (2010) BRS 293: A midseason high-yielding upland cotton cultivar for Brazilian savanna. **Crop Breeding and Applied Biotechnology 10**: 180-182.

Oliveira RL, Pinho RG, Balestre M and Ferreira DV (2010) Evaluation of maize hybrids and environmental stratification by the methods AMMI and GGE biplot. **Crop Breeding and Applied Biotechnology 10**: 247-253.

Ramalho MAP, Santos JB and Zimmermann MJO (1993) **Genética quantitativa em plantas autógamas**: aplicações ao melhoramento do feijoeiro. UFG, Goiânia, 271p.

Silva Filho JL, Morello CL, Farias FJC, Lamas FML, Pedrosa MB and Ribeiro JL (2008) Comparação de métodos para avaliar a adaptabilidade e estabilidade produtiva em algodoeiro. **Pesquisa Agropecuária Brasileira 43**: 349-355.

Toler JE and Burrows PM (1998) Genotype performance over environment arrays: a non-linear grouping protocol. **Journal of Applied Statistics 25**: 131-143.

Yan W and Hunt LA (2001) Interpretation of genotype x environment interaction for winter wheat yield in Ontario. **Crop Science 41**: 19-25.

Yan W, Fetch JM, Frégeau-Reid J, Rossnagel B and Ames N (2011) Genotype x locations interactions patterns and testing strategies for oat in Canadian prairies. **Crop Science 51**: 1903-1914.

Yan W, Hunt LA, Sheng Q and Szlavnics Z (2000) Cultivar evaluation and mega-environment investigation based on the GGE biplot. **Crop Science 40**: 597-605

Yan W, Kang MS, Ma B, Woods S and Cornelius PL (2007) GGE biplot vs. AMMI analysis of genotype-by-environment data. **Crop Science 47**: 643-653.

Yang RC, Crossa J, Cornelius PL and Burgueño J (2009) Biplot analysis of genotype environment interaction: Proceed with caution. **Crop Science 49**: 564-1576.

16

Phenotypic characterization of papaya genotypes to determine powdery mildew resistance

Marcelo Vivas[1*], Silvaldo Felipe da Silveira[2], Janieli Maganha Silva Vivas[2], Pedro Henrique Dias dos Santos[2], Beatriz Murizini Carvalho[2], Rogério Figueiredo Daher[1], Antonio Teixeira do Amaral Júnior[3], and Messias Gonzaga Pereira[3]

Abstract: *In support of breeding of papaya (Carica papaya), the disease incidence and severity of powdery mildew (Ovulariopsis caricicola) were evaluated in papaya genotypes. Two experiments in complete randomized blocks were carried out, one in the field and the other in a greenhouse. In field experiments, the lowest mean disease incidence was observed on the genotypes 'Costa Rica' and 'Baixinho Super', and the lowest mean disease severity on 'Caliman M5', 'GTF', 'SH 11-08', and 'JS 11'. In the greenhouse experiment, the genotypes 'Caliman M5', 'Golden', 'Kapoho Solo', 'Waimanalo', 'Mamão Bené', 'SH 12-07', 'JS 12', and 'GTF' had the lowest mean incidence in at least one evaluation. On the other hand, for severity, the genotypes 'Diva', 'Sunrise Solo 72/12', 'Kapoho Solo PA', 'Waimanalo', 'Maradol', 'Maradol GL', 'SH 15-04', 'FMV', 'JS 12-4', 'SH 12-07' and 'Sekati FLM' had the lowest means. These results indicate these genotypes for a possible use in breeding for reduction of powdery mildew intensity.*

Key words: *Streptopodium caricae, Ovulariopsis caricicola, Carica papaya, genetic resistance.*

*Corresponding author:
E-mail: mrclvivas@hotmail.com

[1] Universidade Estadual do Norte Fluminense Darcy Ribeiro (UENF), Centro de Ciências e Tecnologias Agropecuárias, Laboratório de Engenharia Agrícola, 28.013-602, Campos dos Goytacazes, RJ, Brazil
[2] UENF, Laboratório de Entomologia e Fitopatologia
[3] UENF, Laboratório de Melhoramento Genético Vegetal ,

INTRODUCTION

In Brazil and worldwide, papaya (*Carica papaya* L.) is an important tropical fruit. Today, Brazil ranks second in global production volume and third in acreage. Papaya is cultivated in several Brazilian states, of which Bahia, Espírito Santo, Ceará, Minas Gerais, and Rio Grande do Norte are the largest producers. On the other hand, the states of Espírito Santo, Rio Grande do Norte, Bahia, Ceará, and Paraíba are the largest papaya exporters (Treichel et al. 2016).

Despite the high production and consumption of Brazilian papaya, there are several restrictions to an expansion and even the maintenance of papaya plantations. These limitations are due mainly to the high incidence of pests and diseases (Rezende and Martins 2005, Suzuki et al. 2007, Martelleto et al. 2008). In this context, genebanks play an essential role, particularly with regard to the genetic variability required for plant breeding with a view to developing superior genotypes with different genetic constitutions. Wild species, landraces and obsolete cultivars may contain genes that confer better adaptation to environmental stresses, as well as resistance to numerous pests and diseases (Gepts 2006, Vivas et al. 2010, Vivas et al. 2012, Vivas et al. 2014). However, these genotypes maintained in genebanks should be described and evaluated to deepen the knowledge about the different genotypes or gremplasms, before including them in breeding programs.

With regard to pest and diseases incidence, papaya is described as a host of four powdery mildew species in the world: *Ovulariopsis papayae* Van der Byl in South Africa, *Phyllactinia caricaefolia* Viégas in Brazil, *Oidiopsis haplophylli* (Magnus) Rulamort, anamorph of *Leveillula taurica* (Lév.) G. Arnaud in Australia, India and Portugal, and *Ovulariopsis caricicola* U. Braun, (syn. *Streptopodium caricae* Liberato & R.W. Barreto) in Brazil (Liberato et al. 2004). According to these authors, powdery mildew symptoms of *Streptopodium caricae* species are commonly observed on papaya leaves in the northern region of Espírito Santo. According to Tatagiba et al. (2002), powdery mildew occurs year-round and chemical control is required in certain seasons to reduce symptoms. However, alternatives to the chemical control, e.g., genetic resistance, must also be developed.

The development of papaya varieties with superior agronomic traits, high fruit quality and a high level of disease resistance is a great challenge for crop breeders, because these traits have never been found in a single genotype. Knowledge about the genetic variability of the species available in the germplasm is a prerequisite for the indication of potential parents, to combine alleles related to traits of economic importance and guide crosses in breeding programs (Dias et al. 2011).

In papaya, the estimates of genetic parameters show the possibility of using simple breeding methods, such as mass selection for plant height, stem diameter, internode length, number of leaves, and leaf index (Foltran et al. 1993). With regard to the physiological seed quality, some traits are reported to have high heritability (Cardoso et al. 2009). Estimates of genetic parameters related to agronomic and fruit quality traits also indicate high chances for successful selection in segregating populations, due to the wide genotypic variability and high heritability (Silva et al. 2008).

However, for disease resistance, few such studies are available, in spite of their importance to subsidize the breeding programs targeting disease resistance. In this sense, the purpose of this study was to evaluate the potential of the varieties available in the UENF/CALIMAN germplasm collection for resistance to powdery mildew and to estimate genetic parameters in genotypes related to this disease.

MATERIAL AND METHODS

Two experiments were conducted, one in the field and the other in a greenhouse. The field experiment was conducted on the premises of the company Caliman Agricola S/A, in Linhares, Espírito Santo. The greenhouse experiment was installed on the campus 'Leonael Brizola' of Universidade Estadual do Norte Fluminense Darcy Ribeiro – UENF, in Campos dos Goytacazes, Rio de Janeiro, Brazil.

In the field trial, we evaluated the incidence and severity of powdery mildew on leaves. Disease severity was assessed on leaves with petiole attachment in the axil of newly opened flowers, on a diagrammatic scale for powdery mildew proposed by Santos et al. (2011), in December 2011 and February 2012. The experiment was arranged in a randomized complete block design with two replications, evaluting 59 genotypes. The experimental plots consisted of single rows with 15 plants each. Per unit, three scattered, non-neighboring plants, alternating between rows and excluding border plants were evaluated. For being a germplasm collection and to ensure fruit and seed production, fungicides recommended for the crop were occasionally sprayed. In the 60 days preceding the first evaluation, the active ingredients difenoconazole (23/09) and mancozeb (07/04 and 19/12) were sprayed and mancozeb (02/02) and chlorothalonil (13/02) prior to the second evaluation.

The greenhouse experiment was arranged in a randomized complete block design with five replications per genotype and a total of 27 genotypes. Each experimental unit consisted of one plant per pot. Genotypes were transplanted into 5 L pots containing a 2:1:1 soil/sand/manure mixture, and maintained in the greenhouse. Natural inoculation was evidenced by the symptoms that appeared on the greenhouse plants. The incidence of leaves with powdery mildew symptoms as well as the disease severity on the leaves were evaluated. Severity was assessed in the seventh leaf from the stem base upwards, on a diagrammatic scale (Santos et al. 2011), in August, September and November 2011, and January 2012.

The data were subjected to analysis of variance, considering the genotypes as well as evaluation times, and the interaction between these two factors as source of variation. In case of a significant effect of the genotype x time interaction, separate analyses were performed for each evaluation. When no significant effects of the interaction were detected, analyses were based on the mean of the two assessments. The genotype means were compared by Tukey's test at 0.05 probability.

For each variable, the following estimators and genetic parameters were calculated, considering the model effect as fixed: a) Mean phenotypic variance; b) Mean environmental variance; c) Quadratic component that expresses the mean genetic variability; d) Coefficient of genotypic determination, based on the genotype mean; and, e) Variation index (Cruz et al. 2012). For the genetic analysis, the software package Genes was used (Cruz 2013).

RESULTS AND DISCUSSION

In both experiments, effects of the genotype x evaluation time interaction on disease incidence and severity were detected, so individual analyses were performed for each evaluation time. The significant effects of the interaction suggest differentiated performance of the genotypes in response to environmental changes. According to Cruz et al. (2012), this different behavior is caused by genotype–specific physiological and biochemical factors. In the individual analyses of the field plants, a significant genotype effect for powdery mildew incidence was only observed in the first and for severity in the second evaluation. In the individual analyses in the greenhouse, we observed the effect of powdery mildew severity only in the first and third evaluations, while significant differences between genotypes were observed in all periods for disease incidence.

The above results proved genetic variability among the genotypes of the germplasm collection of UENF/CALIMAN with regard to resistance to *O. caricicola*. However, the variations between experiments and evaluations must be analyzed. For powdery mildew incidence, leaves of greenhouse plants with disease symptoms were evaluated. The

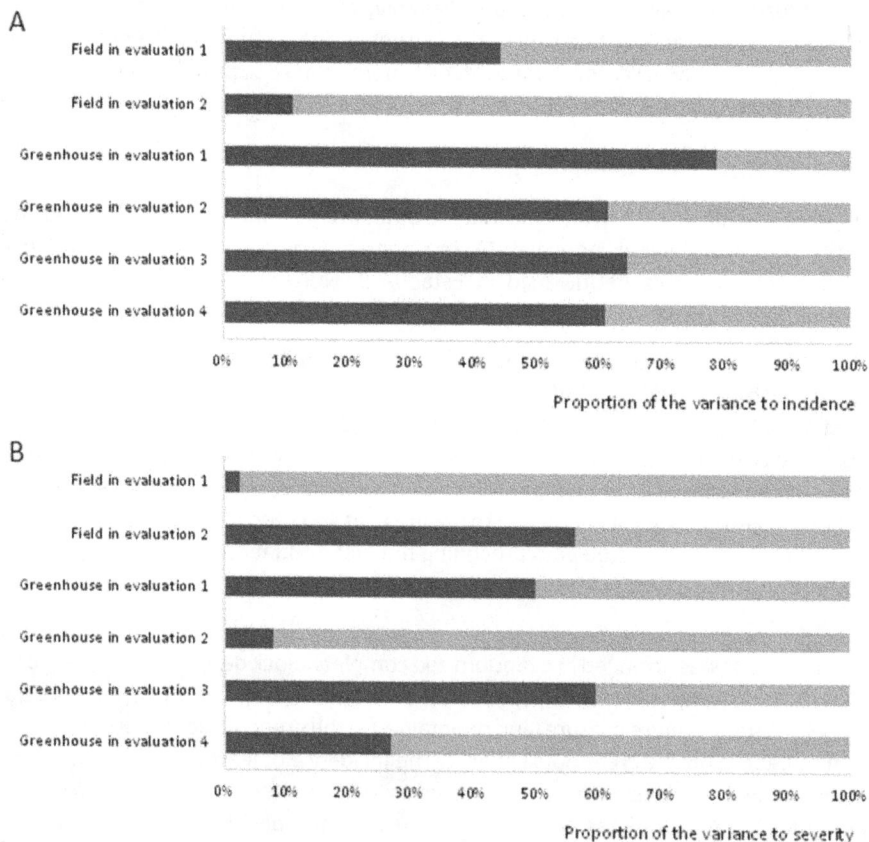

Figure 1. Proportion of the mean environmental variance (light bars) and quadratic component that expresses the mean genetic variability (dark bars) in phenotypic variation, obtained in distinct evaluation times in the field and greenhouse, for powdery mildew incidence (A) and severity (B) on papaya leaves.

genetic variance detected was greater than the environmental variance, i.e., has higher discriminating power (Figure 1). This scenario was repeated for powdery mildew severity on leaves of field plants in the second evaluation, and of greenhouse plants in the first and third evaluations (Figure 1). Aside from information about the genotypic variance, estimates of genetic parameters such as heritability and variation index are highly relevant for breeding, for guiding the choice of the breeding method most indicated for the crop and allowing inferences on estimates of selection gains. When the heritability (h^2) values exceed 80% and the variation index is greater than unity, selection with satisfactory gains is possible (Falconer 1987).

In this study, heritability varied from 11.33 to 78.70 for disease incidence and from 2.27 to 59.58 for severity (Figure 2), indicating the possibility of selecting genotypes with low disease levels. For the genetic parameters related to the physiological seed quality of genotypes of the UENF/CALIMAN papaya germplasm collection, Cardoso et al. (2009) found low heritability estimates and variation indices in a greenhouse evaluation of germination percentage, and germination speed index. On the other hand, significant heritability values (h^2) and index of variation (IV) were found for 1000-seed weight, percentage of germinated seeds, root length, and seedling dry and fresh weight, indicating the possibility of significant gains with selection for seed quality. With regard to resistance to black spot (*Asperisporium caricae*), moderate heritability estimates were found for black spot severity in the tested progenies of a papaya landrace (Vivas et al. 2012).

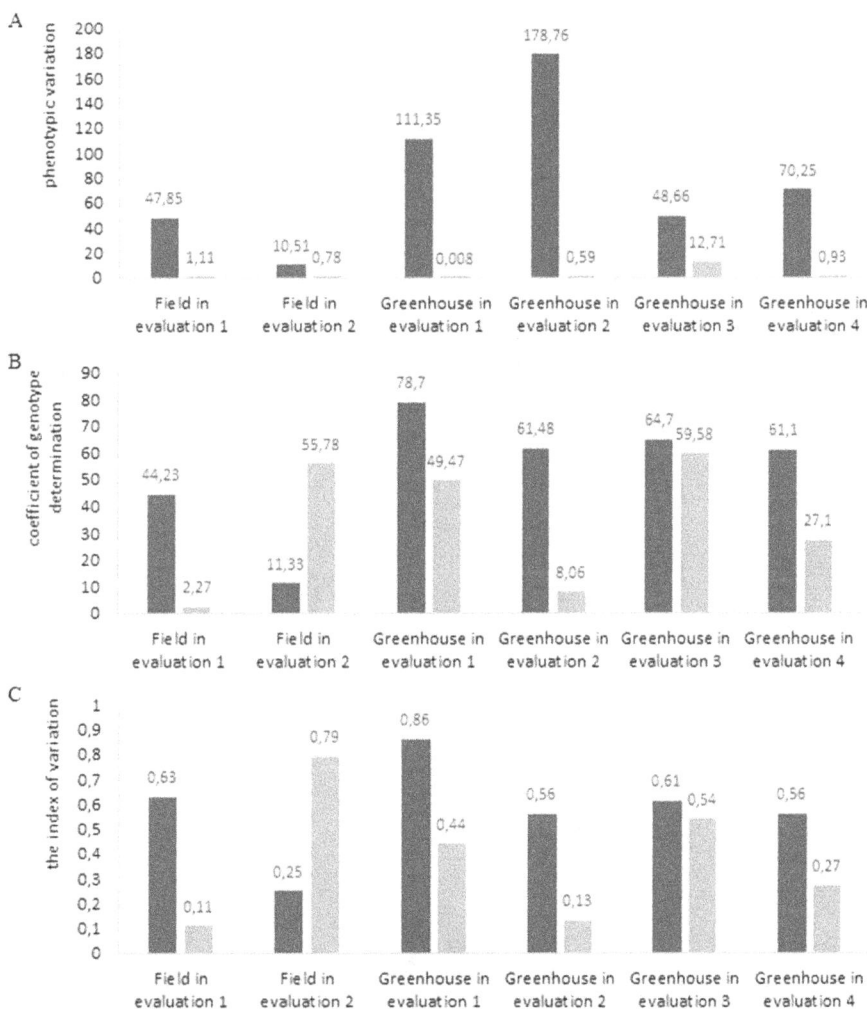

Figure 2. Estimation of the phenotypic variation mean (A), coefficient of genotype determination based on the genotype mean (B) and the index of variation (C). Estimates obtained in field and greenhouse evaluations and distinct periods, for powdery mildew to incidence (dark bars, on the left) and severity (light bars, on the right) on papaya leaves.

The lowest mean variability for leaf severity of powdery mildew symptoms in the field was observed for the genotypes 'Costa Rica' and 'Baixinho Super'. The means of the other 42 genotypes differed from the genotype with lowest mean severity ('Taiwan et') (Table 1). For powdery mildew incidence, the means of the genotypes 'Caliman M5', 'GTF', 'SH

Table 1. Means of powdery mildew severity and incidence on leaves with symptoms of the disease of papaya genotypes of the germplasm collection of UENF/CALIMAN

Genotype		Severity				Incidence			
		Evaluation 1		Evaluation 2		Evaluation 1		Evaluation 2	
1	'Caliman M5'	1.40	a	1.63	bc	5.65	bc	87.15	a
2	'Sunrise Solo 783'	0.57	a	0.74	bc	16.68	a-c	79.67	a
3	'Costa Rica'	2.27	a	0.12	c	17.01	a-c	81.23	a
4	'Taiwan et'	0.45	a	5.25	a	19.10	a-c	88.31	a
5	'Diva'	2.33	a	0.49	bc	10.40	a-c	82.88	a
6	'Grampola'	1.43	a	1.10	bc	15.14	a-c	86.98	a
7	'Sunrise Solo'	2.84	a	0.87	bc	7.99	a-c	83.96	a
8	'Caliman AM'	3.50	a	0.78	bc	26.23	a-c	76.72	a
9	'Caliman GB'	2.45	a	2.00	a-c	11.45	a-c	87.30	a
10	'Caliman SG'	3.90	a	1.87	a-c	9.01	a-c	83.15	a
11	'Golden'	3.50	a	0.87	bc	10.08	a-c	87.61	a
12	'Sunrise Solo 72/12'	1.49	a	0.82	bc	13.67	a-c	83.22	a
13	'Kapoho Solo PA'	2.35	a	1.64	bc	10.62	a-c	87.61	a
14	'Baixinho de Santa Amália'	2.40	a	0.59	bc	37.56	a	80.74	a
15	'Sunrise Solo TJ'	2.25	a	1.25	bc	22.78	a-c	77.76	a
16	'Tailândia'	0.43	a	0.80	bc	15.42	a-c	81.99	a
17	'São Mateus'	2.07	a	1.34	bc	25.60	a-c	84.54	a
18	'Kapoho Solo PV'	1.00	a	0.47	bc	10.89	a-c	82.22	a
19	'Sunrise Solo PT'	3.17	a	0.97	bc	10.33	a-c	83.32	a
20	'Waimanalo'	0.29	a	0.70	bc	8.33	a-c	84.69	a
21	'Mamão Bené'	2.62	a	2.19	a-c	19.33	a-c	86.51	a
22	'Mamão Roxo'	1.35	a	1.85	a-c	21.54	a-c	87.77	a
23	'Maradol'	0.35	a	1.85	a-c	16.67	a-c	87.77	a
24	'Maradol GL'	1.23	a	1.60	bc	13.73	a-c	87.61	a
25	'Sekati'	1.77	a	3.05	a-c	15.01	a-c	77.06	a
26	'Baixinho Super'	3.40	a	0.20	c	26.24	a-c	86.45	a
27	'Americano'	3.67	a	0.88	bc	15.51	a-c	85.32	a
28	'STZ – 51'	1.18	a	1.27	bc	22.80	a-c	84.93	a
30	'Calimosa'	0.67	a	2.15	a-c	14.58	a-c	85.12	a
31	'JS 12'	0.44	a	0.98	bc	12.28	a-c	84.73	a
32	'Cariflora'	2.13	a	1.32	bc	21.45	a-c	86.73	a
33	'GTF'	1.22	a	1.05	bc	1.71	c	82.64	a
34	'STZ – 03'	1.84	a	0.39	bc	21.80	a-c	80.25	a
35	'SH 12-07'	1.84	a	1.50	bc	16.97	a-c	82.98	a
36	'SH 11-08'	0.42	a	1.22	bc	6.00	bc	88.21	a
37	'SH 50-09'	2.97	a	2.07	a-c	21.72	a-c	81.40	a
38	'SH 02-01'	2.14	a	1.10	bc	14.19	a-c	80.94	a
39	'SH 14-05'	3.50	a	1.45	bc	16.77	a-c	83.81	a
40	'SH 15-04'	1.37	a	3.95	ab	18.94	a-c	86.25	a
41	'SH 04-02'	1.14	a	0.63	bc	12.82	a-c	83.72	a
42	'SH 12-06'	0.52	a	0.57	bc	11.95	ab	79.25	a
43	'Papaya 42'	2.17	a	1.87	a-c	7.85	ab	88.17	a
44	'Papaya 45'	0.07	a	2.14	a-c	36.40	ab	86.42	a
45	'Papaya 46'	1.62	a	1.24	bc	15.88	ab	84.37	a

Table 1. (cont)

46	'FMV'	0.38	a	1.37	bc	13.46	ab	87.69	a
47	'Golden R'	2.33	a	1.34	bc	17.08	ab	84.22	a
48	'JS 11'	0.02	a	2.84	a-c	5.81	b	89.00	a
49	'Tainung'	1.68	a	0.83	bc	8.21	ab	82.29	a
50	'STZ 23 PL'	2.27	a	1.34	bc	10.50	ab	85.96	a
51	'STZ – 63'	2.47	a	1.10	bc	8.57	a-c	82.81	a
52	'Sekati FLM'	3.42	a	0.58	bc	13.29	a-c	84.07	a
53	'Gran Golden'	0.48	a	1.10	bc	14.41	a-c	88.21	a
54	'Sunrise Solo PB'	1.09	a	1.75	a-c	17.98	a-c	85.00	a
55	'THB STZ 39'	2.17	a	0.87	bc	7.72	a-c	86.04	a
56	'G39 03-02'	1.69	a	0.42	bc	21.08	a-c	81.02	a
57	'B5'	0.57	a	0.46	bc	18.04	a-c	85.04	a
58	'FG'	0.93	a	1.55	bc	6.29	a-c	77.48	a
59	'FB'	1.80	a	0.48	bc	9.99	a-c	77.42	a
60	'JS 12-4'	0.22	a	1.05	bc	15.54	a-c	87.87	a

* For the same variable, means followed by the same letter represent a statistically homogeneous group by the Tukey test at 0.05 of probability.

Table 2. Comparison of genotype means for leaf incidence of powdery mildew symptoms of greenhouse plants in three evaluations

Genotype		Incidence							
		Evaluation 1		Evaluation 2		Evaluation 3		Evaluation 4	
1	'Caliman M5'	11.09	b	18.86	cd	57.22	a-c	47.14	ab
5	'Diva'	11.50	b	30.16	a-d	58.29	a-c	35.33	ab
11	'Golden'	10.26	b	22.00	b-d	62.73	ab	40.95	ab
12	'Sunrise Solo 72/12'	8.67	b	42.50	a-d	52.39	a-c	49.62	ab
13	'Kapoho Solo PA'	2.00	b	20.55	cd	53.77	a-c	44.00	ab
14	'Baixinho de Santa Amália'	7.67	b	25.85	a-d	67.39	a	44.22	ab
16	'Tailândia'	5.98	b	31.38	a-d	69.12	a	53.33	ab
17	'São Mateus'	10.08	b	29.91	a-d	62.56	ab	47.71	ab
19	'Sunrise Solo PT'	17.68	b	25.28	a-d	53.61	a-c	48.45	ab
20	'Waimanalo'	0.00	b	9.22	d	60.33	ab	43.91	ab
21	'Mamão Bené'	56.24	a	67.74	a	45.00	bc	32.38	b
23	'Maradol'	13.61	b	37.17	a-d	63.12	ab	62.67	a
24	'Maradol GL'	2.22	b	44.06	a-d	66.78	ab	62.24	a
25	'Sekati'	24.45	b	35.90	a-d	59.52	ab	50.57	ab
27	'Americano'	12.89	b	42.22	a-d	57.17	a-c	50.00	ab
30	'Calimosa'	18.64	b	40.24	a-d	62.46	ab	45.33	ab
31	'JS 12'	22.89	b	65.36	ab	51.11	a-c	28.00	b
32	'Cariflora'	8.89	b	31.67	a-d	59.83	ab	53.91	ab
33	'GTF'	16.42	b	36.06	a-d	57.14	a-c	32.86	b
34	'STZ-03'	12.38	b	24.67	a-d	57.43	a-c	44.65	ab
35	'SH 12-07'	10.89	b	34.52	a-d	36.29	c	43.33	ab
40	'SH 15-04'	13.39	b	29.56	a-d	47.38	a-c	35.71	ab
45	'Papaya 46'	17.00	b	35.66	a-d	59.78	ab	55.08	ab
46	'FMV'	13.27	b	24.25	a-d	59.90	ab	50.67	ab
49	'Tainung'	23.61	b	31.94	a-d	57.26	a-c	39.33	ab
52	'Sekati FLM'	12.11	b	54.18	a-c	55.91	a-c	44.19	ab
60	'JS 12-4'	4.86	b	50.67	a-d	58.21	a-c	42.00	ab

* For the same variable, means followed by the same letter belong to a statistically homogeneous group by the Tukey test at 0.05 of probability.

11-08' and 'JS 11' differed from the genotypes with lowest means. For an intermediate group of 46 cultivars, the mean between susceptible and resistant genotypes was not significantly different (Table 1).

In the first evaluation of powdery mildew incidence based on leaf symptoms of greenhouse plants, all genotypes except 'Mamão Bené' had a low mean disease intensity. In the second evaluation, the genotypes 'Caliman M5', 'Golden', 'Kapoho Solo' and 'Waimanalo' had means that differed from the most susceptible genotype ('Mamão Bené'). In the third evaluation, 'Mamão Bené' and 'SH 12-07' had means that differed from 'Thailand' and 'Baixinho de Santa Amália' (genotypes with lowest mean disease intensity). Again, genotype 'Mamão Bené' belonged to the group with the lowest mean, and in addition, 'JS 12' and 'GTF' had means that differed from the genotypes with highest mean disease intensity (Table 2).

Compared with the previous results, different effects of genotypes at different evaluation times can be observed, especially for genotype 'Mamão Bené'. According to Cruz et al. (2012), this different performance is caused by genotype-specific physiological and biochemical factors. The aforementioned genotype 'Mamão Bené' was rather susceptible in the first and second evaluations, however, in the last two it was grouped with the genotypes with lowest disease incidence. This fact allows the assumption that plant age may affect the pathogen resistance/susceptibility of a genotype. However, further studies are required to investigate this presumption. Similar mechanisms were also reported in other pathosystems (Rava et al. 1999, Costa et al. 2006).

In the greenhouse evaluations of powdery mildew severity, a significant genotype effect was observed in the first and third evaluations. The results for genotype 'Mamão Bené' were similar to those mentioned above. In the first

Table 3. Comparison of genotype means for powdery mildew severity on papaya leaves of greenhouse plants in three evaluations

Genotype		Severity							
		Evaluation 1		Evaluation 2		Evaluation 3		Evaluation 4	
1	'Caliman M5'	0.01	ab	0.24	a	13.10	ab	1.10	a
5	'Diva'	0.00	b	0.84	a	8.58	ab	1.49	a
11	'Golden'	0.04	ab	0.18	a	5.80	ab	0.74	a
12	'Sunrise Solo 72/12'	0.00	b	0.17	a	9.00	ab	0.82	a
13	'Kapoho Solo PA'	0.00	b	0.10	a	4.44	ab	4.52	a
14	'Baixinho de Santa Amália'	0.00	b	0.18	a	10.50	ab	1.98	a
16	'Tailândia'	0.02	ab	0.04	a	12.04	ab	1.78	a
17	'São Mateus'	0.02	ab	0.39	a	11.04	ab	1.32	a
19	'Sunrise Solo PT'	0.10	ab	0.02	a	7.30	ab	1.92	a
20	'Waimanalo'	0.00	b	0.12	a	7.50	ab	0.90	a
21	'Mamão Bené'	0.34	a	0.84	a	5.40	ab	1.38	a
23	'Maradol'	0.00	b	0.62	a	8.96	ab	0.64	a
24	'Maradol GL'	0.00	b	0.28	a	3.90	ab	1.66	a
25	'Sekati'	0.02	ab	0.42	a	2.30	ab	1.17	a
27	'Americano'	0.03	ab	0.56	a	3.80	ab	1.10	a
30	'Calimosa'	0.04	ab	2.12	a	10.64	ab	2.04	a
31	'JS 12'	0.28	ab	0.92	a	7.74	ab	1.14	a
32	'Cariflora'	0.06	ab	0.96	a	3.90	ab	1.52	a
33	'GTF'	0.02	ab	0.14	a	5.68	ab	2.94	a
34	'STZ – 03'	0.02	ab	0.04	a	2.60	ab	1.10	a
35	'SH 12-07'	0.02	ab	0.10	a	1.76	b	0.50	a
40	'SH 15-04'	0.00	b	2.36	a	4.96	ab	0.54	a
45	'Papaya 46'	0.10	ab	0.66	a	9.80	ab	1.90	a
46	'FMV'	0.00	b	0.48	a	14.00	a	3.22	a
49	'Tainung'	0.22	ab	0.73	a	6.56	ab	0.32	a
52	'Sekati FLM'	0.10	ab	0.28	a	1.58	b	0.27	a
60	'JS 12-4'	0.00	b	2.66	a	10.80	ab	0.32	a

* For the same variable, means followed by the same letter belong to a statistically homogeneous group by the Tukey test at 0.05 of probability.

assessment, this genotype had the highest mean severity, however, in the third assessment, it formed an intermediate group (Table 3). The group with lowest severity was formed by the genotypes 'Diva', 'Sunrise Solo 72/12', 'Kapoho Solo PA', 'Waimanalo', 'Maradol', 'Maradol GL', 'SH 15-04', 'FMV', and 'JS 12-4' in the first evaluation. In the third evaluation, 'SH 12-07' and 'Sekati FLM' had the lowest disease severity levels (Table 3).

These results are promising for the selection of superior genotypes, with considerable reductions of powdery mildew incidence and/or severity. Of the tested genotypes, 'Maradol', 'Maradol GL', 'Caliman M5', and 'GTF' stood out with lowest mean disease intensity in field and greenhouse. These genotypes can be used in crosses and selection to develop lines and/or hybrids, which aside from the traits of resistance to the above disease, also have agronomic traits of interest such as fruit yield and quality from the agricultural point of view.

ACKNOWLEDGEMENTS

The authors thank the Darcy Ribeiro North Fluminense State University – UENF and the Fundação Carlos Chagas Filho de Amparo à Pesquisa do Estado do Rio de Janeiro (FAPERJ) – FAPERJ, for financial support. The authors also thank the Caliman Agrícola S/A for financial and logistical support.

REFERENCES

Cardoso DL, Silva RF, Pereira MG, Viana AP and Araújo EF (2009) Diversidade genética e parâmetros genéticos relacionados à qualidade fisiológica de sementes em germoplasma de mamoeiro. **Revista Ceres 56**: 572-579.

Costa IFD, Balardin RS, Medeiros LA and Bayer TM (2006) Resistência de seis cultivares de soja ao *Colletotrichum truncatu*m (Schwein) em dois estádios fenológicos. **Ciência Rural 36**: 1684-1688.

Cruz CD (2013) GENES: A software package for analysis in experimental statistics and quantitative genetics. **Acta Scientiarum Agronomy 35**: 271-276.

Cruz CD, Regazzi AJ and Carneiro PC (2012) **Métodos biométricos aplicados ao melhoramento genético**. Editora UFV, Viçosa, 512p.

Dias NLP, Oliveira EJ and Dantas JLL (2011) Avaliação de genótipos de mamoeiro com uso de descritores agronômicos e estimação de parâmetros genéticos. **Pesquisa Agropecuária Brasileira 46**: 1471-1479.

Falconer DS (1987) **Introduction to quantitative genetics**. Longman, London, 340p.

Foltran DE, Gonçalves PS, Sabino JC, Igue T and Vilela RCF (1993) Estimativas de parâmetros genéticos e fenotípicos em mamão. **Bragantia 52**: 7-15.

Gepts P (2006) Plant genetic resources conservation and utilization: the accomplishments and future of a societal insurance policy. **Crop Science 46**: 2278-2296.

Liberato JR, Barreto RW and Louro RP (2004) *Streptopodium caricae* sp. nov, with a discussion on powdery mildew of papaya and an emended description of the genus *Streptopodium* and of *Oidium caricae*. **Mycological Research 108**: 1185-1194.

Martelleto LAP, Ribeiro RLD, Surdo-Martelleto M, Vasconcellos MAS, Marin SLD and Pereira MB (2008) Cultivo orgânico do mamoeiro 'Baixinho de Santa Amália' em diferentes ambientes de proteção.

Revista Brasileira de Fruticultura 30: 662-666.

Rava CA, Costa JGC and Andrade EM (1999) Influência da idade da planta de feijoeiro comum na resistência à antracnose. **Pesquisa Agropecuária Gaúcha 5**: 325-330.

Rezende JAM and Martins MC (2005) Doenças do mamoeiro (*Carica papaya* L.). In Kimati H, Amorim L, Rezende JAM, Bergamin Filho A and Camargo LEA (eds) **Manual de fitopatologia: Doenças das plantas cultivadas**. Agronômica Ceres, São Paulo, p. 435-443.

Santos PHD, Vivas M, Silveira SF, Silva JM and Terra CEPS (2011) Elaboração e validação de escala diagramática para avaliação da severidade de oídio em folhas de mamoeiro. **Summa Phytopathologica 37**: 215-217.

Silva FF, Pereira MG, Ramos HCC, Damasceno Junior PC, Pereira TNS, Gabriel APC, Viana AP, Daher RF and Ferreguetti GA (2008) Estimation of genetic parameters related to morpho-agronomic and fruit quality traits of papaya. **Crop Breeding and Applied Biotechnology 8**: 65-73.

Suzuki MS, Zambolim L and Liberato JR (2007) Progresso de doenças fúngicas e correlação com variáveis climáticas em mamoeiro. **Summa Phytopathologica 33**: 167-177.

Tatagiba JT, Liberato JR, Zambolim L, Costa H and Ventura JA (2002) Controle químico do oídio do mamoeiro. **Fitopatologia Brasileira 27**: 219-222.

Treichel M, Kist BB and Santos CE (2016) **Anuário brasileiro da fruticultura**. Editora Gazeta Santa Cruz, Santa Cruz do Sul, 88p.

Vivas M, Silveira SF, Terra CEPS and Pereira MG (2010) Reação de germoplasma e híbridos de mamoeiro à mancha-de-phoma (*Phoma caricae-papayae*) em condições de campo. **Tropical Plant Pathology 35**: 323-328.

Vivas M, Silveira SF, Vivas JMS and Pereira MG (2012) Patometria, parâmetros genéticos e reação de progênies de mamoeiro à pinta-preta. **Bragantia 71**: 235-238.

Vivas M, Silveira SF, Vivas JMS, Viana AP, Amaral Junior AT and Pereira MG (2014) Seleção de progênies femininas de mamoeiro para resistência a mancha-de-phoma via modelos mistos. **Bragantia 73**: 446-450.

17

BRS AG: first cultivar of irrigated rice used for alcohol production or animal feed

Ariano Martins de Magalhães Júnior[1*], Paulo Ricardo Reis Fagundes[1], Daniel Fernandes Franco[1], Orlando Peixoto de Morais[2], Félix Gonçalves de Siqueira[3], Eduardo Anibele Streck[1], Gabriel Almeida Aguiar[1] and Paulo Henrique Karling Facchinello[1]

Abstract: *BRS AG is an irrigated rice cultivar developed by Embrapa, recommended for cultivation in the state of Rio Grande do Sul. It is intended for grain alcohol production or animal feed, with average thousand-grain weight of 52g (double of the conventional cultivars) and average yield of 8193 kg ha⁻¹.*

Key words: *Oryza sativa L., large grain, crop diversification.*

*Corresponding author:
E-mail: ariano.martins@embrapa.br

[1] Embrapa Clima Temperado, Rodovia BR-392, km 78, 9º Distrito, Monte Bonito, CP 321, 96.010-971, Pelotas, RS, Brazil
[2] Embrapa Arroz e Feijão, Rodovia GO-462, km 12, Fazenda Capivara, Zona Rural, CP 179, 75.375-000, Santo Antônio de Goiás, GO, Brazil
[3] Embrapa Agroenergia, Parque Estação Biológica (pqEB), PqEB s/nº, CP 40.315, 70.770-901, Brasília, DF, Brazil

INTRODUCTION

The economy in the southern half of Brazil is mainly based on agriculture, the irrigated rice and the production of beef and sheep meat being the most prominent sectors. The soybean, maize and sorghum cultures, which are naturally the option to be part of the production system of these areas, though present some problems. The pointed reasons are the climate featuring water restriction in part of the territory, the difficulties of farming in lowlands (hydromorphic, poorly drained soils - ideal for the cultivation of irrigated rice, but challenging for other crops). From the rice production perspective, the RS state meets more than 65% of the Brazilian demand, with approximately 1.1 million hectares being cultivated per year (CONAB 2016).

With the advent of plant breeding, the rate of the genetic progress of rice varieties was intensively accelerated (Breseguello and Coelho 2013), resulting in significant increases in the yield potential.

Thus, Brazil has achieved self-sufficiency in rice production and it seems inevitable that it will continue in this condition of increased production (Adami and Miranda 2011), so there is the need to develop strategies for diversification and development of new ways in the paddy chain. Consequently, thereby avoiding that the surplus production generated reduces the price paid for the product, which would make the activity economically unviable, not covering the production costs.

In this context of the productive chain and the search for renewable energies that are economically viable, BRS AG was developed to use of the rice as a raw material for the production of cereal alcohol or animal feed. It is worth emphasizing that the RS state imports 98% ethanol for consumption with a higher spending to R$ 1 billion per year. Therefore, the aim of this article is to describe agronomically the first cultivar of irrigated rice developed by Embrapa for this purpose.

PEDIGREE AND BREEDING METHOD

The BRS AG is the result of a simple cross between the American irrigated rice cultivar Gulfmont and the SLG1 lineage, of Japanese origin, held in 1994/1995. The F1 seeds were harvested and stored for planting in the 1995/1996 crop, which were sown in a greenhouse and, after reaching 20 cm in height, were transplanted to the experimental field of Embrapa Temperate Climate Lowlands Station. F_1 lines were transplanted alongside their parents for comparison and roguing of selfing. After the rouguing, immature anthers were collected from the effectively hybrid plants that were placed to callus induction "in vitro" in a laboratory and subsequent regeneration of homozygous plants, being acclimatized in a greenhouse for seed production. The chromosome duplication process occurred spontaneously without use of colchicine.

In the 1996/1997 crop, the homozygous seeds were sown on the field originating the CL 485 lineage, where it was observed the agronomic performance of the same. Among the assessed traits, the large size of the grains of the lineage stood out, with average thousand-grain weight around 52 g, almost the double of the mass belonging to the cultivars currently used for the production of irrigated rice in Brazil.

This lineage, by not following the standards required by the national preference, which is a grain of long, thin type, vitreous and loose after cooking, was kept in the program as a source of genetic variability.

In the 2010/11 crop, due to a production surplus in the rice production chain, Embrapa dealt with the demand of a differentiated pattern of rice, seeking to economically enable the paddy activity. In search of a cultivar that would be presented as an alternative of raw material for animal feed or for the production of grain alcohol, removing part of the long fine grain from the consumer market.

In order to satisfy this request, in the 2011/2012 crop, the CL 485 lineage was recovered, which was recoded as AB11047, being part of the trials of Value of Cultivation and Use (VCU), in different parts of Rio Grande do Sul. In 2011/2012, these trials were conducted in Capão do Leão, Santa Vitória do Palmar and Alegrete. In 2012/2013, the trials were conducted in Capão do Leão, Santa Vitória do Palmar and Capivari do Sul.

PERFORMANCE CHARACTERISTICS

The BRS AG is the result of a crossing performed at Embrapa Temperate Climate, which involved genes of the introduced genotype SLG1 (super large grain) (Takita 1983), whose grain dimensions are larger than those of the conventional rice, being nicknamed "Gigante" (Giant). It has an average thousand-grain weight equivalent to 52g, while the majority of irrigated rice cultivars show lower values, as BRS Pampa, showing 25.6 g (Table 1).

The plants of this cultivar (Table 1) have a biological cycle around 126 days, from emergence to maturity. The average plant height is 110 cm, with distinctive architecture (modern-Philippine), however, the thickness of the stem is 5.5 mm, which gives them strong culms, resistant to lodging, despite this high stature of the plants.

The leaves are hairy, the flag leaf being classified as decumbent. It is abscission-resistant, therefore, without risk of becoming a weed plant in the paddy crop. Allied to this fact, the seeds of this lineage lose the germination capacity and the vigor with great ease, helping to reduce their infestation potential, thus they can not be compared to the main weed crop: the red rice.

Through the 35 morphological characters evaluated, it was found that the BRS AG had a high genetic distance from the modern-Philippine cultivars previously released by Embrapa, corroborating with Streck et al. (2017). Thus, this release does not follow mentioned by Rabelo et al. (2015) who found that the irrigated rice Brazilian cultivars have narrow genetic base. This result can be explained mainly by the intrinsic characteristics regarding the type of grain and plant architecture, the BRS AG neither has grains of the *indica* subspecies type nor modern-Philippine architecture. Yet the Arborio genotype, widely used in the Italian cuisine, showed the greatest genetic similarity (Figure 1), being considered the one with best agronomic comparison to the "Gigante".

Regarding the genetic potential of productivity, the BRS AG cultivar stands out greatly in relation to the Arborio genotype, showing yield plateaus around 10,000 kg ha^{-1}. In comparative tests in the municipalities of Capão do Leão and Santa Vitória do Palmar (Figure 2), it showed an average yield of 8,193 kg ha^{-1}, approaching the average yield obtained by the BRS Pampa (9,215 kg ha^{-1}), which is one of the main cultivars in Rio Grande do Sul.

Notwithstanding, the BRS AG "Gigante", by having very large grains and a low amylose-amylopectin ratio (starchy grains), results in a lower quality in cooking, not adapting to the standard Brazilian consumption. It is the first rice cultivar intended for uses other than human consumption, as grain alcohol production or animal feed.

To check the efficiency for these purposes, it were made assessments of the hydrolysis processes, the ethanol fermentation and the bromatological composition. These activities were carried out by the EMBRAPA Agroenergy

Table 1. Comparison of the grain and plant characteristics between the irrigated rice cultivars BRS AG and BRS Pampa

Characteristics	BRS AG	BRS PAMPA
Grain		
Caryopsis shape*	half spindle-shaped	spindle-shaped
Awns	absent	absent
Glumes colour	straw	straw
Colour of apiculus in flowering	white	white
Colour of apiculus in maturaty	white	white
Pubescence grain	present	present
Length of grain (mm)**	7.82	9.82
Width of grain (mm)**	3.64	2.20
Thickness of grain (mm)**	2.60	2.00
Length of caryopsis (mm)**	6.30	7.19
Width of caryopsis (mm)**	2.60	1.96
Thickness of caryopsis (mm)**	1.90	1.76
Length / width $^{-1}$ ratio (mm)*	2.15	3.59
1000-grain weight (g)**	52.00	25.60
Total income (%)**	77.60	68.00
Intact grain (%)****	71.60	62.00
Endosperm amylose content	high	high
Gelatinization temperature	low	low
Potential productivity (t ha^{-1})***	10	10
Plant		
Plant type	intermediate	modern
Days to flowering	96	88
Days to maturity**	126 (medium)	118 (precocious)
Plant height (cm)**	110	96
Culm length (cm)**	88	72
Panicle length (cm)**	23	24
Panicle exsertion*	medium	medium
Leaf colour	green	green
Flag leaf angle	decumbent	erect
Auricle colour	light green	light green
Ligule colour	colorless	colorless
Culm internode color	light green	light green
Anthocyanin colouration on the culms	absent	absent
Panicle type	intermediate	intermediate
Leaf blade pubescence	present	present
Dehiscence*	tolerant	intermediate
Lodging tolerance*	tolerant	tolerant
Tillering*	low	high
Iron indirectly toxicity**	moderate tolerance	moderate tolerance
Rice blast in leaf and panicle**	moderate resistant	moderate resistant
Stain grains**	moderate sensitive	moderate resistant

*Length / width $^{-1}$(withoutshell) ratio
** Can be changed depending on the characteristics of the cultivation environment.
*** Grains in shell, 13% humidity, observed in experiments conducted by Embrapa.
**** The BRS AG cultivar underwent no polishing, while the BRS Pampa was peeled and polished in a Suzuki test device.

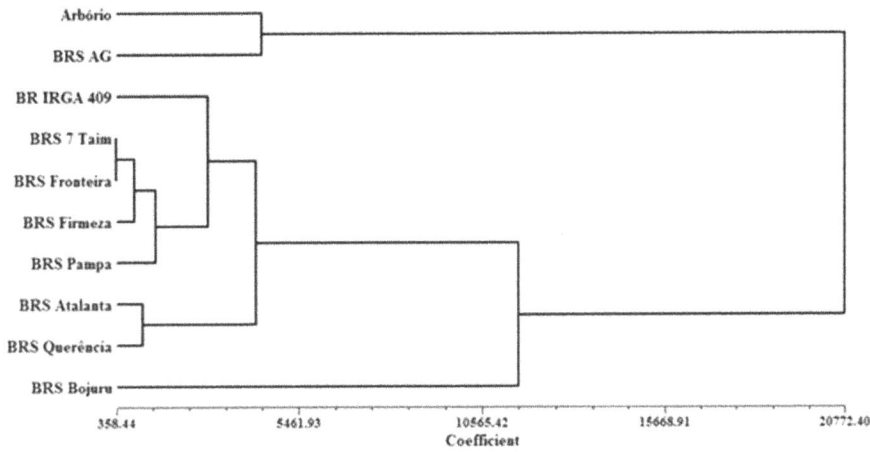

Figure 1. Dendrogram resulting from the genetic distance analysis of 10 irrigated rice genotypes characterized by 35 morphological characters, obtained by the UPGMA grouping method.

laboratories.

The actual yield of the enzymatic hydrolysis of the BRS Pampa and BRS AG grains in the enzyme concentration of 0.1% (m v^{-1}), by the action of the alpha-amylase (dextrinizing) and amyloglucosidase (saccharification) enzymes, was approximately 108.9 g L^{-1} glucose for both genotypes after 24 hours of the procedure. Yet in the enzyme concentration of 0.2%, the yield was 132 g L^{-1} for the BRS Pampa and 155.1 g L^{-1} for the BRS AG. Thus, the efficiencies of the enzymatic hydrolysis for the ground grains of the BRS Pampa and BRS AG cultivars were approximately 89.4% and 104.7%, respectively, for the enzyme concentration of 0.2% (Table 2).

After 24 hours of fermentation for the hydrolysate (with cake) with 0.2% enzyme concentration, it were verified yields of 3.57 g (L h)$^{-1}$ for BRS Pampa and 4.04 g (L h)$^{-1}$ for

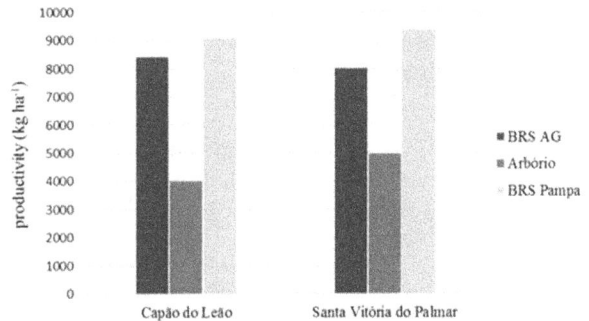

Figure 2. Average yield of grains of the irrigated rice cultivar BRS AG in relation to BRS Pampa and Arborio in the municipalities of Capão do Leão and Santa Vitória do Palmar in the 2011/2012 and 2012/2013 crops.

BRS AG. The ethanol productivity for testing the hydrolysates without the presence of the cake, arising from the process with 0.2% enzyme, showed results of 4.92 and 4.53 g (L h)$^{-1}$ for BRS Pampa and BRS AG, respectively (Table 3).

The bromatological analyses carried out (ground grains) and those regarding the products from the hydrolysis and fermentation processes (Table 4) showed high content of dry matter, with values over 90% for all samples and values less than 6% for the content of mineral matter (ash). The Acid Detergent Fiber (ADF) results were below 16% for most of the samples, while the Neutral Detergent Fiber (NDF) values ranged from 7 to 28.5%, highlighting the sample of the BRS AG rice, analyzed together with the yeast, which showed a percentage of 41.03%. The lignin and ether extract contents were below 8% for most of the analyzed material.

All samples showed a high crude protein content, with values above 30% for most of these materials. The products from fermentation processes, with the presence of yeast, had significant values for total protein, in which the material

Table 2. Efficiency of the enzymatic hydrolysis (%) of samples of the irrigated rice cultivars BRS AG "Gigante" and BRS Pampa, by the alpha-amylase (90 °C for one hour) and amyloglucosidase (60 ºC for 6 hours) enzymes, in the concentration of 0.2% (m v^{-1})

Cultivar	Sample (g)	Volume reactor (L)	Hydrolysis yield (g L^{-1})	Enzymatic efficiency (%)
BRS Pampa	200	1.2	132.4	89.4
BRS AG	200	1.2	155.1	104.7

Table 3. Productivity, substrate yield and efficiency of fermentation of the hydrolysates of samples of the irrigated rice cultivars BRS AG "Gigante" and BRS Pampa, with and without the presence of the residual cake (enzymatic post-hydrolysis, for 24 hours, in the concentration of 0.2%)

Cultivar	Fermentation conditions	Ethanol productivity (L h)$^{-1}$	Substrate yield (gg^{-1})
BRS Pampa	With residual cake	3.57	0.63
	No residual cake	4.92	0.89
BRS AG	With residual cake	4.04	0.83
	No residual cake	4.53	0.79

Table 4. Bromatological analyses of the cakes of the irrigated rice cultivars BRS AG "Gigante" and BRS Pampa, enzymatic post-hydrolysis (without yeast) and post-hydrolysis/fermentation (with yeast), in natural or raw state (ground grain). Analyses made in experiments with the enzyme concentration of 0.2% (v v^{-1})

Cultivar	Sample	Dry matter	Ash	Neutral detergent fiber	Acid detergent fiber	Lignin	Ether extract	Protein
BRS Pampa	Gross	90.06 ± 0.32	3.81 ± 0.05	8.22 ± 0.32	3.57 ± 0.187	0	4.60 ± 0.352	12.22 ± 0.126
	Yeast extract	93.82 ± 1.04	4.27 ± 0.73	0.94 ± 0.26	0.56 ± 0.012	0	1.64 ± 0.12	38.28 ± 0.067
	Cakeno yeast	93.64 ± 0.10	1.64 ± 0.94	18.96 ± 0.23	11.01 ± 0.095	7.60 ± 0.384	5.76 ± 0.18	31.42 ± 0.365
	Cake with yeast	96.00 ± 0.27	2.30 ± 0.04	21.26 ± 2.17	9.99 ± 0.926	7.33 ± 2.58	3.05 ± 0.31	46.45 ± 0.157
BRS AG	Gross	91.26 ± 0.44	0.71 ± 0.26	7.28 ± 0.27	2.96 ± 0.122	1.31 ± 0.312	2.08 ± 0.14	11.99 ± 0.343
	Yeast extract	94.21 ± 0.11	5.80 ± 1.42	0.57 ± 0.17	0.22 ± 0.093	0	1.67 ± 0.29	39.89 ± 0.085
	Cake no yeast	92.64 ± 0.15	0.27 ± 0.05	28.46 ± 0.44	15.31 ± 0.384	5.17 ± 1.04	6.43 ± 0.32	34.44 ± 0.130
	Cake with yeast	95.79 ± 0.20	3.41 ± 0.01	41.03 ± 0.22	14.78 ± 0.228	2.20 ± 2.34	6.77 ± 0.10	51.54 ± 0.193

from the BRS AG cultivar reached the percentage of 51.54%, against 46.45% of the BRS Pampa cultivar. Kunrath et al. (2010) studied, with the same analytical technique, the chemical composition of defatted rice bran for pigs and found values similar to those of this work, showing great potential of the BRS AG cultivar for animal nutrition.

In this context, this type of grain of the BRS AG cultivar, referred to as "DHC" (UHC - unfit for human consumption), has high potential as a feedstock for grain alcohol production and for animal feed. Another advantage of the BRS AG concerns the shell/grain^{-1} ratio, when compared to conventional grain cultivars, of long, thin type, presenting 17.5% shell, while the conventional cultivars present 22%.

BASIC SEED PRODUCTION

The BRS AG is registered in the National Register of Cultivars (RNC) and protected by the Ministry of Agriculture, Livestock and Supply (MAPA). The Business Office of Capão do Leão, of Embrapa Products and Market is responsible for providing the basic seeds of the aforementioned cultivar.

It is indicated the sowing of the cultivar BRS AG "Gigante", following the agricultural zoning for the irrigated rice crop in Rio Grande do Sul. It is recommended that the sowing occurs respecting the cycle of the cultivar in interaction with the cultivation environment, so that the differentiation of primordia occurs until January 1 or as close to that date as possible.

The density of suitable seeds (100% germination capacity) should be about 80 seeds per linear meter (about 200 kg ha^{-1}) for the on-line system, since it does not show high tillering capacity, ensuring a plant population from 200 to 300 plants per square meter (SOSBAI 2014).

The BRS AG shows positive response to different levels of basic and coverage fertilization, without the lodging of plants.

The harvest of this cultivar, aiming to minimize the natural abscission and prevent the grain breakage during the manufacturing process, must be performed when the grain moisture content is between 23% and 18%.

In the homogeneity tests, BRS AG "Gigante" has shown to be uniform, without the presence of atypical plants, demonstrating to be genetically stable, even by the fact that the homozygosity was obtained through the culture of immature anthers.

REFERENCES

Adami ACO and Miranda SHG (2011) Transmissão de preços e cointegração no mercado brasileiro de arroz. **Revista de Economia e Sociologia Rural 49**: 55-80.

Breseguello F and Coelho ASG (2013) Traditional and modern plant breeding methods with examples in rice (*Oryza sativa* L.). **Journal of Agricultural and Food Chemistry 61**: 8277-8286.

CONAB - Companhia Nacional de Abastecimento (2016) **Acompanhamento da safra brasileira: Grãos**. Available at <http://www.conab.gov.br/OlalaCMS/uploads/arquivos/16_06_09_16_49_15_boletim_graos_junho__2016_-_final.pdf>. Accessed on June 21, 2016.

Kunrath MA, Kessler AM, Ribeiro AML, Vieira MM, Silva GL and Peixoto FD'Á (2010) Metodologias de avaliação do valor nutricional do farelo de arroz desengordurado para suínos. **Pesquisa Agropecuária Brasileira 45**: 1172-1179.

Rabelo HO, Guimarães JFR, Pinheiro JB and Silva EF (2015) Genetic base of Brazilian irrigated rice cultivars. **Crop Breeding and Applied Biotechnology 15**: 146-153.

SOSBAI - Sociedade Sul-brasileira de Arroz Irrigado (2014) **Arroz irrigado: recomendações técnicas da pesquisa para o Sul do Brasil**. Editora SOSBAI, Bento Gonçalves, 189p.

Streck EA, Aguiar GA, Magalhães Júnior AM, Facchinello PHK and Oliveira AC (2017) Variabilidade fenotípica de genótipos de arroz irrigado via análise multivariada. **Revista Ciência Agronômica 48**: 101-109.

Takita T (1983) Breeding of a rice line with extraordinarily large grains as a genetic source for high yielding varieties. **Japan Agricultural Research Quarterly 17**: 93-97.

18

RB036088 – a sugarcane cultivar for mechanical planting and harvesting

Edelclaiton Daros[1*], Ricardo Augusto de Oliveira[1], José Luis Camargo Zambon[1], João Carlos Bespalhok Filho[1], Bruno Portela Brasileiro[1*], Oswaldo Teruyo Ido[1], Lucimeris Ruaro[1] and Heroldo Weber[1]

Abstract: *The sugarcane cultivar RB036088 is late-maturing, harvested from September to November in south-central Brazil, and is recommended for soils with medium to high fertility. It stands out with continuously high sugar yield over the harvests, longevity of the ratoon plants, high tillering capacity and can be harvested mechanically.*

Key words: *Saccharum spp., selection, improvement.*

*Corresponding author:
E-mail: brunobiogene@hotmail.com

[1] Universidade Federal do Paraná (UFPR), Departamento de Fitotecnia e Fitossanitarismo, 80.035-050, Curitiba, PR, Brazil

INTRODUCTION

The breeding program of sugarcane (*Saccharum* spp.) of the Federal University of Paraná [PMGCA/UFPR (www.pmgca.ufpr.br)] is part of an inter-university network to foster the development of the sugar-energy sector - RIDESA (www.ridesa.com.br), a framework focused on sugarcane breeding involving 10 Federal Universities (Barbosa et al. 2012). Among the main objectives of RIDESA is the development of sugarcane varieties with different maturation periods and satisfactory yields under specific crop management conditions, e.g., mechanical planting and harvesting (Iaia et al. 2014; Barbosa et al. 2015; Carneiro et al. 2015; Daros et al. 2015). In south-central Brazil, the harvest of cultivars with medium to late maturity and high yields in mechanical cultivation systems is a challenge, due to the damage caused by the currently used sugarcane harvesting machines.

Sugarcane cultivar RB036088 is recommended for planting in medium to high fertility soils, has a high phenotypic stability for the trait tons of sucrose per hectare (TSH), aside from an excellent response to improvements in the production environment. However, the main highlight of this new cultivar is its suitability for mechanical planting and harvesting systems, with continuously high agricultural yields in terms of sugar production per area in the different production cycles. In view thereof, it can be considered an ideotype for mechanical harvesting.

In the central-south region of Brazil, the recommended harvest time for 'RB036088' is the end of the growing season, from September to November. However, when ripeners are applied, it can be cut in the middle of the growing season, from July to August. It has upright, tall growth and excellent suitability for mechanical harvesting, ensuring low levels of mineral impurities for industrial use.

Aside from the above advantages, 'RB036088' can maintain high agricultural productivity over the crop cycles, due to the outstanding tillering capacity, along with the growth of medium-diameter stalks and short to medium internodes, the tall, medium-weight stalks and easily removable leaves.

PEDIGREE AND BREEDING METHOD

In 2003, sugarcane caryopses were obtained from the cross of parent RB855595 with pollen from several other parents of the experimental station Estação de Floração e Cruzamento da Serra do Ouro (lat 9º 13′ S, long 35º 50′ W, alt 450 m asl), in the municipality of Murici, Alagoas, of the Federal University of Alagoas (Figure 1). This parent has an excellent selection rate, since it produces progenies with potential for stalk yield per area (tons of cane per hectare -TCH) and with excellent bud sprouting, upright growth habit, strong tillering, and resistance to the major diseases of the crop. In the same year, the caryopses were germinated in a greenhouse of the experimental station of Paranavaí (lat

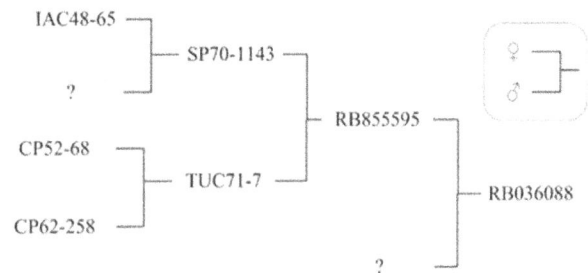

Figure 1. Pedigree of cultivar RB036088.

23º 05′ S, long 52º 27′ W, alt 503 m asl), of the Federal University of Paraná, in the municipality of Paranavaí, Paraná. Plants of the first selection stage (T1) were first planted in the field in November 2003, in two production environments, in the municipalities of Colorado and São Tomé, resulting in approximately 200,000 seedlings derived from hundreds of parents. Individual selection was performed in July 2005 in the sugarcane ratoon cycle. In the first clonal multiplication in 2005 (stage T2), plants were cultivated at two locations of the state of Paraná (Colorado and São Tomé). Each genotype of stage T2 was planted in two 5-m- long rows, spaced 1.4 m apart, in an augmented block design. Clone selection was performed in the experiments in São Tomé in the stages T1 and T2, due to the excellent agronomic performance over three growing seasons, and T2 was named "PRP03088". In 2008, the next stage (stage T3), evaluation and selection were carried out based on data from seven locations in two seasons in Paraná [Mandaguaçú (lat 23º 21′ S, long 52º 05′ W, alt 580 m asl), Bandeirantes (lat 23º 06′ S, long 50º 22′ W, alt 492 m asl), Paranavaí (lat 23º 05′ S, long 52º 27′ W, alt 503 m asl), Colorado (lat 22° 50′ S, long 51° 54′ W, alt 400 m asl), Goioerê (lat 24° 10′ S, long 53° 01′ W, alt 550 m asl), Perobal (lat 23° 54′ S, long 53° 24′ W, alt 410 m asl), Astorga (lat 23° 11′ S, long 51° 09′W, alt 634 m asl), São Pedro do Ivaí (lat 23° 52′ S, long 51° 41′ W, alt 400 m asl). In 2010, the clonal multiplication phase was initiated and in the following year, clone RB036088 was selected for the experimental phase of PMGCA, carried out at 10 locations in Paraná. At this stage, the agronomic traits were evaluated, i.e., tons of cane and sucrose per hectare, as well as adaptability and yield stability in different soil-climatic areas of the northern and northwestern regions of the state of Paraná (Figure 2a, b). This stage was monitored for four growing seasons. In this period, the resistance/tolerance to the major diseases relevant for the south-central sugarcane region was also evaluated. Between 2011 and 2012, experiments were conducted at nine locations in Paraná to evaluate the maturation period of cultivar RB036088. Prior to the market release, data of 54 growing seasons were compiled, covering from the first cut (10 growing seasons) to the fourth cut (6 growing seasons), which allowed a description of the main qualities of the cultivar, particularly the high yield of ratoon crops, associated with high yield stability and wide adaptability to soils, with medium to high agricultural productivity (Oliveira et al. 2015).

Since 2011, when the preliminary positive results were repeated in various environments in Paraná, sugar mills and distilleries indicated an interest in planting 'RB036088' in multiplication areas for performance evaluation under the management conditions of the production units. This resulted in the confirmation of the excellent performance of the cultivar under different conditions of commercial management. During this period, the cultivar stood out with an excellent performance in areas with mechanical harvesting, due to its upright growth habit and tall height, with no lodging, even in the final stage of the crop cycle of sugarcane plantations. This information motivated further propagation of the cultivar and in 2015, when 'RB036088' was officially released, over 10% of the sugarcane acreage was already destined for this cultivar in commercial areas in Paraná, due to all qualities described above.

In June 2016, the Federal University of Paraná requested the protection of cultivar RB036088 by the National Plant Variety Protection (SNPC) and National Register of Cultivars of the Ministry of Agriculture, Livestock and Supply (MAPA).

PERFORMANCE

The experimental results obtained in sugar mills and distilleries of Paraná demonstrated the superior performance of cultivar RB036088 over the standard cultivars (RB867515 and RB855536), mainly in medium to high fertility soils, as shown by the results obtained by the method of stability and adaptability proposed by Eberhart and Russell (1966) (Figure 2). Yield stability was high in various environments, indicating superior performance of yield-related traits in in medium to high fertility soils.

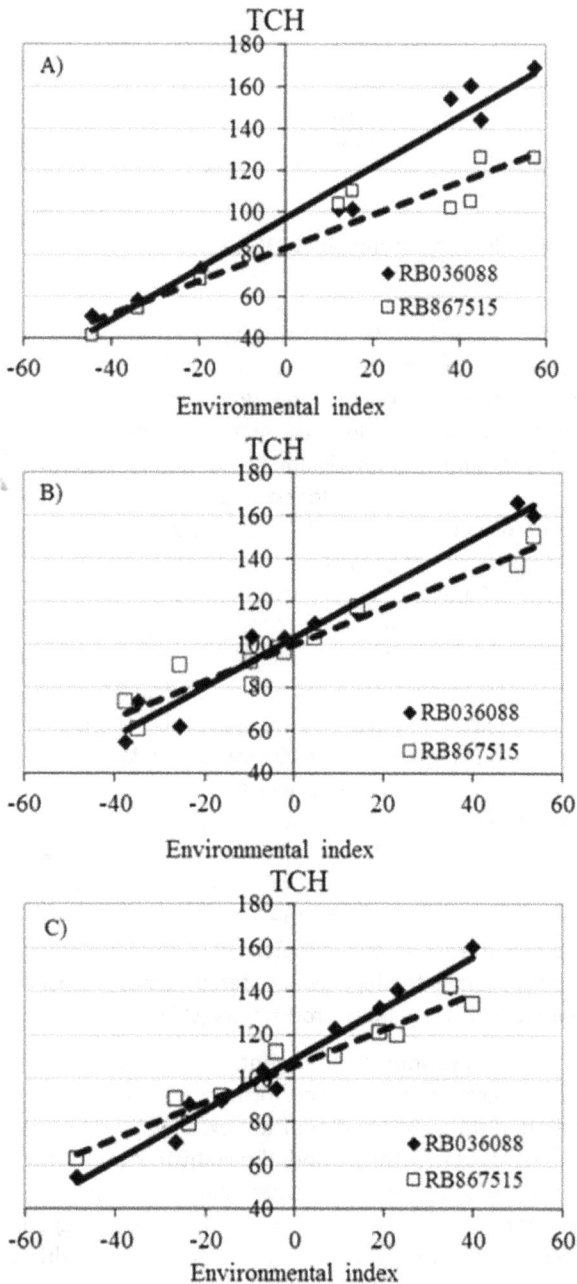

The phenotypic adaptability of cultivar RB036088 was also high, indicating excellent response in yield for cultivation in high fertility soils. This performance was observed both for plant cane as well as ratoon cane crops (Figure 2). In a comparison of cultivar RB036088 with the standard cultivars, there was a 14% increase in TCH yield in the mean of four cycles (Table 1). This characteristic of high agricultural productivity (111.06 Mg ha^{-1}) associated with a mean sucrose content enabled an increase in sucrose yield.

The maturation curve of cultivar RB036088 was constructed with data from eight evaluation locations in Paraná. The data of sucrose percentage in cane juice (SPC) was determined according to the method described by Fernandes (2003). The data of SPC indicated late maturation of the cultivar, suggesting harvest from September onwards in the center-south of Brazil (Figure 3). When comparing the maturation curve of 'RB036088' with another late-maturing cultivar (RB867515), sucrose levels were similar from August to September. In comparison with the intermediate-maturing cultivar RB855536 on high fertility soils (Bandeirantes and São Pedro do Ivai), it was found that from September onwards the SPC increased, reaching the standard of early-maturing cultivars. This new cultivar is therefore an excellent alternative for cultivation in the south-central region, in view of the high stalk yield and sugar content for harvest between September and November. In years with high flowering induction in sugarcane plantations, frequent flowering of cultivar RB036088 was observed, but in evaluations of stalk density and stalk sucrose content, the performance was 2 to 5% below that of other cultivars commonly used as standards in experimental trials. In contrast, the application of ripeners in the management of cultivar RB036088 shifted the harvest period to July and August, with high agricultural yields, expanding the period of adequacy for industrial use of this new sugarcane cultivar.

Cultivar RB036088 has a high level of plant health, and is resistant to brown rust (*Puccinia melanocephala* H. and P. Sydow), resistant to orange leaf rust (*Puccinia kuehnii* H. and P. Sydow, resistant to sugarcane smut (*Sporisorium scitamineum* (Syd.) M. Piepenbr., M. Stoll & Oberw), according to evaluations in field experiments with natural infection.

Figure 2. Phenotypic performance of RB036088 and RB867515 in 10 environments in (a) plant-cane, (b) first ratoon and (c) second ratoon crop, Paraná. * TCH – tons of cane per hectare.

Table 1. Comparison of RB036088 with mean yields in crop cycles of important cultivars tested in Paraná, from 2010 to 2014

Crop cycle	Cultivar	Cane yield t ha⁻¹	(%)*	Sugar yield t ha⁻¹	(%)	SPC** (%)	(%)	Fiber (%)	(%)	Apparent sucrose g kg⁻¹	(%)
	RB867515	98.56	100	11.69	100	11.69	100	12.04	100	139.45	100
Plant-cane	RB855536	84.54	86	10.61	91	12.37	106	9.31	77	142.64	102
	RB036088	114.43	116	13.14	112	11.32	97	12.76	106	139.41	100
	RB867515	112.09	100	13.42	100	13.42	100	11.70	100	134.55	100
First-ratoon	RB855536	111.26	99	13.66	102	13.76	103	11.07	95	138.70	103
	RB036088	122.75	110	13.89	104	12.69	95	11.94	102	133.89	100
	RB867515	90.91	100	15.23	100	15.23	100	12.44	100	150.19	100
Second-ratoon	RB855536	100.89	111	17.00	112	15.32	101	11.05	89	155.94	104
	RB036088	110.18	121	17.36	114	14.32	94	12.81	103	149.37	99
	RB867515	106.40	100	16.85	100	16.85	100	13.42	100	142.00	100
Third-ratoon	RB855536	98.65	93	14.73	87	15.89	94	10.80	81	149.64	105
	RB036088	111.06	104	15.39	91	14.74	87	12.34	92	143.34	101

* Relative yield, considering cultivar RB867515 as reference.
** SPC - Pol in juice (sucrose percentage in cane juice).

OTHER TRAITS

Based on the official descriptors for sugarcane (SNPC/MAPA), cultivar RB036088 has an upright growth habit, a purple sugarcane heart with a low amount of wax, semi-open, minor part of visible internodes, and medium to easy husking. Regular amount of leaves with upright leaf architecture and closed capitula with short green sugarcane heart. The stalks have conoidal-shaped internodes, a round section, arranged in gentle zigzag, short to medium length and medium diameter, greenish - yellow and yellowish-green color when exposed to the sun, with mottled appearance without cracks and little wax.

The yellowish-green growth ring has a medium width and bud prominence. The root region has medium width and

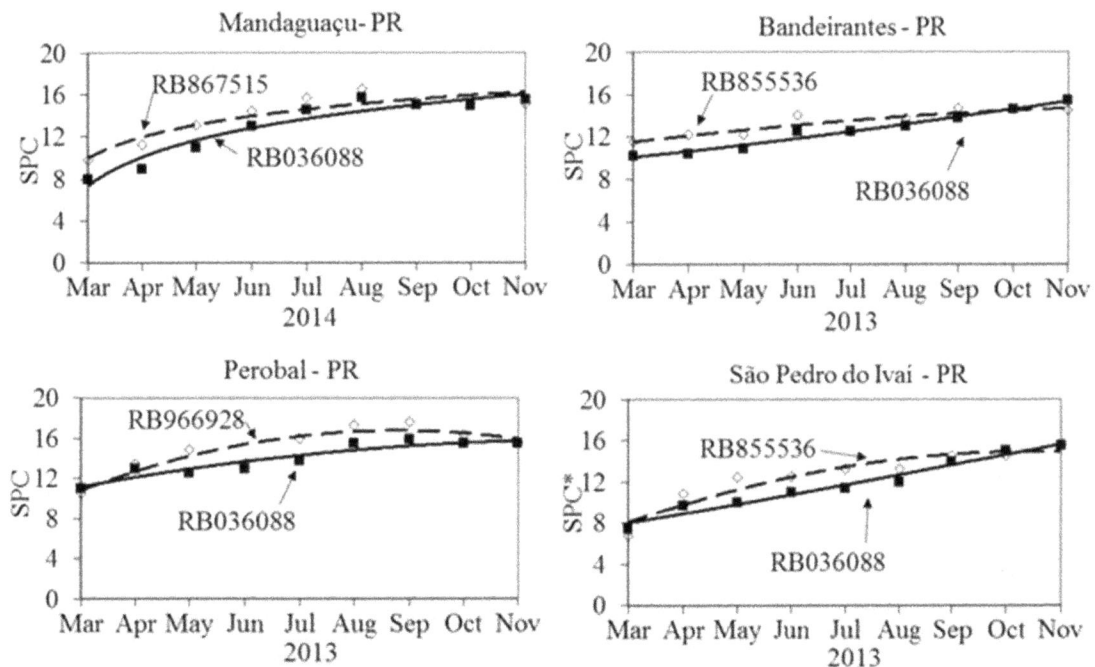

Figure 3. Maturation curves of 'RB036088' and other important early-maturing sugarcane cultivars used in the center-south of Brazil, at four test locations. *SPC - Pol in juice (sucrose percentage in cane juice).

bud prominence, greenish-yellow color, abundant primordial roots and absent bud pubescence. Clear presence of wax in the root node region. The medium-sized bud is oval, slightly prominent, occasionally touching the growth ring, has no flower cushion, apical position of the germ pore and haired apex. Light to intermediate green leaves, with medium length, narrow width and regular volume in the canopy. The auricle is small, asymmetric and deltoid-shaped. Green dewlap of the normal type. Short, green to purple sheaths, in a straightforward arrangement, with weak presence of wax, without fine hairs.

SEEDLING MAINTENANCE AND DISTRIBUTION

Seedlings of cultivar RB036088 are maintained and distributed by the Sugarcane Breeding Program of the Department of Plant Science and Plant Health, Sector of Agricultural Sciences, Federal University of Paraná, 80035-050, Curitiba, Brazil.

REFERENCES

Barbosa GVS, Oliveira RA, Cruz MM, Santos JM, Silva PP, Viveiros AJA, Sousa AJR, Ribeiro CAG, Soares L, Teodoro I, Sampaio Filho F, Diniz CA and Torres VLD (2015) RB99395: Sugarcane cultivar with high sucrose contente. **Crop Breeding and Applied Biotechnology 15**: 187-190.

Barbosa MHP, Resende MDV, Dias LAS, Barbosa GVS, Oliveira RA, Peternelli LA and Daros E (2012) Genetic improvement of sugar cane for bioenergy: the Brazilian experience in network research with RIDESA. **Crop Breeding and Applied Biotechnology S2**: 87-98.

Carneiro MS, Chapola RG, Fernandes Júnior AR, Cursi DE, Barreto FZ, Balsalobre TWA and Hoffmann HP (2015) RB975952 – Early maturing sugarcane cultivar. **Crop Breeding and Applied Biotechnology 15**: 193-196.

Daros E, Oliveira RA and Barbosa GVS (eds) (2015) **45 anos de variedades RB de cana-de-açúcar: 25 anos de Ridesa**. Graciosa, Curitiba, 156p.

Eberhart SA and Hussell WA (1966) Stability parameters for comparing varieties. **Crop Science 6**: 36-40.

Fernandes AC (2003) **Cálculos na agroindústria da cana-de-açúcar**. 2nd edn, EME, Piracicaba, 240p.

Iaia AM, Oliveira RA, Melo LJOT, Daros E, Simões Neto DE, Bastos GQ, Oliveira FJ, Chaves A and Melo TTAT (2014) RB002504 – New early-maturing sugarcane cultivar. **Crop Breeding and Applied Biotechnology 14**: 45-47.

Oliveira RA, Daros E and Hoffmann HP (eds) (2015) **Liberação nacional de variedades RB de cana-de-açúcar**. Graciosa, Curitiba, 72p.

19

Marker-assisted screening of breeding populations of an apomictic grass *Cenchrus ciliaris* L. segregating for the mode of reproduction

Suresh Kumar[1, 2*], Sheena Saxena[1] and Madan G. Gupta[1]

Abstract: *Cenchrus ciliaris L. is an apomictic forage grass grown in pastures and rangelands of the semi-arid tropics. It reproduces predominantly through apomixis; rarely, obligate sexual plants have also been reported. Absence of sexual reproduction limits the possibility of genetic improvement through hybridization. This study reports on hybridization of an obligate sexual, self-incompatible buffelgrass with pollen from apomictic plants towards development of an F_2 population segregating for mode of reproduction and use of sequence characterized amplified region (SCAR) markers for screening the population. The segregation ratio of 3:1 (facultative: apomictic) was observed in the F_1 generation, whereas it was 1:2:1 (apomictic: facultative: sexual) in the F_2 generation. A number of obligate sexual F_2 progenies with desirable agronomic traits were obtained. The SCAR markers were able to screen out apomictic plants from sexual ones, but failed to discriminate between facultative and sexual. Marker-assisted screening could be useful for introgression of desirable trait(s) in the apomictic genotype through hybridization.*

Key words: *SCAR, apomixis, molecular marker, sexual reproduction.*

*Corresponding author:
E-mail: sureshkumar3_in@yahoo.co.uk

[1] ICAR - Indian Grassland and Fodder Research Institute, Division of Crop Improvement, Jhansi-284003, India
[2] ICAR - Indian Agricultural Research Institute, Division of Biochemistry, Pusa Campus, New Delhi, Delhi-110012, India

INTRODUCTION

Buffelgrass (*Cenchrus ciliaris* L.) is one of the most important perennial forage grasses grown throughout the tropical and subtropical regions of the world. It is an apomictic, polyploid grass suited to pastures and rangelands of Australia, South Africa, and India (Bhat et al. 2001). It is drought tolerant and well adapted to arid and semi-arid areas. A serious disease in buffelgrass is leaf blight caused by *Magnaporte grisea*, which reduces forage quality and yield (Rodriguez et al. 1999). A natural source of resistance to this disease is known (Diaz-Franco and Mendez-Rodríguez 2005), but lack of sexual reproduction restricts genetic improvement of this species through hybridization. Apomixis not only makes genetic improvement of the species difficult and time consuming, but restricts it to selection of elite lines from the natural variants (Kumar and Bhat 2012). On the other hand, apomixis provides a means of clonal propagation through seeds because the progenies produced through apomixis are genetically identical to the female parent. Apomictic *Cenchrus* species reproduce by apospory, characterized by apomeiosis and parthenogenesis. Recently, the ASGR-BABY BOOM-like (*PsASGR-BBML*) gene from *Pennisetum squamulatum* (L.) R.Br. has been reported to express in egg cells before fertilization and induce parthenogenesis, as well as produce haploid offspring in transgenic sexual pearl millet (Conner et al. 2015).

Induction of parthenogenesis by PsASGR-BBML can be of value for inducing parthenogenesis to synthesize apomixis in crop plants and may be applied to haploid induction to rapidly obtain homozygous lines for breeding. However, success in transfer of apomixis to a crop species has not yet been reported, mainly because the gene(s) for the components of apomixis has (have) not yet been identified (Spillane et al. 2004, Kandemir and Saygili 2015).

Though apomixis may be introgressed into the crop plant through conventional breeding, the process is slow and laborious and requires embryological/progeny analysis of huge breeding populations for selection of apomictic genotypes after each round of backcrossing (Albertini et al. 2001). Moreover, there are certain breeding constraints in this, including the availability of desired parental lines and efficient techniques for screening the segregating populations. In most of the species, apomixis shows dominance over the sexual mode; hence, the occurrence of an obligate sexual plant is rare and, over time, apomictic individuals outnumber sexual ones. Nevertheless, obligate sexual plants of buffelgrass have been identified (Bray 1978, Kumar et al. 2010b). A natural variant of an Indian accession of buffelgrass was reported to be short in stature, protogynous, and obligate sexual in nature (Kumar et al. 2013). The first genetic linkage map for the sexual mode of reproduction in C. ciliaris was reported based on recombining and closely linked AFLP markers by Yadav et al. (2012). Lack of sexual reproduction limits genetic improvement of this species through hybridization (Kumar and Bhat 2012). The only successful method of varietal development in this species has been selection of elite lines from the natural variants. With the identification of an obligate sexual plant of buffelgrass (Kumar et al. 2010b), it was possible to conduct hybridization experiments not only for basic studies (genetic analysis for the apomixis trait) but also for applied research (creation of genetic diversity and selection of elite genotypes) towards variety development.

To discriminate between apomictic and sexual modes of reproduction, embryological analysis of developing ovules using paraffin and resin embedded sectioning have been successful, but they are very time-consuming and cumbersome methods. The benzyl benzoate-$4^1/_2$ technique was first used by Herr (1971) for this purpose; later on, several modifications of the original method were adopted to study ovule development in angiosperms (Farence and Smith 1975, Shealy 1980). Subsequently, Young et al. (1979) used a pistil-clearing technique as an alternative to the time-consuming and cumbersome embedding and sectioning methods. The pistil-clearing technique requires less (one-tenth) time compared to that of the sectioning method, yet it involves nine subsequent changes through ethanol and methyl salicylate series. Moreover, it requires fixing of florets at an appropriate developmental stage after initiation of flowering. Since then, the most commonly used and reliable methods for investigating the apomictic mode of reproduction in plants have been either embryological analysis of the mother plant (Mazzucato et al. 1996) or its progeny analysis (Barcaccia et al. 1997). Since the pistil- clearing technique is still time consuming, labor intensive, and reliant on the developmental stage (flowering) of the plant, it may prove not to be an efficient technique for screening larger segregating populations.

Marker-assisted screening (MAS) is an efficient technique to minimize the time and labor required in a breeding program. Some of the requirements for MAS include (i) simple genetic inheritance of the trait under selection and (ii) availability of markers tightly linked with the trait (Albertini et al. 2001). The data available on the apomictic mode of reproduction indicate that apomixis is under simple genetic control not only in C. ciliaris (Goel et al. 2003, Dwivedi et al. 2007, Yadav et al. 2012) but also in Poa pratensis (Barcaccia et al. 1998), Panicum maximum (Savidan 1983), Pennisetum squamulatum (Gustine et al. 1997, Ozias-Akins et al. 1998, Roche et al. 1999), Brachiaria decumbens (Pessino et al. 1997), Paspalum notatum (Martinez et al. 2003), and Tripsacum dactyloides (Leblanc et al. 1995). Molecular analyses have revealed that only a few dominant genes are required for genetic transmission of apomixis (Barcaccia et al. 1998, Goel et al. 2003, Yadav et al. 2012). Moreover, the role of transposable elements and their epigenetic controls (Wang et al. 2016) are also being investigated in regulation of the genes associated with apomixis (our unpublished data). The molecular markers linked with sexual and apomictic modes of reproduction (Kumar et al. 2010c, Kumar and Saxena 2016) would be very useful for screening of the segregating population, mapping of genes responsible for the mode of reproduction, and their characterization using a reverse genetics approach (Kumar 2014).

The present study was undertaken to create genetic diversity in apomictic buffelgrass through hybridization and to demonstrate the utility of the SCAR markers (Kumar et al. 2010c, Kumar and Saxena 2016) in breeding procedures. The SCAR marker-assisted screening of the segregating population would be very useful for genetic/molecular analyses of apomixis and marker-assisted breeding of buffelgrass.

MATERIAL AND METHODS

Plant material

A tetraploid obligate sexual *C. ciliaris* (IGFRI-CcSx -08/1) plant (Kumar et al. 2013) was grown in a greenhouse in isolation, as well as cross pollinated with an apomictic *C. ciliaris* plant (IG-693108). Seeds were collected from the cross-pollinated obligate sexual plant and the obligate apomictic plant. Progenies of both the plants were raised and analyzed using RAPD and SCAR markers. A facultative F_1 progeny of the sexual plant was selfed by bagging an individual panicle before emergence of the stigma. A total of 287 F_2 progenies were raised from the seeds collected from the selfed F_1 and screened using four apomixis-specific SCAR markers (Kumar and Saxena 2016) and one sexual SCAR marker (Kumar et al. 2010c).

Genomic DNA isolation

Genomic DNA was isolated from leaf tissues of the parents and F_1 and F_2 progenies following a simplified protocol reported elsewhere (Kumar et al. 2010a). Young leaf tissues were ground in 200 µL of AP1 Lysis buffer (QIAGEN GmbH). The ground sample was incubated at 65 °C for 15 min, and then 65 µL of P3 Neutralization buffer (QIAGEN GmbH) was added, followed by incubation on ice for 5 min. The content was centrifuged for 5 min at 9000 g, and a 0.6 volume of isopropanol was mixed with the supernatant. The mixture was centrifuged and the DNA pellet was washed with 70% ethanol, air-dried, and finally dissolved in 30 µL of sterilized double-distilled water. The quality of the genomic DNA was checked by agarose gel (0.8% w/v) electrophoresis.

Progeny analysis

Progenies of the sexual and the apomictic plants were subjected to comparative evaluation for morphological variation (Figure 1), and analyzed for genetic variability using PCR-based RAPD markers with reactive decamer primers (Unpublished data). Morphological diversity among individuals was assessed based on the data collected (with respect to plant height, leaf length and width, panicle size, and fresh biomass production) in triplicate at the flowering stage over three years. PCR amplification for RAPD analysis was performed in a 20 µL reaction volume containing 100 ng genomic DNA, 400 mM of each dNTP, 30 pmol primer, 2.5 mM $MgCl_2$, 1x Taq buffer, and 3 U Taq DNA polymerase on a PTC-100° Peltier thermal cycler (MJ Research). PCR conditions were 94 °C for 5 min, followed by 40 cycles of 94 °C for 60 s, 37 °C for 60 s, 72 °C for 2 min, and a final extension at 72 °C for 10 min. The amplification products were visualized on 1.5% agarose gel.

Marker-assisted screening (MAS) of progenies

The mode of reproduction in the progenies of the sexual and the apomictic plants was first determined using a sexual SCAR marker (CcSex-260: Forward 5'-GAGCAGGGGTTAGAGGTAA-3', Reverse 5'-ACATTCAGCCTACGGAGTG-3') (Kumar et al. 2010c) and four apomixis-specific SCAR markers (Apo-C270, Apo-C470, Apo-C730, Apo-C930) (Kumar and Saxena 2016). The apomixis-specific markers can detect the obligate apomictic mode of reproduction in *Cenchrus* spp. The sexual SCAR marker detects sexuality in *Cenchrus ciliaris*; however, it

Figure 1. Morphological diversity among F_1 hybrids of the sexual *C. ciliaris* plant. A - Apomictic F_1 hybrid with longer, broader leaves and better regrowth potential, B - Apomictic F_1 hybrid showing heterosis for leaf size, C - Facultative F_1 hybrid with distinct morphological features, D - Another facultative F_1 hybrid morphologically similar to the male parent.

does not differentiate between sexual and facultative modes of reproduction. Screening of 287 F_2 progenies of *C. ciliaris* for mode of reproduction was performed using the sexual and apomixis-specific SCAR markers. PCR-based MAS was carried out using 100 ng genomic DNA, 100 mM of each dNTP, 10 pmol of each of the primers, 2 mM $MgCl_2$, 1x Taq buffer, and 3 U Taq DNA polymerase. PCR conditions were 94 °C for 5 min, followed by 38 cycles of DNA amplification (94 °C for 60 s, 60 °C for 60 s, and 72 °C for 30 s) and final incubation at 72 °C for 5 min. The PCR products were visualized by 1.4% agarose gel electrophoresis.

Embryo sac analysis

To confirm the mode of reproduction detected by the molecular markers and to validate fidelity of the MAS, embryo sac analysis was performed as described elsewhere (Kumar et al. 2015). All of the F_1 hybrids and selected F_2 individuals were subjected to the pistil-clearing technique (Young et al. 1979) by analyzing 25 cleared pistils from each plant. Based on the presence or absence of antipodal cells, the embryo sac was categorized as sexual or apomictic. The presence of both sexual and apomictic embryo sacs on the same inflorescence categorized the plant as facultative.

Statistical analysis

Statistical analysis of the data collected from a minimum of three replications was performed by analysis of variance. Duncan's multiple range test (DMRT) was used to compare the means. The χ^2 test was used to determine goodness of fit (at $P < 0.05$) between the observed and expected number of genotypes for the segregation ratio of either 3:1 or 1:2:1.

RESULTS AND DISCUSSION

When the obligate sexual *C. ciliaris* plant (IGFRI-CcSx -08/1) was grown in isolation, normal flowering was observed, but no seed setting was found. On growing the sexual plant along with apomictic plants in a greenhouse, seed setting on the sexual plant was observed. Thus, cross-pollination of the sexual plant was necessary for seed setting. The self-incompatible nature of the sexual plant bearing viable pollens was confirmed over three years of experimentation. This report on the self-incompatibility of the sexual parent is unique and has not previously been reported in buffelgrass. The reported self-incompatibility of the sexual plant would be a desirable feature so as to minimize the efforts required for cross-hybridization (hand-pollination).

Marker-assisted screening of segregating populations

SCAR markers known to be linked with sexual and apomictic modes of reproduction (Kumar et al. 2010c, Kumar and Saxena 2016) were utilized for screening of segregating populations developed by crossing the obligate sexual plant with pollen from the tetraploid obligate apomictic *C. ciliaris* plant (IG-693108). The sexual mode of reproduction and cross-fertilization of the tetraploid obligate sexual plant was confirmed by F_1 progeny analyses wherein morphologically diverse (Figure 1), facultative, and apomictic progenies were observed. SCAR marker-assisted screening of F_1 hybrids resulted in identification of 10 hybrids (out of 34 F_1 progenies) as apomictic. Screening with the sexual SCAR marker, followed by embryo sac analysis, confirmed that the remaining 24 hybrids were facultative. Facultative sexual progenies are supposed to be heterozygous, and they were found to be self-compatible since selfing of the facultative F_1 progeny produced sexual, apomictic, and facultative F_2 individuals. The F_1 progenies were found to be either apomictic or facultative, but no sexual progeny was observed. This may be due to the tetraploid nature of the species, and this observation is in agreement with earlier reports in buffelgrass (Dwivedi et al. 2007, Yadav et al. 2012) and *Brachiaria* (Pessino et al. 1997). Selfing of a facultative F_1 plant by bagging, followed by collection of seeds to raise the F_2 generation, resulted in a segregating population of 287 individuals. Screening of the F_2 individuals using the sexual and apomixis-specific SCAR markers resulted in identification of 71 apomictic individuals, showing amplification with the apomixis-specific SCAR markers but no amplification with the sexual SCAR marker (Figure 2A and B).

Use of more than one SCAR markers (four apomixis-specific and one sexual SCAR marker) rendered sufficient fidelity to MAS for screening out apomictic individuals. In fact, the absence of the sexual SCAR marker itself indicated the apomictic mode of reproduction in the plant, but amplification with apomixis-specific SCAR markers confirmed the result. The sexual SCAR marker was detected in the remaining 216 F_2 individuals (Figure 2B), with no band for any of the apomixis-specific SCAR markers (Figure 2A). However, further examination of these individuals by the pistil-clearing technique resulted in identification of 158 individuals as facultative. The test of segregation ratio for the mode of reproduction using the c^2 test indicated a 1:3 ratio (apomictic: facultative) in the F_1 generation, whereas a 1:2:1 (apomictic: facultative: sexual) ratio was observed in the F_2 generation (Table 1). The observed distortion in the F_1 segregation ratio (3:1 instead of 1:1) might be due to the polyploid nature of the species. In F_2, a dosage effect can be assumed, with sexual being aaaa, facultative being Aaaa, and apomictic being AAaa. After selfing a facultative (Aaaa), we expect a sexual: facultative: apomictic segregation ratio of 1:2:1 (71:155:58).

Molecular markers provide a reliable and sensitive technique for molecular plant breeding (Albertini et al. 2001, Santana et al. 2014). Their use is particularly important when the selectable phenotype is manifested late in the plant life cycle (Albertini et al. 2001). Since the apomixis-specific SCAR markers were identified from conserved apomixis-specific loci in four apomictic Cenchrus species (Kumar and Saxena 2016), they were able to unequivocally distinguish apomictic genotypes from sexual/facultative. Therefore, these SCAR markers can be used for MAS of segregating populations, including those from inter-specific hybridization carried out for introgression of desirable trait(s) from other Cenchrus species.

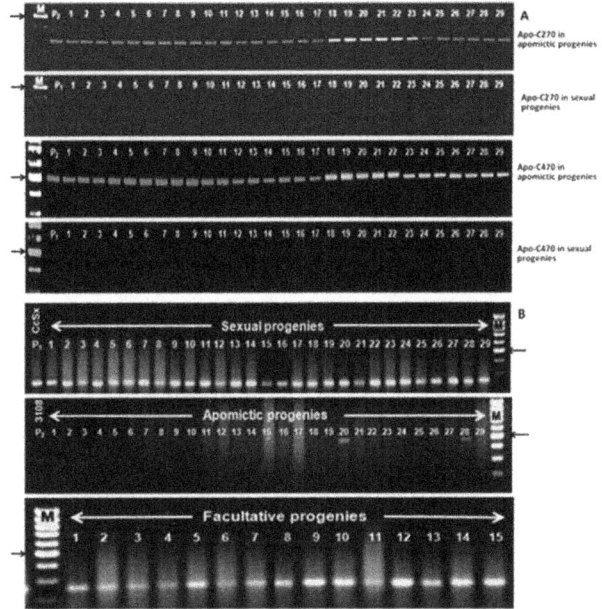

Figure 2. Marker-assisted screening of F_2 progenies using A - apomixis-specific SCAR markers (Apo-C270, Apo-C470) and B - sexual SCAR marker (CcSx-260). P_1 = mother sexual plant, P_2 = apomictic parent, CcSx= sexual parent, 3108= apomictic parent. Arrows (-->) indicate 500 bp band in the 100 bp DNA size marker.

Embryological analysis to detect mode of reproduction

Based on the presence of antipodal cells, the embryo was categorized as sexual, whereas the absence of antipodal cells led to an apomictic classification. To validate the fidelity of MAS, we performed embryological analysis of the F_1 and F_2 progenies and looked for the presence or absence of antipodal cells in the cleared pistil under a DIC microscope. Only the individual that was positive for the sexual SCAR marker, bearing an eight-nucleate embryo sac, was categorized as sexual, whereas those bearing both sexual and apomictic embryos were categorized as facultative. Since embryological and progeny analyses are laborious and time consuming, there are limitations to their application in screening of breeding populations. Using PCR-based (SCAR) markers linked with the mode of reproduction, we could overcome some of these limitations.

Diversity among the progenies

Whereas the mother sexual plant was short in stature, with poor growth and development, its F_1 progenies showed considerable morphological variation with respect to plant height (56 to 127 cm), leaf length (11 to 30.8 cm), width (4.3 to 15.2 mm), panicle structure and size (5.8 to 9.8 cm), and fresh biomass production (756 to 3458 g per plant) at flowering. RAPD analysis of randomly selected F_1 hybrids of the sexual plant showed significant variation in their DNA profiles, whereas the randomly selected progenies of an apomictic plant showed similar DNA profiles (Figure 3). F_2 progenies of the sexual plant also showed considerable morphological diversity with respect to plant height (26 to

Table 1. Trait segregation (apomictic, sexual, and facultative) in F_1 and F_2 progenies of Cenchrus ciliaris

Progeny	Observed ratio	Expected ratio	Segregation ratio	χ^2 value	P-value (P = 0.05)
F_1	24(F): 10(A)	25(F): 9(A)	3: 1	0.151*	3.84
F_2	71(A): 158(F): 58(S)	72(A): 143(F): 72(S)	1: 2: 1	4.309*	5.99

A= apomictic, F= facultative, S= sexual
* Non-significant at P = 0.05

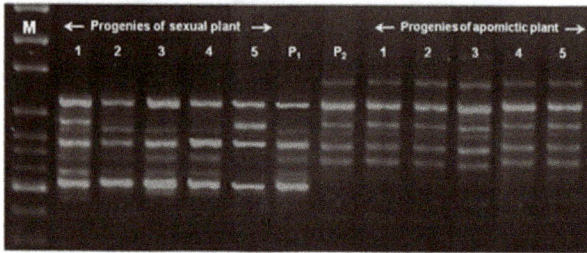

Figure 3. RAPD profile of five (1-5) randomly selected progenies of the sexual (P_1) and an apomictic (P_2) *C. ciliaris* plant. M = 100 bp DNA size marker.

132 cm), leaf morphology, panicle characteristics, mode of reproduction (apomictic/sexual/facultative), regrowth potential, and biomass production (356 to 3656 g per plant) at the 50% flowering stage. The morphological variations observed among the progenies of the sexual plant confirmed its utility in creating the genetic diversity required for genetic improvement and varietal development in this apomictic species. The observed morphological diversity among the apomictic F_1 hybrids (Figure 1A and B) demonstrated that apomixis can be successfully utilized to fix heterosis in a hybrid.

Selection of elite genotypes

The mother sexual plant showed meager growth and development over the years (Figure 4A) compared to the normally occurring apomictic buffelgrass (Figure 4B). While most of the progenies surpassed the mother sexual plant for several agronomic traits, a number of sexual (Figure 4C)

Figure 4. Morphological diversity among the parents and F_2 progenies of A - sexual *C. ciliaris* plant, and B - apomictic plant. C - An obligate sexual F_2 progeny with desirable agronomic features, D - an apomictic F_2 progeny with distinct morphological features.

and apomictic (Figure 4D) F_2 progenies with desirable agronomic features could also be identified. Diversity among the F_2 progenies resulted in selection of 17 genotypes on the basis of their plant height, leaf structure, growth potential, and biomass production for station trials toward varietal development. Most of the selected progenies were on par with the apomictic parent for many of the agronomic traits, while exceeding it for one or two features, particularly for biomass production and regrowth potential.

Although there are reports of hybridization in buffelgrass (Dwivedi et al. 2007, Yadav et al. 2012), the main objective of those studies was to develop an F_2 mapping population for genetic analysis of the apomixis trait and identification of molecular markers linked to it. No morphological assessment of parents and progenies for agronomic traits has been reported before. While the mother sexual plant (IGFRI-CcSx -08/1) was the only obligate sexual genotype (showing meager growth and development), a number (8) of obligate sexual F_2 progenies with desirable agronomic traits (Figure 4C) was able to be obtained through hybridization. Several apomictic F_2 progenies with desirable attributes (leaf structure, regrowth potential, and biomass production) led to the selection of 9 promising genotypes for station trials toward varietal development (Figure 4D).

These findings show that hybridization can be used for creating genetic diversity in apomictic species and fixing heterosis through introgression of apomixis. Whenever a desirable trait has to be introgressed into an apomictic genetic background, an apomictic parent may be used as a male parent, and sexual progenies with the desirable trait(s) will need to be selected in each backcross generation. Once the recurrent parental genotype is recovered, selection will need to be directed to select apomictic individuals with the introgressed trait. Therefore, the molecular markers must have maximum fidelity in selection of apomictic and sexual genotypes. The sexual and apomixis-specific SCAR markers used in the present study were successful in MAS of segregating populations with sufficient fidelity. Although the sexual SCAR marker could not discriminate between sexual and facultative individuals, the four apomixis-specific SCAR markers unambiguously identified the apomictic genotypes. Most of the earlier reports on molecular markers linked with the apomictic or sexual mode of reproduction (Albertini et al. 2001, Yadav et al. 2012) did not describe their status in facultative genotypes. Although a facultative-specific marker may not be so important in a breeding program, it may be useful for genetic/molecular analyses of apomixis. Therefore, it would be desirable to identify a marker linked with the facultative mode of reproduction to minimize dependence on embryological/progeny analyses.

REFERENCES

Albertini E, Barcaccia G, Porceddu A, Sorbolini S and Falcinelli M (2001) Mode of reproduction is detected by Parth1 and Sex1 SCAR markers in a wide range of facultative apomictic Kentucky bluegrass varieties. **Molecular Breeding 7**: 293-300.

Barcaccia G, Mazzucato A, Albertini E, Zethof J, Pezzotti M, Gerats A and Falcinelli M (1998) Inheritance of parthenogenesis in *Poa pratensis* L.: auxin test and AFLP linkage analyses support monogenic control. **Theoretical and Applied Genetics 97**: 74-82.

Barcaccia G, Mazzucato A, Belardinelli A, Pezzotti M, Lucretti S and Falcinelli M (1997) Inheritance of parental genomes in progenies of *Poa pratensis* L. from sexual and apomictic genotypes as assessed by RAPD markers and flow cytometry. **Theoretical and Applied Genetics 95**: 516-524.

Bhat V, Dalton SJ, Kumar S, Bhat BV, Gupta MG and Morris P (2001) Particle-inflow gun-mediated genetic transformation of buffelgrass (*Cenchrus ciliaris* L.): optimizing biological and physical parameters. **Journal of Applied Genetics 42**: 405-412.

Bray RA (1978) Evidence for facultative apomixis in *Cenchrus ciliaris*. **Euphytica 27**: 801-804.

Conner JA, Mookkan M, Huo H, Chae K and Ozias-Akins P (2015) A parthenogenesis gene of apomict origin elicits embryo formation from unfertilized eggs in a sexual plant. **Proceedings of National Academy of Science of USA 112**: 11205-11210.

Diaz-Franco A and Mendez-Rodriguez A (2005) Leaf blight [*Pyricularia grisea* (Cooke) Sacc] in buffelgrass (*Cenchrus ciliaris* L.) meadows and reaction of genotypes in North Tamaulipas, Mexico. **Revista Mexicana de Fitopatologia 23**: 232-237.

Dwivedi KK, Bhat SR, Bhat V, Baht BV and Gupta MG (2007) Identification of a SCAR marker linked to apomixis in buffelgrass (*Cenchrus ciliaris* L.). **Plant Science 172**: 788-795.

Farence DR and Smith BB (1975) Effect of chemical pretreatments on *Ludwigia alternifolia* L. ovules prior to immersion in Herr's clearing fluid. **Proceedings of Pennsylvania Academy of Science 29**: 89-91.

Goel S, Chen Z, Conner JA, Akiyama Y, Hanna WW and Ozias-Akins P (2003) Physical evidence that a single hemizygous chromosomal region is sufficient to confer aposporous embryo sac formation in *Pennisetum squamulatum* and *Cenchrus ciliaris*. **Genetics 163**: 1069-1082.

Gustine DL, Sherwood RT and Huff DR (1997) Apospory-linked molecular markers in buffelgrass. **Crop Science 37**: 947-951.

Herr JMJr (1971) A new clearing squash technique for the study of ovule development in Angiosperms. In Radford AE, Dickson WC, Massey JR and Bell CR (eds) **Vascular plant systematic**. Harer and Row, New York, p. 230-235.

Kandemir N and Saygili I (2015) Apomixis: new horizons in plant breeding. **Turkish Journal of Agriculture and Forestry 39**: 1-8.

Kumar S (2014) RNAi (RNA interference) vectors for functional genomics study in plants. **National Academy of Science Letters 37**: 289-294.

Kumar S and Bhat V (2012) High-frequency direct plant regeneration via multiple shoot induction in the apomictic forage grass *Cenchrus ciliaris* L. **In Vitro Cellular & Developmental Biology-Plant 48**: 241-248.

Kumar S and Saxena S (2016) Sequence characterized amplified regions linked with apomictic mode of reproduction in four different apomictic *Cenchrus* species. **Molecular Plant Breeding 7**: 1-14.

Kumar S, Arul L and Talwar D (2010a) Generation of marker-free Bt transgenic indica rice and evaluation of its yellow stem borer resistance. **Journal of Applied Genetics 51**: 243-257.

Kumar S, Chandra A, Gupta MG and Shukla GP (2010b) Molecular and embryological analyses of rare sexual plant in buffelgrass (*Cenchrus ciliaris* L.). **Range Management & Agroforestry 31**: 36-40.

Kumar S, Chandra A, Gupta MG and Shukla GP (2010c) SCAR marker linked to sexuality in *Cenchrus ciliaris* L. **Range Management & Agroforestry 31**: 149-150.

Kumar S, Chandra A, Gupta MG and Shukla GP (2013) IGFRI-CcSx-08/1 (IC0590889; INGR11062), an Anjan grass (*Cenchrus ciliaris* L.) germplasm with a rare obligate sexual plant. **Indian Journal of Plant Genetic Resources 26**: 99-100.

Kumar S, Sahu N and Singh A (2015) High-frequency in vitro plant regeneration via callus induction in a rare sexual plant of *Cenchrus ciliaris* L. **In Vitro Cellular & Developmental Biology—Plant 51**: 28-34.

Leblanc O, Grimanelli D, de Leon DG and Savidan Y (1995) Detection of the apomictic mode of reproduction in maize-*Tripsacum* hybrids using maize RFLP markers. **Theoretical and Applied Genetics 90**: 1198-1203.

Martinez EJ, Hopp HE, Stein J, Ortiz JPA and Quarin CL (2003) Genetic characterization of apospory in tetraploid *Paspalum notatum* based on the identification of linked molecular markers. **Molecular Breeding 12**: 319-327.

Mazzucato A, Falcinelli M and den Nijs APM (1996) Estimation of parthenogenesis frequency in Kentucky bluegrass with auxin induced parthenocarpic seeds. **Crop Science 36**: 9-16.

Ozias-Akins P, Roche D and Hanna WW (1998) Tight clustering and hemizygosity of apomixis-linked molecular markers in *Pennisetum squamulatum* implies genetic control of apospory by a divergent locus which may have no allelic form in sexual genotypes. **Proceedings of National Academy of Science of USA 95**: 5127-5132.

Pessino SC, Ortiz J, Leblanc O, do Valle CB and Hayward MD (1997) Identification of a maize linkage group related to apomixis in *Brachiaria*. **Theoretical and Applied Genetics 94**: 439-444.

Roche D, Peisheng C, Zhenbang C, Hanna WW, Gustine DL, Sherwood RT and Ozias-Akins P (1999) An apospory-specific genomic region is conserved between buffelgrass (*Cenchrus ciliaris* L.) and *Pennisetum squamulatum* Fresen. **Plant Journal 19**: 203-208.

Rodriguez O, Gonzalez-Dominguez J, Krausz JP, Odvody GN, Wilson JP, Hanna WW and Lew M (1999) First report and epidemics of buffelgrass blight caused by *Pyricularia grisea* in South Texas. **Plant Diseases 83**: 398.

Santana FA, Silva MF, Guimarães JKF, Silva MF, Pereira WD, Piovesan WD and Barros WG (2014) Marker-assisted selection strategies for developing resistant soybean plants to cyst nematode. **Crop Breeding and Applied Biotechnology 14**: 180-186.

Savidan Y (1983) Genetics and utilization of apomixis for the improvement of guinea grass (*Panicum maximum* Jacq.). In Smith JA and Hays VW (eds) **Proceedings of international grassland congress**. Lexington, KY, p. 182-184.

Shealy HEJr (1980) Treatment of dense ovule wall with monoethanolamine used in conjunction with clearing technique. **Association of Southeastern Biologists Bulletin 27**: 62.

Spillane C, Curtis MD and Grossniklaus U (2004) Apomixis technology development– virgin births in farmers' fields? **Nature Biotechnology 22**: 687-691.

Wang X, Li Q, Yuan W, Cao Z, Qi B, Kumar S, Li Y and Qian W (2016). The cytosolic Fe-S cluster assembly component MET18 is required for the full enzymatic activity of ROS1 in active DNA demethylation. **Scientific Reports 6**: 26443 doi: 10.1038/srep26443.

Yadav CB, Anuj, Kumar S, Gupta MG and Bhat V (2012) Genetic linkage maps of the chromosomal regions associated with apomictic and sexual modes of reproduction in *Cenchrus ciliaris*. **Molecular Breeding 30**: 239-250.

Young BA, Sherwood RT and Bashaw EC (1979) Cleared pistil and thick sectioning techniques for detecting aposporous apomixis in grasses. **Canadian Journal of Botany 57**: 1668-1672.

20

Combining ability of sugarcane genotypes based on the selection rates of single cross families

Priscilla Neves de Santana[1], Américo José dos Santos Reis[2] and Lázaro José Chaves[2*]

*Corresponding author:
E-mail: lchaves@ufg.br

[1] Faculdade Integrada Aparício Carvalho (FIMCA), 78.912-640, Porto Velho, RO, Brazil
[2] Universidade Federal de Goiás, Escola de Agronomia, Campus Samambaia, 74.690-900, Goiânia, GO, Brazil

Abstract: *This study evaluated the genetic potential of parents used in sugarcane genetic breeding programs based on the performance of previously conducted single crosses. The average selection rate of each family, predicted using Best Linear Unbiased Prediction (BLUP) procedure, was used as a surrogate to the cross performance in the initial evaluation phase. Data analysis was performed using Griffing's method IV adapted for the available set of crosses to detail the general combining ability (GCA) and specific combining ability (SCA) effects. Significant GCA effects were detected, which demonstrated the possibility of selecting parents based on this parameter. SCA had a higher coefficient of determination than GCA. In conclusion, the selection rate is an effective indicator for evaluation of the combining ability of parents in the first selection stage of a sugarcane breeding program.*

Key words: *Saccharum, breeding, general combining ability, specific combining ability.*

INTRODUCTION

Sugarcane genetic breeding has become a decisive factor in sugar alcohol industry development in Brazil, with considerable selection gains over the years and regular release of new cultivars (Barbosa et al. 2012, Ramalho et al. 2012, Daros 2014, Iaia et al. 2014, Melo et al. 2014, Barbosa et al. 2015, Carneiro et al. 2015).

New sugarcane cultivars are obtained by vegetative propagation of selected genotypes, which are obtained by sexual reproduction of suitable parents. Selection is applied to all breeding stages, from parental choice to the final evaluation phase of the network trials. Individual selection is inefficient during the early stages due to the low heritability coefficients of most traits. Nevertheless, selection based on phenotypic evaluations of individual plants is commonly performed in the early stages of genetic breeding programs (Barbosa et al. 2005, Matsuoka et al. 2005).

Parental selection should fall on genotypes with the characteristics of interest and good cross performances to obtain populations with good performances and containing genes of interest. Parental crosses should be carefully planned, and crosses between related individuals should be avoided to reduce the occurrence of inbreeding depression and the narrowing of the genetic base (Matsuoka et al. 2005).

The correct choice of parents for crosses depends on the goals and objectives of the genetic breeders, the characteristics of the parents in morphological and agronomic variables, and the progeny performance in previous crosses (Badaloo et al. 1999). The previous performances of the parents may be inferred by calculating the selection rate, which is the ratio between the genotypes selected from each cross and the number of genotypes in that cross. The average selection rate of a parent may be used as an indicator of its general combining ability (Badaloo et al. 1999).

The general combining ability (GCA) is used to estimate the average performance of genotypes in various hybrid combinations and is associated with additive allele effects and additive-by-additive epistatic effects. Conversely, the specific combining ability (SCA) is used to identify specific hybrid combinations that are better or worse than the expected abilities based on the GCA and is associated with gene dominance effects (Halauer et al. 2010).

Genetic variability of the germplasm collection is a basic need for a genetic breeding program. Knowledge of the genetic divergence among genotypes determines the success of hybridization from parental selection. Divergence between genotypes may be indirectly inferred using genealogy data, morphological and agronomic variables, or molecular markers. The SCA of a cross may also be used as an indicator of genetic divergence between parents for a given trait provided that the trait-controlling loci in question show dominance effects (Halauer et al. 2010). This study evaluated the genetic potential of sugarcane parents based on the GCAs estimates using the selection rates of previously generated single-cross families.

MATERIAL AND METHODS

Populations derived from 3043 single crosses (biparental crosses) involving 541 parents from the sugarcane genetic breeding program of the Inter-University Network for the Development of the Sugar Alcohol Industry (*Rede Interuniversitária de Desenvolvimento do Setor Sucroalcooleiro* – RIDESA) were used. The single cross progenies were assumed to be full-sib families, although contaminants may occur (Santos et al. 2014). The breeding program crosses were routinely conducted at the Serra do Ouro Flowering and Crossing Station, Murici, Alagoas state, Brazil. The seeds were sent to the federal universities (IFES) that composed the RIDESA (Barbosa et al. 2012). Sowing was performed in each IFES, and seedlings were produced. After a nursery growth phase, the seedlings were used to outline the first selection stage (termed stage T1) and installed in their own areas and experimental areas of associated sugar alcohol industries. In this stage, the best genotypes were selected by visual evaluation and measurements of some basic traits; then, the selected clones advanced to the next stage (termed T2). The average number of replications per family in experimental fields was 2.64.

The selection rate within each full-sib family evaluated in stage T1 was the basic variable used to evaluate the genetic value of a single cross in the present study. The contribution of each breeding program to the whole set of data used in this study (8045 data) is shown in Table 1. These data refer to crosses series RB 94 to RB 10 (1994 to 2010). The number of crosses series per breeding program varied from five (UFG and UFMT) to 17 (UFV). The variable corresponds to the ratios in percentages between the number of genotypes selected for stage T2 and the number of genotypes evaluated

Table 1. Number of original data per RIDESA Brazil breeding program and mean effects of selection rates predicted using the BLUP procedure (Intercept = 1.7702)

Breeding program	Number of data	BLUP – selection rate
Universidade Federal de Goiás – UFG	345	1.2817
Universidade Federal de São Carlos – UFSCar	1628	0.8788
Universidade Federal de Alagoas – UFAL	2497	0.2524
Universidade Federal do Paraná – UFPR	1033	-0.1072
Universidade Federal do Mato Grosso – UFMT	190	-0.1675
Universidade Federal de Pernambuco – UFRPE	124	-0.3334
Universidade Federal de Viçosa – UFV	1878	-0.8276
Universidade Federal Rural do Rio de Janeiro – UFRRJ	350	-0.9772
Total	8045	-

in stage T1. Only parents that participated in two or more crosses were analyzed to allow GCA estimation.

The selection rates of each cross were predicted using the Best Linear Unbiased Predictor (BLUP) method according to the model: $Y_{ijk} = \mu + \tau_i + \alpha_j + \rho_k + \varepsilon_{ijk}$; where, Y_{ijk}: selection rate (%) from the cross k, in the year j, in IFES i; μ: effect of the mean (intercept), fixed, with $E(\mu) = \mu$ and $E(\mu^2) = \mu^2$; τ_i: effect of IFES (breeding program), random, with $E(\tau_i) = 0$ and $E(\tau_i^2) = \sigma_\tau^2$; α_j: effect of year j, random, with $E(\alpha_j) = 0$ and $E(\alpha_j^2) = \sigma_\alpha^2$; ρ_k: effect of cross k, random, with $E(\rho_k) = 0$ and $E(\rho_k^2) = \sigma_\rho^2$; ε_{ijk}: error (deviation) associated with the observation Y_{ijk}, random, with $E(\varepsilon_{ijk}) = 0$ and $E(\varepsilon_{ijk}^2) = \sigma_\varepsilon^2$. This procedure was adopted to minimize differences in the selection criteria of each program and other unmeasured effects between experiments. The predicted value of a specific cross (ρ_k, $k = 1,2,...,3043$) representing the adjusted average performance of its progeny in stage T1 was used to outline a partial diallel table. The use of the Restricted Maximum Likelihood (REML) mixed models method in sugarcane genetic breeding has been used for several authors (Barbosa et al. 2005, Barbosa et al. 2012, Silva et al. 2015).

Model I: Method IV proposed by Griffing (1956) was used for the combining ability analysis as follows: $Y_{ij} = \mu + g_i + g_j + s_{ij}$, wherein Y_{ij} is the BLUP value regarding the cross between parents i and j, μ is the overall mean, g_i and g_j are the general combining ability (GCA) effects of the ith and j^{th} parents, respectively ($i < j = 2,3...,541$), and s_{ij} s_{ij} is the specific combining ability (SCA) effect of the cross between the i- and j-order parents.

The estimates of the effects and their respective sums of squares were obtained using the least squares method. The model is described using matrix notation as follows: $Y = X\beta + \varepsilon$, wherein Y is the vector of observed means, X is the incidence matrix, β is the vector of parameters and ε is the vector of deviations from the model. Some restrictions, like the sum of estimates equal to zero, were added to obtain unique solutions for the vector β estimate.

The matrix X was constructed using the reduced model $Y_{ij} = \mu + g_i + g_j$. The parameter estimates were obtained by $\hat{\beta} = (X'X)^{-1} (X'Y)$. The sum of squares of the model was obtained by $SS_{Model} = \hat{\beta}(X'Y)$. SCA estimates were obtained by the differences between the observed value of each cross and predicted values by model based on GCA ($\hat{g}_i$ and $\hat{g}_j$) (i.e., $S = Y - \hat{X}\beta$, wherein S is the vector of s_{ij} estimates). The sums of squares of the diallel analysis were obtained by $SS_{SCA} = Y'Y - SS_{Model}$ and $SS_{GCA} = SS_{Model} - C$; C is the sum of squares regarding the constant of the model. The analysis was performed using a specific script in the R environment (R Core Team 2014).

RESULTS AND DISCUSSION

Selection rate

The selection rate varied considerably between RIDESA member institutions depending on the purpose of each program, the experimental area, the material availability, the subjective evaluation of the breeder, and the year and site in which the materials were evaluated in stage T1. The average selection rate calculated from 8045 data points resulting from 3043 different crosses was 1.84% of genotypes selected per family, with a range from 0% to 90% and a 142.04% coefficient of variation, which shows a wide variation of selection rates among crosses and among experiments. Most values ranged from 0% to 5%, thereby generating a frequency distribution skewed to near-zero values; values higher than 10% were rare (Figure 1a). Applying the BLUP procedure to predict the genetic values of crosses proved efficient and generated a frequency distribution similar to the standard normal curve, with the higher frequency classes close to zero (Figure 1b).

The selection rate used in the present study to evaluate the genetic potential of the parents was a composite variable derived from different primary variables used in the T1 selection stage of the RIDESA sugarcane genetic breeding program. Conceptually, the selection rate may be considered a selection index because it encompasses several variables, including vigor, health, height, and stalk diameter, depending on the criteria adopted in each IFES. High selection rates are generated in crosses (families) with higher means and phenotypic variability when considering a single selection criterion and the same sample size. Populations with low means and variances generate low or null selection rates, whereas average rates occur in populations with a low mean and high variance, high mean and low variance or with both parameters with intermediate values.

The average selection rate per cross predicted using the BLUP method had values ranging from -0.692 to 4.734, with

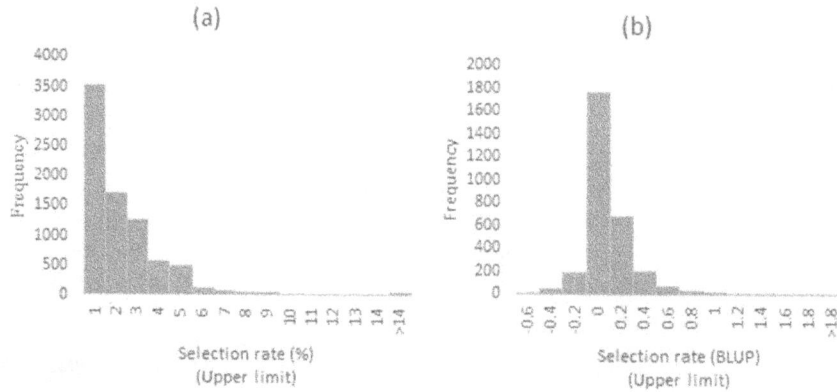

Figure 1. Histograms of the frequencies of selection rate values: a) original data (8045 values); b) mean values of 3043 crosses predicted using the BLUP procedure.

a 1.7702 intercept (Figure 1b). The effects of institutions ranged from -0.977 (Federal Rural University of Rio de Janeiro – UFRRJ) to 1.281 (Federal University of Goiás – UFG). These values showed the different selection intensities in T1 stage between institutions (Table 1).

Analysis of variance

The results from the analysis of variance performed according to the model by Griffing (1956) adapted for the available set of crosses are outlined in Table 2. A mean squared error on the same scale as the other mean squares was not available because the BLUP values used in the

Table 2. Analysis of variance using the model by Griffing (1956) for the selection rate (BLUP) in 3043 hybrid combinations involving 541 parents of the sugarcane genetic breeding program of RIDESA Brazil

Sources of variation	df	MS	R²
GCA	540	0.0747***	0.231
SCA	2502	0.0537	0.769
Total	3042	-	-

*** Significant at 0.1% probability level by F test.

analysis were obtained at the level of means of crosses. Thus, the significance of the general combining ability (GCA) was tested using the F test with the mean square of the specific combining ability (SCA) as a denominator, which is a conservative test that is equivalent to considering the parental effects as random. The F test was significant at a 0.1% probability level for the GCA effect. The significance of the SCA was not tested using the F test for the aforementioned reason.

Table 2 also outlines the coefficients of determination for GCA (0.231) and SCA (0.769). These values indicate that 76.9% of the total data variation results from SCA and 23.1% from GCA. Although the significance of the mean squared SCA was not tested, the R² value demonstrated the importance of this effect for the determination of data variation compared with the GCA. These results demonstrate the possibility of parental selection for future crosses based on the GCA and the existence of specific complementation shown in particular crosses.

The results from the GCA and SCA analysis using diallelic models are usually interpreted regarding genetic effects; the GCA is predominantly associated with the additive effects and the SCA with the non-additive effects of the genes (Silva et al. 2002, Camacho et al. 2015). This interpretation seemed inappropriate in the present case because the variable was a composite variable and a large number of diverse primary variables were analyzed in the selection. The SCA shown using this model may at least partially result from the interaction between genotypes and the test environments.

General combining ability

The present study focused more on the GCA effects for both the reliability and its statistical significance; the SCA effects were estimated for a limited number of crosses compared to the total crosses possible if a full diallel was available.

Estimates of the general combining ability effects for the 541 parents ranged from 0.454 to -0.353. These values showed that the best parent added 0.45 percentage points to the selection rate (predicted using the BLUP procedure) of crosses in which the best parent participated. Considering that the prediction intercept was 1.770, the best parent was 25.6% higher than the average of all parents. On the opposite end, the parent that contributed least to the crosses was 19.9% lower than the parental average. The GCA estimation method reported herein consists in modeling the concept introduced by Badaloo et al. (1999) in which the average selection rate of a parent is indicative of its GCA.

The 50 best genotypes regarding the GCA are outlined in Table 3. These parents are potentially the best based on the contribution of the highest positive values of the general combining ability effects. For these, the number of evaluated crosses per parent varied from two to 110 (Table 3). Among the 50 best parents, 29 participated of four or more crosses, allowing an acceptable accuracy for the estimates of GCA. Simulated study on the efficiency of partial circulant diallels showed good coincidence of the estimates of GCA with complete diallels, even with small number of crossings per parent (Veiga et al. 2000). A total of 1225 single crosses would be possible with these parents. The choice of crosses to be performed could consider other criteria, including genetic diversity and complementarity between parents, and specific characteristics according to the demands of each program. Some of the parents with the best GCAs outlined in Table 3 are commercial varieties cropped in Brazil, including RB855453, RB987935, SP832847, SP801816 and SP803280 (Chapola et al. 2013, Daros 2014). Only RB855453 was among the eight most planted varieties in 2011 (Barbosa et al. 2012).

Table 3. Estimates of general combining ability effects () and number of crosses (NC) for the 50 best sugarcane parents

Rank	Genotype	NC	GCA	Rank	Genotype	NC	GCA
1	RB7893	2	0.4538	26	RB945957	3	0.1761
2	RB945956	9	0.4227	27	RB855589	22	0.1742
3	RB977666	3	0.4206	28	SP775181	76	0.1734
4	RB951558	3	0.4179	29	RB00509	11	0.1725
5	B70710	4	0.3712	30	RB991555	8	0.1723
6	RB008304	6	0.3454	31	RB961	7	0.1722
7	RB99710	2	0.3299	32	SP801816	90	0.1703
8	JA6420	2	0.3118	33	RB957712	17	0.1669
9	RB865513	8	0.2922	34	RB00512	6	0.1657
10	CP82550	4	0.2889	35	RB951521	2	0.1640
11	RB855063	14	0.2781	36	RB971739	5	0.1611
12	IAC873396	18	0.2664	37	RB835487	4	0.1572
13	RB987965	2	0.2642	38	RB965908	3	0.1564
14	RB97327	2	0.2529	39	RB95549	5	0.1518
15	CB3822	8	0.2245	40	RB931546	2	0.1483
16	RB987935	9	0.2090	41	SP831483	3	0.1465
17	RB735200	11	0.2044	42	RB997810	2	0.1431
18	RB758516	4	0.2036	43	RB855453	26	0.1415
19	HJ5741	2	0.2008	44	CB654	2	0.1328
20	SP853877	16	0.1918	45	SP811663	9	0.1325
21	RB9358	2	0.1880	46	H566724	4	0.1320
22	CB4176	2	0.1869	47	RB997627	7	0.1295
23	SP832847	110	0.1839	48	RB961552	2	0.1293
24	RB971741		0.1783	49	SP803280	60	0.1280
25	RB975948		0.1762	50	RB865084	2	0.1235

Heterogeneity for GCA is of great relevance because the existence of genetic diversity among genotypes is important to generate genetic variability and obtain genetic gains on the variables selected in the T1 stage of the genetic breeding programs.

Specific combining ability

The estimates of SCA effects relative to the 3043 hybrid combinations ranged from 4.080 to -0.948 according to the model by Griffing (1956). This estimate had a noticeably higher value (4.080) and thus might be considered an outlier regarding the distribution of values concentrated below 2.0. This cross between parents RB945956 and IAC873396 had only 10 genotypes in the first selection stage in one of the environments; nine genotypes were selected for the following stages generating a selection rate of 90%.

As shown by the coefficient of determination (R^2), the SCA was high, with approximately 77% of the total variation of the BLUPs of selection (Table 2). Therefore, the extreme SCA values were noticeably higher in absolute values than the GCA values. Thus, selection may also be performed based on the SCA estimates.

A total of 146,070 hybrid combinations could be obtained using the 541 parents available. Only 3043 of these combinations were evaluated and used to perform the SCA predictions, which was a small percentage of the total that could be studied. Nevertheless, the number of crosses performed was high and might be considered a representative sample of the possible crosses. The genetic potential of the non-evaluated crosses could be estimated based on a reduced model that only analyzed the GCA of those parents (Reis et al. 2005).

SCA estimates enabled the identification of the most promising hybrid combinations to obtain families for genetic breeding programs. From a practical standpoint, crosses with high potential that were previously underexploited could be exploited using a larger sample. The following were the best hybrid combinations in terms of SCA: RB945956 with IAC87-3396, SP80-1816 with RB855063, RB72454 with RB721012, RB855063 with RB855127, RB008304 with RB92579, and RB955970 with SP91-1049.

Table 3 shows that some of the best parents in GCA are involved in the crosses with the highest SCAs. Parent SP83-2847 is noteworthy; this parent had a high GCA and participated in seven of the 45 best crosses. Similarly, parent SP77-5181 ranked 28th regarding the GCA values (Table 3) and participated in five crosses with good specific combining ability. Parent SP80-1816, which participated in six crosses with the best SCA, was also noteworthy.

The effect of SCA is associated with deviations from the mean cross compared to the expected outcome based on the GCA, which results from genetic complementation between parents. Thus, the selection of hybrid combinations with more favorable SCA estimates involving at least one parent with favorable GCA effects is recommended (Bressiani et al. 2002).

The genetic variability that exists between the parents noticeably indicates the possibility of successful selection. However, the selection rate variable, which is composed of other variables, hinders the comparison of the results obtained in this study with other studies because they are based on primary variables (i.e., each trait is individually evaluated), such as soluble solids (Brix), polarizable sugars (POL), tons of sugarcane per hectare (TCH) and stalk number (Bressiani et al. 2002, Silva et al. 2002, Bastos et al. 2003).

GCA has been mostly prioritized in parental selection complemented by SCA, which is also significant for various traits. Thus, it is best to use analyses based on primary variables when selecting individual traits to enable the identification of genotypes or hybrid combinations for the trait of interest. Conversely, the present study demonstrates the viability of using a composite variable to evaluate both parents and hybrid combinations. This procedure allows the use of data from previous cross performances of the parents in a single analysis, thereby enabling the planning of future crosses without conducting diallel crosses, which are labor-intensive and limit the number of parents that may be evaluated. In this case, GCA-based selection should be prioritized as an indicator of the genetic value of the parents. SCA-based selection can be performed within the evaluated set of crosses, and those with the greatest potential may be repeated using a more appropriate sample.

ACKNOWLEDGEMENTS

The authors thank the RIDESA, Brazil for financial support and the coordinators of sugarcane breeding programs Antônio Marcos Iaia (UFMT), Djalma Euzébio Simões Neto (UFRPE), Edelclaiton Daros (UFPR), Geraldo Veríssimo de Souza Barbosa (UFAL), Hermann Paulo Roffmann (UFSCar), Jair Felipe Garcia Pereira Ramalho (UFRRJ) and Márcio Henrique Pereira Barbosa (UFV), for sharing the data used in this study. The authors also thank the National Council for Scientific and Technological Development (CNPq, Brazil), for research grant to LJ Chaves and scholarship grant to PN Santana.

REFERENCES

Badaloo GH, Domaingue R and Ramdoyal KA (1999) Critical review of parental choice and cross prediction techniques in the MSIRI sugar cane breeding program. In Lalouette JA, Bachraz DY and Sukurdeep N (eds) **Proceedings third annual meeting of agricultural scientists.** Food and Agricultural Research Council Reduit, Mauritius, p. 47-54.

Barbosa MHP, Resende MDV, Bressiani JA, Silveira LCI and Peternelli LA (2005) Selection of sugarcane families and parents by Reml/Blup. **Crop Breeding and Applied Biotechnology 5**: 443-450.

Barbosa MHP, Resende MDV, Dias LAS, Barbosa GVS, Oliveira RA, Peternelli LA and Daros E (2012) Genetic improvement of sugar cane for bioenergy: The Brazilian experience in network research with RIDESA. **Crop Breeding and Applied Biotechnology S2**: 87-98.

Barbosa GVS, Oliveira RA, Cruz MM, Santos JM, Silva PP, Viveiros AJA, Sousa AJR, Ribeiro CAG, Soares L, Teodoro I, Sampaio Filho F, Diniz CA and Torres VLD (2015) RB99395: Sugarcane cultivar with high sucrose content. **Crop Breeding and Applied Biotechnology 15**: 187-190.

Bastos IT, Barbosa MHP, Cruz CD, Burnquist W, Bressiani JA and Silva F L (2003) Análise dialélica em clones de cana-de-açúcar. **Bragantia 62**: 199-206.

Bressiani JA, Burnquist WL, Fuzatto SR, Bonato ALV and Geraldi IO (2002) Combining ability in eight selected clones of sugarcane (*Saccharum* sp). **Crop Breeding and Applied Biotechnology 2**: 411-416.

Camacho LRS, Scapim CA, Senhorinho HJC and Conrado TV (2015) Diallel analysis of popcorn lines and hybrids for baby corn production. **Crop Breeding and Applied Biotechnology 15**: 33-39.

Carneiro MS, Chapola RG, Fernandes Junior AR, Cursi DE, Barreto FZ, Balsalobre TWA and Hoffmann HP (2015) RB975952 – Early maturing sugarcane cultivar. **Crop Breeding and Applied Biotechnology 15**: 193-196.

Chapola RG, Cruz JÁ, Nunes IK and Fernandes JAR (2013) **Censo varietal 2012.** Universidade Federal de São Carlos, Centro de Ciências Agrárias, São Carlos, 55p.

Daros E (ed) (2014) **Clones RB de cana-de-açúcar.** Graciosa, Curitiba, 112p.

Griffing B (1956) Concept of general and specific combining ability in relation to diallel crossing systems. **Australian Journal of Biological Science 9**: 463-493.

Halauer AR, Carena MJ and Miranda Filho JB (2010) **Quantitative genetics in maize breeding.** Springer, New York, 663p.

Iaia AM, Oliveira RA, Melo LJOT, Daros E, Simões Neto DE, Bastos GQ, Oliveira FJ, Chaves A and Melo TTAT (2014) RB002504 – New early-maturing sugarcane cultivar. **Crop Breeding and Applied Biotechnology 14**: 45-47.

Matsuoka S, Garcia AAF and Arizono H (2005) Melhoramento da cana-de-açúcar. In Borém A (ed) **Melhoramento de espécies cultivadas.** 2nd edn, Editora UFV, Viçosa, p. 205-251.

Melo LJOT, Daros E, Simões Neto DE, Chaves A, Silva LJ, Silva AEP and Melo TTAT (2014) RB962962, a sugarcane cultivar for late harvest. **Crop Breeding and Applied Biotechnology 14**: 132-135.

R Core Team (2014) **R: A language and environment for statistical computing.** R Foundation for Statistical Computing, Vienna. Available at <http://www.R-project.org/>. Accessed in 15 Jun, 2013

Ramalho MAP, Dias LAS and Carvalho BL (2012) Contributions of plant breeding in Brazil – progress and perspectives. **Crop Breeding and Applied Biotechnology 2**: 107-112.

Reis AJS, Chaves LJ, Duarte JB and Brasil EM (2005) Prediction of hybrid means from a partial circulant diallel table using the ordinary least square and the mixed model methods. **Genetics and Molecular Biology 28**: 314-320.

Santos JM, Barbosa GVS, Ramalho Neto CE and Almeida C (2014) Efficiency of biparental crossing in sugarcane analyzed by SSR markers. **Crop Breeding and Applied Biotechnology 14**:102-107.

Silva MA, Landell MGA, Gonçalves PS, Bressiani JA and Campana MP (2002) Estimates of general and specific combining ability for yield components in a partial sugarcane diallel cross. **Crop Breeding and Applied Biotechnology S2**: 111-120.

Silva FL, Barbosa MHP, Resende MDV, Peternelli LA and Pedrozo CA (2015) Efficiency of selection within sugarcane families via simulated individual BLUP. **Crop Breeding and Applied Biotechnology 15**: 1-9.

Veiga RD, Ferreira DF and Ramalho MAP (2000) Eficiência dos dialelos circulantes na escolha de genitores. **Pesquisa Agropecuária Brasileira 35**: 1395-1406.

21

BRS Kurumi and BRS Capiaçu - New elephant grass cultivars for grazing and cut-and-carry system

Antônio Vander Pereira[1], Francisco José da Silva Lédo[1] and Juarez Campolina Machado[1]*

Abstract: *Cultivar BRS Kurumi is characterized by short plant height, short internodes, high dry matter production (29.25 t ha^{-1}yr^{-1}), has a high nutritional value and is recommended for the cut-and-carry system or grazing. Cultivar BRS Capiaçu, indicated for silage and the cut-and-carry system, has tall upright growth, high dry matter production (49.75 t ha^{-1}yr^{-1}), good forage quality, is lodging-resistant and suitable for mechanical harvesting. Both cultivars are vegetatively propagated by stem cuttings.*

Key words: *Pennisetum purpureum, plant breeding, cultivars, forage traits.*

*Corresponding author:
E-mail: juarez.machado@embrapa.br

[1] Embrapa Gado de Leite, Rua Eugênio do Nascimento, 610, Dom Bosco, 36.038-330, Juiz de Fora, MG, Brazil

INTRODUCTION

Elephant grass (*Pennisetum purpureum* Schum.) is one of the most important forages, grown in almost all tropical and subtropical regions of the world. This forage stands out for its high dry matter production potential, forage quality, palatability, vigor, and persistence. It is mainly used in cut-and-carry system, and can also be used for silage and rotational grazing (Pereira et al. 2001).

Milk and meat production in Brazil are based on the use of pastures. However, these are strongly influenced by seasonal variations in forage supply, with negative consequences for animal production. Elephant grass is a low cost alternative for roughage supply, with positive effects on pasture stocking rate (Cóser et al. 2000).

In 1991, Embrapa Gado de Leite initiated a breeding program of this forage, addressing the development of cultivars for cut-and-carry system and grazing. To this end, the Germplasm bank of Elephantgrass (BAGCE) was established, with 110 accessions introduced from different regions of Brazil and abroad. The accessions were described based on morphological, cytogenetic and molecular descriptors (Shimoya et al. 2001, Techio et al. 2006, Pereira et al. 2008), and agronomic evaluations were carried out, underlying the selection of genetically divergent accession with high forage value for the breeding program. New genotypes were developed by controlled crosses and evaluation, selection and cloning of the best progenies, followed by clonal performance tests in different environments.

Cultivar BRS Kurumi, destined for grazing, and BRS Capiaçu, for silage and cut-and-carry feeding, represent the most recent cultivars of this breeding program.

CULTIVAR BRS KURUMI

Breeding Methods

To breed short cultivars appropriate for grazing, several crosses were carried out between the normal-sized accessions selected from BAGCE and accessions carrying the dwarf gene. This recessive gene, in homozygous condition, causes a shortening of the stem internodes, thus reducing plant height (Sollenberger et al. 1988). Consequently, the leaf/stem ratio increases significantly and the forage quality is improved, since the nutritional value of the leaves is higher than that of the stems.

The BAGCE accessions carrying the recessive dwarf gene (Merkeron de Pinda and Mott) were used as parents in crosses to breed short-height progenies. The accession Merkeron de Pinda has a normal height, since its genotypic background is heterozygous for the dwarf gene, while the homozygous accession Mott has a short plant height.

Cultivar BRS Kurumi was derived from a cross between the accession Merkeron de Pinda (BAGCE 19) and Roxo (BAGCE 57). The F_1 progenies of this cross had a tall size and the best were selected and recombined by polycross. The resulting progeny segregated for tall and short-height green and purple plants. Cultivar BRS Kurumi was obtained by selection and cloning of one of the short and green progenies.

Tests for the Value for Cultivation and Use (VCU) were performed for the Atlantic Forest, Amazon and Cerrado biomes, where this cultivar is recommended for cultivation. Cultivar BRS Kurumi was released by Embrapa in partnership with the Empresa de Pesquisa Agropecuária de Santa Catarina - Epagri, Agência Paulista de Tecnologia dos Agronegócios - Apta and Universidade Estadual do Norte Fluminense Darcy Ribeiro - Uenf.

Figure 1. BRS Kurumi, low plant height cultivar for rotational grazing.

Cultivar characteristics

Cultivar BRS Kurumi is a vegetatively propagated, perennial, short-height clone, recommended for the cut-and-carry system and for grazing. This cultivar is characterized by semi-open clumps, green leaves and stems, short internodes (mean of 4.8 cm) and mean height of 70 cm during the growing season (Figures 1 and 2, Table 1).

This cultivar has vigorous vegetative growth, rapid leaf expansion and intense basal and axillary tillering. Flowering occurs from June to July, and at this stage the stem is elongated and the plant can grow up to 3 m tall. The cultivar is propagated by vegetative cuttings resulting from stem subdivision, and the gems have excellent germination capacity.

According to Gomide et al. (2011), cultivar BRS Kurumi has a potential dry matter (DM) production of 29.25 t ha^{-1}yr^{-1}, of which 70% is concentrated in the rainy season. Compared to the short-height cultivar Mott, BRS Kurumi had a higher dry matter production of forage and leaves, as well as more axillary and basal tillers. In an experiment conducted over five grazing periods, from November 2001

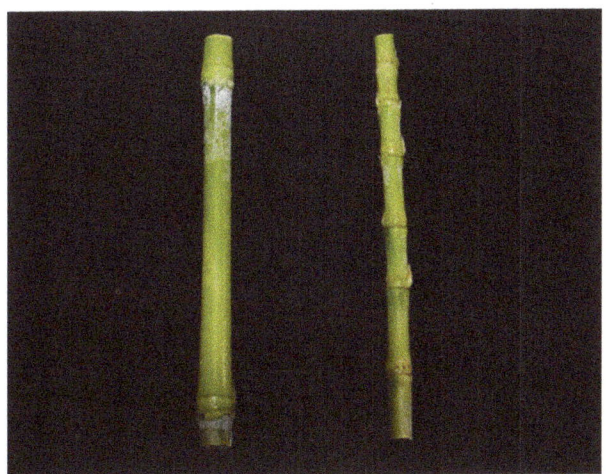

Figure 2. Comparison of internode length of BRS Capiaçu (tall plant) and BRS Kurumi (dwarf plant).

to June 2002, BRS Kurumi produced 16.2 t ha^{-1} forage DM and 11.1 t ha^{-1} leaf DM; while cultivar Mott produced 7.7 t ha^{-1} and 7.1 t ha^{-1}, respectively. Almeida et al. (2004) observed growth of 36 and 24 basal tillers per m^2 and 96 and 66 axillary tillers per m^2 for the cultivars BRS Kurumi and Mott, respectively. Similarly, Gomide et al. (2015) observed that cultivar BRS Kurumi had higher volumetric forage and leaf density than the tall cultivar Napier.

Aside from the ease of management for grazing compared to normal-sized cultivars, BRS Kurumi has a high production of forage with high nutritional value. Crude protein (CP) content in the forage reaches 18-20% and the *in vitro* dry matter digestibility (IVDMD) varies around 70% (Gomide et al. 2015).

Cultivar BRS Kurumi is susceptible to *Mahanarva spectabilis*, and is not recommended for cultivation in areas with a history of high pasture spittlebugs infestation. It was registered by the Ministry of Agriculture, Livestock and Supply (MAPA) on 04/17/2012 (No. 28690), and was licensed by the plant variety protection certificate on 02/02/2012 (No. 20120164), while Embrapa holds the property rights of this cultivar. Embrapa Gado de Leite is responsible for the maintenance of the genetic stock of this cultivar and information about propagation material can be requested from Embrapa Produtos e Mercado, agency in Brasilia-DF (www.embrapa.br/produtos-e-mercado).

CULTIVAR BRS CAPIAÇU

Breeding methods

Cultivar BRS Capiaçu was developed from a cross made in 1992, between the accessions Guaco (BAGCE 60) and Roxo (BAGCE 57). The full-sib seeds from this cross were sown in beds, since strong segregation for forage-related traits had been observed. The 10 best progenies of this family were cloned for comparative tests with clones selected from other crosses. The selected clones were evaluated by the national network of elephant grass evaluation (RENACE) in 17 Brazilian states, from 1999 to 2008 (Pereira and Lédo 2008). From 2009 to 2011, clone CNPGL 92-79-2 was tested for the value of cultivation and use (VCU), and registered and protected as BRS Capiaçu.

Cultivar characteristics

Cultivar BRS Capiaçu was developed by the Embrapa breeding program of elephant grass. This cultivar is characterized by late flowering; tall size; upright clumps, leaves with wide, long and green blades; yellowish green leaf sheath and; stem with thick diameter and yellowish internodes (Table 2 and Figure 3).

This cultivar is distinguished by high yield and forage quality (Table 3); high lodging resistance and by excellent adaptation to mechanical harvest.

BRS Capiaçu also produces good quality silage, representing a cheaper alternative than corn, for being a

Table 1. Main traits of cultivar BRS Kurumi

Traits	Description
Plant material	Clone
Ploidy level	Tetraploid (2n=4x=28)
Cultivation	Vegetative propagation (stem cuttings)
Flowering period	June - July
Growth habit	Clumps (semi-open clump)
Basal tiller density	High (36 tillers m^{-2})
Axillary tiller density	High (96 tillers m^{-2})
Plant height	70 cm
Stem diameter	1.2 cm
Internode length	4.8 cm
Leaf/stem ratio[1]	11.8
Leaf width	3.4 cm
Leaf length	69 cm
Leaf color	green
Midrib color	white
Total dry matter production	20.2 t ha^{-1}yr^{-1}
Crude protein (CP)[1]	20.6%
Digestibility (IVDMD)[1]	67.6

[1] Summer (harvest after 20 days).

Table 2. Main traits of cultivar BRS Capiaçu

Trait	Description
Plant material	Clone
Ploidy level	Tetraploid (2n=4x=28)
Cultivation	Vegetative propagation (cuttings)
Flowering period	Late (June - July)
Growth habit	Clumps (upright clump)
Basal tiller density	Medium (30 tillers m^{-2})
Plant height	Tall (4.20 m)
Stem diameter	Thick (1.6 cm)
Internode length	16 cm
Leaf/stem ratio	0.75
Leaf width	5.17 cm
Leaf length	106 cm
Leaf color	Green
Midrib color	White
Total forage yield (t ha^{-1}yr^{-1})	49.75
Leaf yield (t ha^{-1}yr^{-1})	21.60
Crude protein (%) – Whole-plant content	9.10
IVDMD (%) – Whole-plant content	54.76
Fiber NDF (%) – Whole-plant content	71.5

Table 3. Total dry matter production (TDMP), leaf dry matter production (LDMP) and crude protein (CP) content

Cultivars	Annual TDMP (t ha^{-1})	Annual LDMP (t ha^{-1})	Whole-plant CP content (%)
BRS Capiaçu	49.75	21.60	9.10
Mineiro	36.79	16.16	6.94
Cameroon	29.87	14.32	7.17

perennial crop that does not require annual seed purchase and with higher productivity.

The cultivar has vegetative propagation by cuttings and is indicated for forage cutting, destined for roughage supply in the form of silage or cut-and-carry system. The elephant grass cultivar BRS Capiaçu was registered by MAPA on 01/08/2015 (No. 33503), and licensed by a cultivar protection certificate on 01/23/2015 (No.20150009), whereas Embrapa maintains the property rights of the cultivar. Based on results of VCU tests, this cultivar is recommended for cultivation in the Atlantic Forest biome. The genetic stock of BRS Capiaçu is maintained by Embrapa Gado de Leite and information on propagation material can be obtained at Embrapa Produtos e Mercado, agency in Brasília-DF (www.embrapa.br/produtos-e-mercado).

Figure 3. BRS Capiaçu – elephant grass cultivar for production of silage or cut-and-carry system.

REFERENCES

Almeida EX, Baad EAS and Pereira AV (2004) Avaliação de novos genótipos de capim-elefante sob pastejo. **Agropecuária Catarinense 17**: 75-78.

Cóser AC, Martins CE and Deresz F (2000) Capim-elefante: formas de uso na alimentação animal. Embrapa, Juiz de Fora, 27p. (Circular Técnica, 57).

Gomide CA, Paciullo DSC, Ledo FJ S, Pereira AV, Morenz MJF and Brighenti AM (2015) Informações sobre a cultivar BRS Kurumi. Embrapa, Juiz de Fora, 4p. (Comunicado Técnico, 75).

Gomide CAM, Paciullo DSC, Ledo FS, Castro CRT and Morenz MJF (2011) Produção de forragem e valor nutritivo de clones de capim-elefante anão sob estratégias de desfolha intermitente. Embrapa, Juiz de Fora, 23p. (Boletim de Pesquisa e Desenvolvimento, 31).

Pereira AV and Lédo FJS (2008) Melhoramento genético de *Pennisetum purpureum*. In Resende RMS, Valle CB and Jank L (eds) **Melhoramento de forrageiras tropicais**. Embrapa, Campo Grande, p. 89-116.

Pereira AV, Machado MA, Azevedo ALS, Nascimento CS, Campos AL and Ledo FJS (2008) Diversidade genética entre acessos de capim-elefante obtida com marcadores moleculares. **Revista Brasileira de Zootecnia 37**: 1216-1221.

Pereira AV, Valle CB, Ferreira RP and Miles JW (2001) Melhoramento de forrageiras tropicais. In Nass LL, Valois ACC, Melo IS and Valadares Inglis MC (eds) **Recursos genéticos e melhoramento.** Fundação MT, Rondonópolis, p. 549-602.

Shimoya A, Ferreira RP, Pereira AV, Cruz CD and Carneiro PCS (2001) Comportamento morfo-agronomico de genótipos de capim-elefante. **Revista Ceres 48**: 1-19.

Sollenberger LE, Prine GN, Ocumpaugh WR, Hanna WW, Jones Jr. CS, Schank SC and Kalmbacher RS (1988) Mott elephantgrass: a high quality forage for the subtropics and tropics. (s.p.): Florida Agricultural Experimental Station, 18p. (Circular, 5-356).

Techio VH, Davide LC and Pereira AV (2006) Meiosis in *Pennisetum purpureum, P. glaucum* and interspecific hybrids (Poaceae, Poales). **Genetics and Molecular Biology 29**: 253-262.

22

BRS FC402: high-yielding common bean cultivar with carioca grain, resistance to anthracnose and fusarium wilt

Leonardo Cunha Melo[1*], **Helton Santos Pereira**[1], **Luís Cláu-dio de Faria**[1], **Thiago Lívio Pessoa Oliveira de Souza**[1], **Adriane Wendland**[1], **José Luis Cabrera Díaz**[1], **Hélio Wilson Lemos de Carvalho**[2], **Carlos Lásaro Pereira de Melo**[3], **Antônio Félix da Costa**[4], **Mariana Cruzick de Souza Magaldi**[1] and **Joaquim Geraldo Cáprio da Costa**[1]

Abstract: *BRS FC402 is a common bean cultivar of the carioca-grain group with commercial grain quality, suitable for cultivation in 21 Brazilian states. Cultivar has a normal cycle (85-94 days), high yield potential (4479 kg ha⁻¹), 10.1% higher mean yield than the controls (2462 kg ha⁻¹) and resistance to fusarium wilt and anthracnose.*

Key words: *Phaseolus vulgaris, crop breeding, disease resistance, yield stability.*

*Corresponding author:
E-mail: leonardo.melo@embrapa.br

[1] Embrapa Arroz e Feijão, Rod. GO 462, km 12, CP 179, 75.375-000, Santo Antônio de Goiás, GO, Brazil
[2] Embrapa Tabuleiros Costeiros, Avenida beira Mar, 3250, Bairro Jardins, CP 44, 49.025-040, Aracaju, SE, Brazil
[3] Embrapa Agropecuária Oeste, Rod. BR 163, km 253,6, CP 449, 79.804-790, Dourados, MS, Brazil
[4] Instituto Agronômico de Pernambuco, Avenida General San Martin, 1371, Bairro Bongi, 50.761-000, Recife, PE, Brazil

INTRODUCTION

Brazil is one of the world's leading producers and consumers of common bean (*Phaseolus vulgaris* L.) (FAO 2015). Nationwide, the crop has a high socioeconomic value, for being part of small, medium and large-scale production systems, widely distributed across the country.

In view of the key importance of the crop, common bean breeding programs in Brazil conducted by public and private research institutions have continuously supplied the domestic market with new cultivars (Ramalho et al. 2014, Barili et al. 2016). Primarily, these programs sought the association of desirable traits such as disease resistance, low loss in mechanical harvesting, earliness, higher yield and yield stability, thus contributing to increase crop yields from 810 kg ha⁻¹ in 2000 to 1353 kg ha⁻¹ in 2013, i.e., a mean yield gain of 67.0%. The mean gains in grain yield obtained by the national common bean breeding program conducted by the Brazilian Agricultural Research Corporation (Embrapa) and partners are in the order of 0.72% per year for the carioca market class (Faria et al. 2013) and 1.1% for the black market class (Faria et al. 2014).

The recommendation of region- and season-specific cultivars has been a challenge due to the low seed use rate. The reasons were the low seed demand and, consequently, low production level, which failed to generate a satisfactory economic return for the seed industry. Thus, a commonly used strategy for the development of new common bean cultivars is to select genotypes with broad adaptation, allowing the recommendation of cultivars that maintain their competitiveness under the most diverse growth conditions and production systems.

Embrapa has established agreements with the main common bean breeding programs in Brazil and the world, at different levels of cooperation. These partnerships allow a wide network for the improvement of the crop in Brazil, generating technical/scientific benefits for all partner institutions, but mainly for producers and the Brazilian society. Between 1984 and 2015, Embrapa and partners developed 55 cultivars of different market classes, with an annual mean of 1.8 cultivars, of which 32 were released after the plant variety protection law came into effect. Estimates of Embrapa of the cost-benefit relation of the above common bean breeding program indicated an average return of about 10 dollars in benefits to the Brazilian society, for every dollar invested in the development of cultivars (Alves et al. 2002).

BREEDING METHODS

Cultivar BRS FC402 resulted from the cross between the common bean elite lines LM 96200246 and LP 9632, performed in 2000 by Embrapa Rice and Beans, in Santo Antônio de Goiás (GO). In 2001, the F_1 generation of this population was sown in a greenhouse during the dry growing season. In the same year, generations F_2 and F_3 were also sown in a greenhouse, in the winter and rainy growing season, respectively. In the dry season of 2002, the F_4 generation was grown in Ponta Grossa (PR), in bulk, with selection for upright plant architecture and resistance to anthracnose, rust, common bacterial blight and angular leaf spot. In the rainy season of 2002, the F_5 generation was grown in bulk, also in Ponta Grossa, and individual plants were selected for upright plant architecture and resistance to anthracnose, rust, common bacterial blight and angular leaf spot. In the winter of 2003, $F_{5:6}$ lines were evaluated in Santo Antônio de Goiás, and selected for carioca grain with commercial standard, upright plant architecture, and for resistance to anthracnose, rust, common bacterial blight, and angular leaf spot. One of the selected lines was LM 203200638.

After this stage, line LM 203200638 was evaluated in field trials with replications for grain yield and other important traits, e.g., reaction to diseases and plant architecture. In 2004, LM 203200638 was evaluated together with 159 other elite lines and 9 control cultivars (IPR Juriti, FTS Magnífico, Pérola, Carioca Eté, BRSMG Talismã, BRS Pontal, Iapar 81, Aporé, and BRS Requinte) in field trials with a triple lattice design in Ponta Grossa (dry growing season) and Santo Antônio de Goiás (winter), for resistance to anthracnose, angular leaf spot and for plant architecture and yield.

In 2005, this line was evaluated in the Preliminary Tests in a triple lattice design with plots consisting of two 4.0-m rows, along with 59 other elite lines and 4 control cultivars (BRS Pontal, Pérola, FTS Magnífico, and IPR Juriti). The trials were conducted in four environments: Santo Antônio de Goiás (GO), winter; Lavras (MG), dry season; and Ponta Grossa (PR), rainy and dry growing seasons.

In 2007, LM 203200638 was labeled with the pre-commercial name CNFC 11948 and evaluated in the Intermediate Tests with 29 other elite lines and 5 control cultivars (Pérola, BRS Pontal, BRS Requinte, BRS Cometa, and IPR Juriti) in a randomized block design with three replications and four 4.0-m rows, in eight environments: Santo Antônio de Goiás (GO), Sete Lagoas (MG) and Uberlandia (MG), in the winter growing season; Ponta Grossa (PR), in the rainy and dry seasons; Ijaci (MG), in the dry season; Simão Dias (SE) and Frei Paulo (SE), in the rainy season. The combined analysis of data for grain yield and other agronomic traits qualified the common bean elite line CNFC 11948 for the final field tests, the Tests of Value for Cultivation and Use (VCU).

In 2008, seeds were multiplied for the VCU tests. In 2009 and 2010, line CNFC 11948 was evaluated in 78 field trials with four control cultivars (BRS 9435 Cometa, BRS Estilo, IPR Juriti, and Pérola), in a randomized block design with three replications and four 4.0-m rows, applying the recommended cultivation techniques for the different environments and cropping systems. The two middle rows of each plot were harvested and the grains evaluated for yield, percentage of commercial grain, 100-grain weight, cooking time, and protein content.

Grain yield was measured in kg ha^{-1}, corrected to 13% moisture. The percentage of commercial grain size (sieve yield) was determined in grain samples of 300 g per plot, which were sieved through sieves with oblong holes (width 4.25 mm). The seeds retained on the sieve were weighed and this weight divided by the initial sample weight, to obtain the percentage of seeds with commercial size. From the seeds retained on the sieve, a new 100-seed sample was taken to determine 100-grain weight. In the tests with best results (highest mean yield and lowest variation coefficient), samples were taken from each plot to analyze cooking time and protein content. For cooking time, the grains were soaked in distilled water at a 1:4 (w/v) ratio at room temperature. After 16 hours, the water was poured off and the grains placed in a Mattson cooker. The cooking time was determined from water boiling until the point when the cooker needles penetrated 50% + 1 of the grains, according to the methodology adapted from Proctor and Watts (1987). The protein content was analyzed in grain flour (grain ground in a ball mill), according to the micro-Kjeldahl method.

In addition, the following agronomic traits were evaluated: lodging tolerance, plant architecture, and disease reaction to: common bacterial blight (*Xanthomonas axonopodis* pv *phaseoli* and *Xanthomonas fuscans pv. fuscans*); bacterial wilt (*Curtobacterium flaccumfaciens* pv *flaccumfaciens.*); angular leaf spot (*Pseudocercospora griseola*); anthracnose (*Colletotrichum lindemutianum*); rust (*Uromyces appendiculatus*), fusarium wilt (*Fusarium oxysporum* f. sp. *phaseoli*); *Bean common mosaic virus* (BCMV), and *Bean golden mosaic virus* (BGMV). For these assessments, rating scales of disease severity were used, as described by Melo (2009), ranging from 1 (absence of pathogen symptoms and signs) to 9 (100% disease severity or plant death).

GRAIN YIELD AND YIELD POTENTIAL

In 78 final field trials (VCU tests) conducted in 2009 and 2010 during the rainy growing season in Sergipe, Alagoas, Pernambuco, Bahia, and São Paulo, at sowing in the dry season in Mato Grosso do Sul and Rio Grande do Sul, at sowing in the rainy and dry seasons in Santa Catarina and Paraná, and at sowing in the rainy, dry and winter growing seasons in Goiás and the Federal District, the cultivar BRS FC402 (CNFC 11948) had a 10.1% higher grain yield than the mean of the control cultivars BRS Estilo and Pérola. The superiority of the mean performance of BRS FC 402, in the three main regions of cultivar recommendation for common bean (Pereira et al. 2009), was 4.5% in Region 3 (Sergipe, Alagoas, Pernambuco, Paraíba, Rio Grande do Norte, Ceará, and Piauí), 15.0% in Region 1 (São Paulo, Mato Grosso do Sul, Paraná, Santa Catarina, and Rio Grande do Sul) and 6.7% in Region 2 (Mato Grosso, Goiás/Federal District, Minas Gerais, Rio de Janeiro, Espírito Santo, Bahia, Tocantins, and Maranhão) (Table 1).

The overall mean yield of BRS FC402 was 2462 kg ha⁻¹, versus 2238 kg ha⁻¹ of the control cultivars Pérola and BRS Estilo. Considering the data for each region for official recommendation of common bean cultivars, BRS FC402 exceeded the yield of the control cultivars by 10% in the rainy and dry growing seasons in Region 1, and by 17.7% in the rainy season. In Region 2, BRS FC402 reached a 6.7% higher mean and up to 23% higher mean yield in the dry season (Table 1). In Region 3, the superiority was 4.5%, indicating that BRS FC402 is a broadly adapted cultivar, which can be grown advantageously in the main bean-producing areas of Brazil.

The yield potential of BRS FC402, calculated from the mean of the five best yield test results of this cultivar, was 4,479 kg ha⁻¹. This estimate indicates the high genetic potential of this cultivar and that high yields can be achieved in favorable environments and under good growing conditions.

Based on the agronomic performance, cultivar BRS FC402 was registered as suitable for the rainy and winter growing seasons in the states of Bahia, Mato Grosso and Tocantins; rainy, dry and winter seasons in Goiás, Federal District, Espírito Santo, Rio de Janeiro; dry season in Mato Grosso do Sul; rainy and dry seasons in Paraná, Santa Catarina, São

Table 1. Mean grain yield of the common bean cultivar BRS FC402 compared to the means of the control cultivars Pérola and BRS Estilo in region and season-specific final field trials (Tests of Value for Cultivation and Use - VCU tests) in 2009 and 2010

Region	Season	BRS FC402 (kg ha⁻¹)	Mean yield of control cultivars (kg ha⁻¹)	Relative mean yield (%)	Number of environments
I	Rainy	2955	2575	117.7	23
	Dry	2119	1946	110.2	13
	Overall	2653	2348	115.0	36
II	Rainy	2594	2467	108.4	16
	Dry	1770	1437	123.0	4
	Winter	2196	2410	93.5	7
	Overall	2368	2299	106.7	27
III	Rainy	2171	2096	104.5	15
Overall	-	2462	2283	110.1	78

Region I – RS, SC, PR, MS, and SP; Region II – MG, ES, RJ, GO, DF, MT, TO, BA, and MA; Region III – SE, AL, PE, PB, CE, RN, and PB.

Paulo and Rio Grande do Sul; and rainy season in Maranhão, Sergipe, Alagoas, Pernambuco, Rio Grande do Norte, Piauí, Ceará and Paraíba.

OTHER AGRONOMIC TRAITS

With regard to the grain quality traits, BRS FC402 has high nutritional value with regular and standard grain size (Table 2). In field trials without fungicide spraying, BRS FC402 reached 87% of the sieve yield, and the mean 100-grain weight was 26 g, similar to the control cultivars Pérola and BRS Estilo (Table 2), indicating the high commercial value of grains. For this reason, it is expected that BRS FC402 presents grain quality parameters higher than those observed in the present work when grown in commercial fields with disease control through fungicide spraying. The mean cooking time of BRS FC402 was 32 min, i.e., slightly higher than that of Pérola and BRS Estilo (31 and 28 min, respectively). The mean grain protein content of BRS FC402 was practically identical to that of the control cultivars (Table 2). Cultivar BRS FC402 has grains with a 7% higher zinc content than the two control cultivars and a 4.5% and 19% higher iron content than Pérola and BRS Estilo, respectively, indicating its strong potential to be used in governmental nutrition programs using biofortificated foods, addressing the nutritional complementation of low-income populations.

Under artificial inoculation, BRS FC402 is resistant to Bean common mosaic virus. In field trials, it proved moderately resistant to anthracnose, rust and fusarium wilt. However, it was susceptible to angular leaf spot, common bacterial blight, bacterial wilt, and BGMV (Table 3).

Cultivar BRS FC402 has a normal cycle (85-94 days from seedling emergence to physiological seed maturity), similar to that of Pérola and BRS Estilo (Table 3). The plants have a semi-upright architecture and an indeterminate growth habit (Type II). This new cultivar is moderately tolerant to lodging and can be used for mechanical and even direct harvesting. The flowers are white and the physiologically mature pods reddish yellow. At harvest maturity, the pods turn sand-yellow. The grains are beige with light brown stripes, a nearly full elliptical shape, not shiny.

Compared to BRS Estilo (Melo et al. 2010) and BRSMG Uai (Ramalho et al. 2016), the most promising and modern cultivars with carioca grain released by Embrapa and partners, BRS FC402 has a higher yield, resistance to anthracnose and wilt fusarium, but the same commercial grain quality as BRS Estilo. For this reason, BRS FC402 is expected to be adopted as a new technical solution for common bean growers throughout Brazil. This cultivar can contribute efficiently to the sustainability of common bean crop in the Brazilian agribusiness. The greatest impact of the introduction of this cultivar is expected mainly for winter cultivation under pivot irrigation and in areas of long-standing and intensive use, for being resistant to fusarium wilt, as well as in the rainy season in high altitude regions and throughout South-central Brazil, for being resistant to anthracnose.

Table 2. Grain traits of the common bean cultivar BRS FC402 compared to the control cultivars Pérola and BRS Estilo in Tests of Value for Cultivation and Use (VCU tests) in 2009 and 2010

Cultivar	Iron content (mg kg⁻¹)	Zinc content (mg kg⁻¹)	Cooking time (minutes)	Protein content (%)	Sieve yield (%)*	100-grain weight (g)*
BRS FC402	61.8	30.8	32.0	21.3	87.3	26.0
BRS Estilo	51.9	28.7	30.9	21.8	86.0	24.2
Pérola	59.1	28.8	27.8	21.1	84.0	25.6

* Estimates determined in final field trials without disease control, using mesh 11 (4.25 mm).

Table 3. Agronomic traits and disease reaction of the common bean cultivar BRS FC402 compared to the control cultivars Pérola and BRS Estilo

Cultivar	Cycle	PAR	AN	CBB	RU	ALS	BCMV	BGMV	FW	BW
BRS FC 402	N	Semi-upright	MR	S	MR	S	R	S	MR	S
BRS Estilo	N	Upright	MS	S	MR	S	R	S	S	S
Pérola	N	Semi-prostrate	S	S	S	S	R	S	MS	S

N- Normal cycle (85-94 days); PAR- plant architecture; AN- Anthracnose; CBB- common bacterial blight; RU- Rust; ALS- angular leaf spot; BCMV- *Bean common mosaic virus*; BGMV- *Bean golden mosaic virus*; FW- fusarium wilt; BW- bacterial wilt. R-Resistant (grade 1); MR- Moderately resistant (grades 2 and 3); MS- Moderately susceptible (grades 4-6); and S Susceptible (grades 7-9).

SEED PRODUCTION

BRS FC402 was registered as a new common bean cultivar by the Ministry of Agriculture, Livestock and Supply (MAPA) on September 30, 2015, under number 34531, and protected in May 06, 2016 under number 20160087. "Embrapa Produtos e Mercado" is in charge of the basic seed production.

CONCLUSIONS

The common bean cultivar BRS FC402 with carioca grain has a normal cycle (85-94 days), high yield, high commercial grain quality and yield stability, and effective resistance to anthracnose and fusarium wilt.

This cultivar is indicated for cultivation in the following States and growing seasons: winter and rainy season in Bahia, Mato Grosso and Tocantins; rainy, dry and winter seasons in Goiás, Distrito Federal, Espírito Santo and Rio de Janeiro; dry season in Mato Grosso do Sul; rainy and dry seasons in Paraná, Santa Catarina, São Paulo and Rio Grande do Sul; and rainy season in Maranhão, Sergipe, Alagoas, Pernambuco, Rio Grande do Norte, Piauí, Ceará and Paraíba.

ACKNOWLEDGEMENTS

The following partner institutions contributed to the agronomic evaluation of the cultivar BRS FC402: Embrapa Arroz e Feijão; Embrapa Transferência de Tecnologia; Embrapa Tabuleiros Costeiros; Embrapa Agropecuária Oeste; Instituto Agronômico de Pernambucano; Emater Alagoas; Emater Goiás; Fundação de Ensino Superior de Rio Verde; Instituto Federal de Goiás, Universidade Estadual de Goiás, Uni-Anhanguera; Embrapa Cerrados; Empresa de Pesquisa Agropecuária e Extensão Rural de Santa Catarina; Embrapa Milho e Sorgo; Universidade Federal de Lavras; Universidade Federal de Uberlândia; Universidade de Cruz Alta; Embrapa Soja; Universidade Estadual Paulista; Universidade Federal do Mato Grosso do Sul; and Universidade Estadual do Mato Grosso do Sul.

REFERENCES

Alves ERA, Magalhães MC and Guedes PP (2002) **Calculando e atribuindo os benefícios da pesquisa de melhoramento de variedades: o caso da Embrapa**. Embrapa, Brasília, 248p.

Barili LD, Vale NM, Prado AL, Carneiro JES, Silva FF and Nascimento M (2015) Genotype-enviroment interaction in common bean cultivars with carioca grain cultivated in Brazil in the last 40 years. **Crop Breeding and Applied Biotechnology 15**: 244-250.

FAO (2015) **Faostat**. Available at <http://faostat3.fao.org/browse/Q/QC/E>. Accessed Jan 06, 2015.

Faria LC, Melo PGS, Pereira, HS, Del Peloso MJ, Brás AJBP, Moreira, JAA, Carvalho HWL and Melo LC (2013) Genetic progress during 22 years of improvement of carioca-type common bean in Brazil. **Field Crops Research 142**: 68-74.

Faria LC, Melo PGS, Pereira, HS, Del Peloso MJ, Wendland A, Borges SF, Pereira Filho IA, Diaz J LC, Calgaro M and Melo LC (2014) Genetic progress during 22 years of black bean improvement. **Euphytica 199**: 261-272.

Melo LC, Del Peloso MJ, Pereira HS, Faria LC, Costa JGC, Díaz, JLC, Paiva CA, Wendland A and Abreu AFB (2010) BRS Estilo - Common bean cultivar with Carioca grain, upright growth and high yield potential. **Crop Breeding and Applied Biotechnology 10**: 377-379.

Melo LC (2009) **Procedimentos para condução de ensaios de valor de cultivo e uso em feijoeiro-comum**. Embrapa Arroz e Feijão, Santo Antônio de Goiás, 104p. (Documentos, 239).

Pereira HS, Melo LC, Silva SC, Del Peloso MJ, Faria LC, Costa JGC, Magaldi MCS and Wendland A (2009) **Regionalização de áreas produtoras de feijão comum para recomendação de cultivares no Brasil**. Embrapa Arroz e Feijão, Santo Antônio de Goiás, 6p. (Comunicado técnico, 187).

Proctor JR and Watts BM (1987) Development of a modified Mattson bean cooker procedure based on sensory panel cookability evaluation. **Canadian Institute of Food Science and Technology 20**: 9-14.

Ramalho MAP, Abreu AFB and Guilherme SR (2014) **Informações técnicas para o cultivo do feijoeiro-comum na região central brasileira: 2015-2017**. Universidade Federal de Lavras, Lavras, 168p.

Ramalho MAP, Abreu AFB, Carneiro JES, Melo LC, Paula Júnior TJ, Pereira HS, Del Peloso MJ, Pereira Filho IA, Martins M, Del Giúdice MP and Vieira RF (2016) BRSMG Uai: common bean cultivar with carioca grain type and upright plant architecture. **Crop Breeding and Applied Biotechnology 16**: 261-264.

23

BRS Pampeira: new irrigated rice cultivar with high yield potential

Ariano Martins de Magalhães Júnior[1]*, Orlando Peixoto de Morais2, Paulo Ricardo Reis Fagundes1, José Manoel Colombari Filho2, Daniel Fernandes Franco1, Antônio Carlos Centeno Cordeiro3, José Almeida Pereira4, Paulo Hideo Nakano Rangel2, Francisco Pereira Moura Neto2, Eduardo Anibele Streck1, Gabriel Almeida Aguiar1 and Paulo Henrique Karling Facchinello1

Abstract: *BRS Pampeira is a rice cultivar developed by Embrapa, recommended for irrigated cultivation in Brazil. It shows modern architecture, with high tillering and tolerance to lodging. It stands out for its high yield potential, medium cycle and good grain quality.*

Key words: *Oryza sativa L., productivity, plant breeding.*

*Corresponding author:
E-mail: ariano.martins@embrapa.br

[1] Embrapa Clima Temperado, Rodovia BR-392, km 78, 9º Distrito, Monte Bonito, CP 321, 96.010-971, Pelotas, RS, Brazil
[2] Embrapa Arroz e Feijão, Rodovia GO-462, km 12, Fazenda Capivara, Zona Rural, CP 179, 75.375-000, Santo Antônio de Goiás, GO, Brazil
[3] Embrapa Roraima, Rodovia BR 174, km 8, Distrito Industrial, CP 133, 69.301-970, Boa Vista, RR, Brazil
[4] Embrapa Meio Norte, Av. Duque de Caxias, 5.650, Bairro Buenos Aires, CP 001, 64.006-220, Teresina, PI, Brazil

INTRODUCTION

Rice (*Oryza sativa* L.) is the basis of the diet and the main source of proteins and carbohydrates for more than half the world's population, its relative importance is even more evident in poor and developing countries (Lee et al. 2011).

Currently, the increase in yield is a major challenge for the genetic breeding of irrigated rice, because, in addition to the difficulties arising from the complexity of this trait, the improvement must meet the industrial and culinary standards on the Brazilian market. Thus, the global food security will continue to depend on the ability to sustain high production yields (Zeigler and Barclay 2008).

Embrapa's Irrigated Rice Breeding Program has the challenge to develop cultivars that have a high stability and adaptability to the different environments in which they were grown and that express high yield, associated with the appropriate agronomical and industrial characteristics.

In this sense, the rice cultivar BRS Pampeira was developed to meet a gap of medium maturity cultivars, with grain quality and high yield potential. Therefore, the goal is to present and agronomically describe the irrigated rice cultivar BRS Pampeira developed by Embrapa, recommended for cultivation under flood irrigation, initially for the state of Rio Grande do Sul, but with the perspective of adoption also throughout the Brazilian tropical region.

PEDIGREE AND BREEDING METHOD

The cultivar BRS Pampeira was originated through a cross single, having as female parent the cultivarIR-22, introduced from the International Rice Research Institute (IRRI), and, as male parent, the CNA 8502 line. The aim was to gather in this new cultivar good agronomical traits, such as better blast resistance,

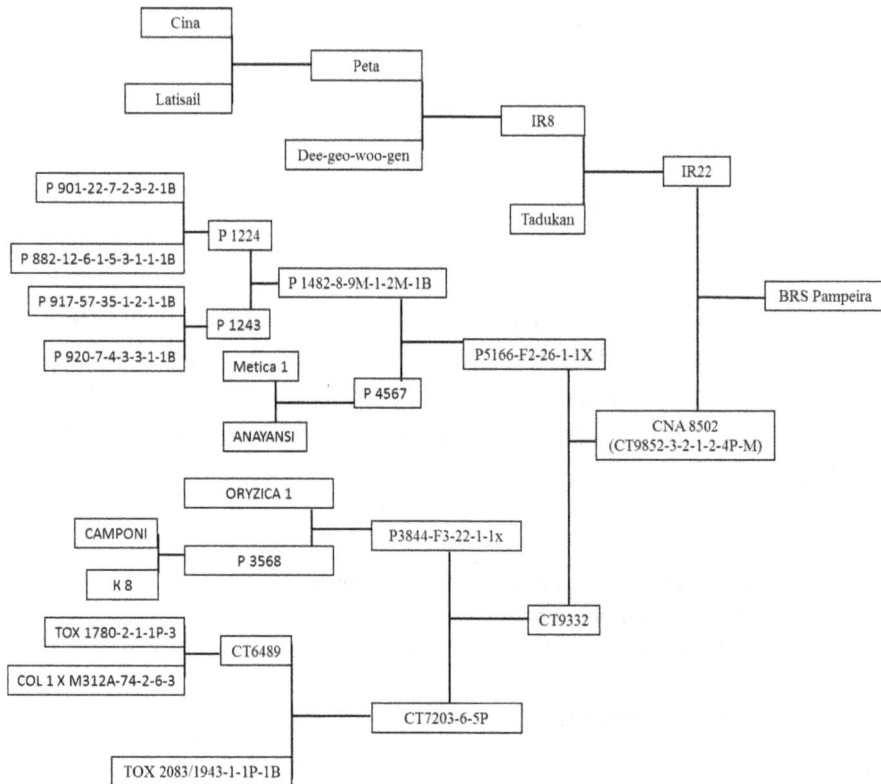

Figure 1. Genealogy of the irrigated rice cultivar BRS Pampeira.

hardiness, yield potential and grain quality (Figure 1).

In 2000, the cross between these parents was done, and it progenies were coded by the breeding program as CNAx8133. In the first half of 2001, in Goianira, in the state of Goiás (GO), after the multiplication of F_1 seeds, the F_2 generation was sown in the nursery 1 (SN1) for the selection of individual plants, resulting in the F_3 progenies. These progenies, in the 2001/2002 crop season, were evaluated in the Test of Observation of Families (EOF) in the city of Goianira. The progenies selected in this experiment ($F_{2:4}$) were re-evaluated in 2002/2003, in the Tropical Families Yield Test (ERFT) in the counties of Goianira (GO), Formoso do Araguaia (TO) and Boa Vista (RR).

In the pooled analysis of these trials, it was possible to identify the CNAx8133-B-4-B-B family, originated from the fourth plant selected in the 2001 field trial, producing in average 6,594 kg ha^{-1}, and considered to be promising with respect to the other features. This family was used as a source of lines in the VS2, in the year 2003/2004, when 15 plants were selected from its scope.

In Goianira, in the 2004/2005 growing season, its progenies ($F_{5:6}$) were evaluated in the Lines Assessment Test (EOL), where the line derived from the first selected plant in the former family stood up. So, in the following year, it was included in the preliminary test of irrigated rice yield in the tropical region (EPT), with the identification BRA051108. In this trial, its yield was similar to that of Metica 1, until then one of the most productive tropical cultivars, and presented low temperature of grain gelatinization, similar to the IR-22, one of the desired targets during the selection process.

The BRA051108 line, in the 2006/2007 crop season, was part of the Tropical Yield Regional Tests (ERT), held in five locations in the tropical region of Brazil (Formoso do Araguaia – Tocantins, Belém – Pará, Bragança – Pará, Salvaterra – Pará, Cantá – Roraima). In these field trials, the BRA051108 line showed good agronomic performance and yield.

In the crop seasons from 2007/2008 to 2012/2013, the BRA051108 was included on tests required by the Ministry of Agriculture, Livestock and Supply (MAPA) for the release of new cultivars, i.e., the tests of Value of Cultivation and Use

(VCU). These were conducted in all the irrigated rice-producing regions of Brazil. In the joint analysis of the VCU tests, the BRA051108 line reached higher yields than the controls BR IRGA 409, BRS 7 Taim and BRS Tropical. The BRA051108 has good quality grain, tolerance to lodging and to the major diseases of the culture (rice blast, leaf scald), in addition to high genetic potential for productivity. Thus, it is an excellent the new cultivar to be release and recommended for cultivation in the different rice-producing regions in Brazil.

PERFORMANCE CHARACTERISTICS

The cultivar BRS Pampeira has biological cycle around 133 days, from emergence to maturity in the Rio Grande do Sul state, being classified as medium cycle. The plants have modern-Philippine height, with hairy leaves and erect leaf flags. The average plant height is 91.5 cm, which can vary depending on the cultural management and the environmental conditions found in other states of Brazil. This cultivar has high tillering, strong stems and resistance to plant lodging (Table 1).

The grains are long and thin, vitreous, with low incidence of white center, the average 1000-grain weight being around 27 g. The hulls of the grains have straw yellow color is hairy and has no awns. The average panicle length is 23.9 cm. The industrial yield of grains, under normal conditions of environment and crop management, is higher than 62% whole polished grains, with a total yield of 68% (Table 2).

It has excellent cooking attributes compared to the best cultivars highlighted by the industry. In indirect tests of cooking quality, the grain has amylose content (AC) ranked as high and low gelatinization temperature (GT), as expected for a cultivar with good characteristics of cooking, giving loose and soft pattern after baking. These attributes found follow the line of the cultivar release obtained by Schiocchet et al. (2015), being a standard currently sought by irrigated rice breeding in Brazil.

As for the response to biotic stresses, BRS Pampeira shows a reaction ranging from intermediate to moderately resistant to rice blast (*Pyricularia grisea*) in the leaf and in the panicle. The level of resistance refers to what was observed in the

Table 1. Comparison of plant traits between the irrigated rice cultivars BRS Pampeira and BRS Pampa

Characteristics	Cultivars	
	BRS Pampeira	BRS Pampa
Plant type	modern	modern
Days to flowering*	103	88
Days to maturity**	133 (medium)	118 (precocious)
Plant height (cm)**	91.5	96
Culm length (cm)**	67.6	72
Panicle length (cm)**	23.9	24
Panicle exsertion*	medium	medium
Leaf colour	dark green	green
Flag leaf angle	erect	erect
Auricle colour	light green	light green
Ligule colour	colorless	colorless
Culm internode color	light green	light green
Anthocyanin colouration on the culms	absent	absent
Panicletype	intermediate	intermediate
Leaf blade pubescence	present	present
Dehiscence*	intermediate	intermediate
Lodging tolerance*	tolerant	tolerant
Tillering*	high	high
Iron indirectly toxicity**	moderate tolerance	moderate tolerance
Rice blast in leaf and panicle**	moderate resistant	moderate resistant
Stain grains**	moderate sensitive	moderate resistant

* Atypical plants may arise due to the occurrence of natural crosses.
** Can undergo changes depending on the characteristics of the environment in which it is grown.

Table 2. Comparison of grain-related characteristics between the irrigated rice cultivars BRS Pampeira and BRS Pampa

Grain characteristics	Cultivars	
	BRS Pampeira	**BRS Pampa**
Caryopsis shape*	spindle-shaped	spindle-shaped
Awns	absent	absent
Glumes colour	straw	straw
Colour of apiculus in flowering	white	white
Colour of apiculus in maturaty	white	white
Pubescence grain	present	present
Lengthof grain (mm)**	9.2	9.82
Widthof grain (mm)**	2.1	2.2
Thicknessof grain (mm) **	2	2
Length of caryopsis (mm)**	7.15	7.19
Width of caryopsis (mm)**	1.93	1.96
Thickness of caryopsis (mm)**	1.75	1.76
Length / width $^{-1}$ ratio (mm)*	3.4	3.59
1000-grain weight (g)**	27	25.6
Total income (%)**	68.2	68
Intact grain (%)****	62	62
Endosperm amylose content	high	high
Gelatinization temperature	low	low
Potential productivity (t ha^{-1})***	12	10

* Length/width (without shell) ratio
** Can be changed depending on the characteristics of the environment in which it is grown.
*** Grains in shell, 13% humidity, observed in experiments conducted by Embrapa.
**** Grains peeled and polished in a Suzuki test device..

average of the VCU tests and can undergo changes in view of the different races, which alter with changing environments (interaction between places and years). It also displays average resistance to scald and medium susceptibility to brown spot and grain stain.

In relation to abiotic stresses such as toxicity to excess of iron in the soil, the cultivar is classified as moderately tolerant. It has a slow initial development after emergence, characteristic of cultivars with longer cycles such as the IRGA 424.

The cultivar BRS Pampeira meets a demand of medium cycle cultivars tending to long cycle with high yield potential and grain quality, being an option to producers that use the cultivar IRGA 424 (3rd most sown cultivar in RS). It has shown high yield levels in the main rice-producing areas, especially in the west border of the RS state, where it showed an average yield of 14 t ha^{-1} in the city of Uruguaiana (Figure 2).

In studies of estimates of the phenotypic stability and adaptability parameters, by the coefficient and the regression deviations from the average yield of rice grains in ten different RS environments, consisting of five environments in two crop seasons, the cultivar BRS Pampeira presented specific adaptability to the favorable environments.

The yield of the cultivar BRS Pampeira in other states of Brazil, compared to the standard cultivars of irrigated rice for each region, is presented in Table 3. It can be observed the yield potential of the cultivar and its wide adaptation. In the homogeneity tests, the BRS Pampeira has shown to be uniform, without the presence of atypical plants, proving to be genetically stable.

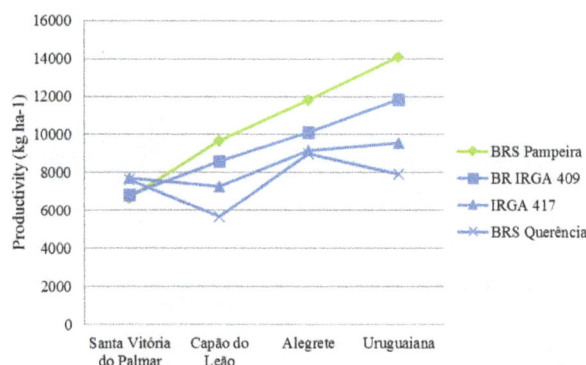

Figure 2. Yield of the cultivar BRS Pampeira compared with control cultivars at the VCU experiment, in the 2011/2012 and 2012/2013 season crops, in different regions of the RS state.

BASIC SEED PRODUCTION

The cultivar BRS Pampeira is registered in the National Register of Cultivars (RNC) and protected by the Ministry of Agriculture, Livestock and Supply (MAPA - Brazil). The Business Office of Capão do Leão, of Embrapa Products and Market is responsible for providing the basic seeds of the aforementioned cultivar.

Table 3. Yield assessment of the cultivar BRS Pampeira in kg ha^{-1} of paddy rice, at 13% humidity, for each location and year, in different regions of Brazil

Region	Year	County	Cultivars					
			BRS Pampeira	BRS Jaçanã	BR IRGA 409	BRS Tropical	BRS Fronteira	Average
Midwest	2007	Goianira, GO	7,263	6,989	5,965	.	.	6,477
	2008	Goianira, GO	10,437	8,620	7,018	.	.	7,819
	2009	Goianira, GO	11,244	7,976	.	8,945	.	8,461
		Miranda, MS	8,122	.	.	6,605	.	6,605
		Dourados, MS	7,566	.	.	8,558	.	8,558
		Rio Brilhante, MS	10,372	.	.	10,153	.	10,153
	2010	Goianira, GO	5,717	4,773	.	5,361	.	5,067
		Miranda, MS	7,198	6,545	.	4,676	.	5,611
		Dourados, MS	8,552	8,926	.	6,668	.	7,797
		Rio Brilhante, MS	7,298	5,231	.	6,270	.	5,751
North	2007	Formoso do Araguaia, TO	6,030	4,977	4,558	.	.	4,767
		Belém, PA	7,512	6,078	5,350	.	.	5,714
		Bragança, PA	5,645	5,208	3,048	.	.	4,128
		Salvaterra, PA	6,067	6,191	5,651	.	.	5,921
		Cantá, RR	8,171	8,257	7,367	.	.	7,813
	2008	Formoso do Araguaia, TO	7,492	6,160	6,565	.	.	6,346
		Lagoa da Confusão, TO	7,195	5,516	5,625	.	.	5,571
		Belém, PA	5,750	4,438	4,604	.	.	4,521
		Bragança, PA	5,481	7,331	7,323	.	.	7,327
		Salvaterra	4,702	4,699	3,302	.	.	4,001
		Cantá, RR	6,664	5,555	.	6,871	.	6,213
	2009	Formoso do Araguaia, TO	7,151	6,963	.	7,062	.	7,013
		Belterra, PA	9,707	8,500	.	9,148	.	8,824
		Cantá, RR	6,664	5,555	.	6,871	.	6,213
	2010	Formoso do Araguaia, TO	5,125	4,531	.	4,335	4,398	4,421
		Lagoa da Condusão, TO	7,047	8,797	.	6,422	9,047	8,089
		Belém, PA	8710	6588	.	5604	6666	6286
		Bragança, PA	3883	4500	.	3430	3726.	3885
Northeast	2008	Arari, MA	5,672	.	.	6,697	.	6,697
		Buriti dos Lopes, PI	8,861	.	.	8,518	.	8,518
		Teresina, PI	9,848	.	.	9,942	.	9,942
		Iguatu, CE	6,520	.	.	6,978	.	6,978
	2009	Arari, MA	8,668	.	.	8,488	.	8,488
		Buriti dos Lopes, PI	8,052	.	.	7,754	.	7,754
		Teresina, PI	11,708	.	.	11,399	.	11,399
	2010	Arari, MA	9,451	.	.	7,940	.	7,940
		Buriti dos Lopes, PI	10,513	.	.	9,156	.	9,156
		Teresina, PI	8,056	.	.	8,631	.	8,631
		Iguatu, CE	11,758	.	.	11,334	.	11,334
		Limoeiro do Norte, CE	10,986	.	.	10,624	.	10,624
	2012	Buriti dos Lopes, PI	6,340	6,282	.	7,020	5,734	6,345
		Teresina, PI	9,133	8,724	.	8,264	8,466	8,485
		Igreja Nova, AL	5,254	5,048	.	5,503	4,567	5,039

The sowing of the cultivar BRS Pampeira should follow the agricultural zoning for irrigated rice in Rio Grande do Sul and other states of the union. In the RS it is recommended that the sowing occurs respecting the cultivar cycle in interaction with the cultivation environment in such a way that the panicle differentiation occurs until early January or as close to that date as possible. In this case, it is recommended the early sowing season, which in the RS corresponds to the first half of October, so that it can express its maximum yield potential.

The cultivar BRS Pampeira is recommended for sowing in six rice-producing regions of the RS state, the west frontier being the preferred region, where it showed greater adaptability to the favorable environment. In registration with the Ministry of Agriculture, Livestock and Supply (MAPA), the cultivar was also recommended to the states of Goiás and Mato Grosso do Sul (midwest region); Tocantins, Pará and Roraima (north region); Maranhão, Piauí, Ceará, Rio Grande do Norte, Paraíba, Pernambuco, Alagoas and Sergipe (northeast region).

The density of suitable seeds (100% GC) should be about 60 seeds per linear meter (approximately 100 kg ha^{-1}) for the online system, to ensure a plant population from 200 to 300 plants per square meter (SOSBAI 2014). In the germination and seedling emergence tests performed at low temperatures, the cultivar showed intermediate response to cold.

The cultivar BRS Pampeira shows positive response to different levels of basic and coverage fertilization, without lodging of plants.

The harvest of this cultivar, aiming to minimize the natural abscission and prevent the grain breakage during the manufacturing process, should be performed when the grain moisture content is between 23% and 18%.

REFERENCES

Lee I, Seo YS, Coltrane D, Hwang S, Oh T, Marcotte EM and Ronald PC (2011) Genetic dissection of the biotic stress response using a genome-scale gene network for rice. **Proceedings of the National Academy of Sciences 108**: 18548-18553.

Schiocchet MA, Noldin JA, Marschalek R, Wickert E, Martins GN, Eberhardt DS, Hickel E, Knoblauch R, Scheuermann KK, Raimondi JV and

Andrade A (2015) SCS121 CL: Rice cultivar resistant to herbicides of imidazolinone chemical group. **Crop Breeding and Applied Biotechnology 15**: 282-284.

SOSBAI - Sociedade Sul-brasileira de Arroz Irrigado (2014) **Arroz irrigado: recomendações técnicas da pesquisa para o Sul do Brasil**. Editora SOSBAI, Bento Gonçalves, 189p.

Zeigler RS and Barclay A (2008) The relevance of rice. **Rice 1**: 3.

24

Field and laboratory assessments of sugarcane mutants selected in vitro for resistance to imazapyr herbicide

RS Rutherford[1,2], KZ Maphalala[1,2], AC Koch[1,2], SJ Snyman[1,2] and MP Watt[1*]

Abstract: Seven imazapyr-tolerant mutant sugarcane plants, previously generated by in vitro mutagenesis, were studied. The imazapyr concentrations required to inhibit their acetolactate synthase (ALS basal activity) (IC_{50} as µmoles acetoin h^{-1} mg^{-1} protein) were 0.77 – 5.36 times greater than that of the N12 'parent'. The basal ALS activities of Mut1 and Mut6 were 1.4-fold higher than that of N12. When the mutants were sprayed with Arsenal® GEN 2 (312 and 624 g a.i. imazapyr ha^{-1}), 2 months after field planting, and evaluated 9 months later, live stalk height and number were significantly lowest in Mut2, Mut3 and the control N12. No differences in sucrose, fibre and estimated yield were observed amongst lines in untreated plots. Mutant plants germinated and grew in soil treated with the herbicide (at the lethal dose of 1248 g a.i. ha^{-1}). The Mut lines tested in this study offer improved options for weed control.

Key words: *Acetolactate synthase, ethyl methanesulfonate, imidazolinone, mutation breeding.*

*Corresponding author:
E-mail: wattm@ukzn.co.za

[1] School of Life Sciences, University of KwaZulu-Natal, Westville campus, Private Bag X54001, Durban, 4000, South Africa
[2] South African Sugarcane Research Institute, Private Bag X02, Mount Edgecombe, KwaZulu-Natal, 4300, South Africa

INTRODUCTION

Weeds can drastically reduce cane and sugar yields (Millhollon 1992). The application of herbicides is therefore a well-established necessity and is most crucial during plant cane establishment and subsequent ratoon crop regeneration (Campbell 2008). Herbicides must be carefully selected and applied as they disrupt essential processes (e.g. photosynthesis, amino acid biosynthesis) shared by crops and weeds. In sugarcane, this is especially difficult, as a many of the weeds are also graminaceous species, e.g. *Cynodon dactylon* and *Digitaria longifolia* (Campbell 2008).

One approach to herbicide phytotoxicity is the development of cultivars resistant to broad-spectrum herbicides. In sugarcane, as in most crops, this can be achieved by conventional plant breeding, genetic transformation (Leibbrandt and Snyman 2003) and induced mutagenesis (Rutherford et al. 2014). Because of the lengthy plant breeding and selection process in sugarcane, legislative restrictions, licensing costs and public opposition to transgenesis, our preferred approach is the generation of herbicide-resistant variants of proven elite genotypes using mutagenesis. Hence, cultivar N12 was selected for generating variants resistant to the herbicide imazapyr (Koch et al. 2012). N12 is known to be hardy (McIntyre and Nuss 1998) and is a favored cultivar of emerging small-scale farmers in South Africa, who operate under low input conditions, on predominantly strongly acid (pH<5) soils, with weed pressure as a major production constraint (Cockburn et al. 2012).

The active ingredient imazapyr, an imidazolinone (IMI) compound, is registered in South Africa [no. L8817; Arsenal® GEN 2; BASF South Africa (Pty) Ltd] for the control of grass and broadleaf weeds prior to sugarcane re-planting. Imidazolinone herbicides are active against the enzyme acetolactate synthase (ALS; EC 2.2.1.6), also known as acetohydroxyacid synthase (AHAS; EC 4.1.3.18), which catalyses the first step in the biosynthesis of isoleucine, leucine and valine.

Seven putatively imazapyr-resistant sugarcane mutant plants (Mut1-Mut7) were generated from N12 by callus exposure to 16 mM ethyl methanesulfonate, and selection on imazapyr-containing medium (Koch et al. 2012). This report describes subsequent studies on these mutants grown in the field addressing their imazapyr resistance, basal activity of the ALS enzyme and the imazapyr concentration required to inhibit it by 50 % (IC_{50}), and plant characteristics and yield, compared with the 'parent' N12.

The mutant genotypes were also tested for their tolerance to the persistent herbicide residual activity in the soil. For other crops (Santos et al. 2014), and as per herbicide label instructions, a 4-month waiting period and at least 600 mm of precipitation are necessary before planting to avoid suppression of sugarcane sett germination and growth. However, the ability to replant during the herbicide soil residual period would afford the sugarcane farmer improved weed control, as the crop would attain full canopy before any significant weed pressure. Setts of the mutant genotypes were, therefore, also tested for germination and growth shortly after the soil was sprayed with imazapyr.

MATERIAL AND METHODS

Plant material, field trials and imazapyr application

The Mut1-7 obtained by Koch et al. (2012) and N12 control plants were bulked-up *in vitro* (Ramgareeb et al. 2010). After 3 months (30 - 40 cm in height), they were planted in field areas A, B and C. Stalks from 11-month-old plants from area A were cut into 3-budded setts and planted in area D. The field (lat 29° 42′ 24.5585″ S, long 31° 02′ 45.1735 E″; long-term mean annual rainfall 1023 mm) was divided into four areas (A-D); A, B and C were subdivided into three replicated plots (1 x 3.5 m row, 10 plants per row) in a randomized complete block design. Two months after planting (4-6 leaf stage), Arsenal® GEN 2 [240 g a.i. ha⁻¹ imazapyr; BASF, Ago BV Arnhem, Switzerland] was applied directly over the top of the plants at 312 and 624 g a.i. ha⁻¹ in areas B and C, i.e. ¼ and ½ of the lethal dose, with a gas-regulated sprayer and flat-fan nozzle (Albuz APE 110°) at 194.2 l ha⁻¹ and 1.515 l min⁻¹. Area A was unsprayed. Area D was halved (8 x 9.5 m each) and the soil of one half was treated with Arsenal® GEN 2 at 1248 g a.i. ha⁻¹ (lethal dose), 3 weeks prior to planting. Both halves were planted with 3-budded setts from mutant (Mut1-Mut7) and N12 control plants, as 90 - 100 buds per 9.5 m. Rainfall was 78 mm between herbicide application and planting. Germination (per genotype in the treated section as % germination in the corresponding untreated one) was determined after 3 weeks and shoot length after 12 weeks.

Yield component and quality of field-grown plants

Stalk number per plot and stalk height and diameter were determined (on 20 randomly chosen stalks per plot) for mutant and N12 plants in areas A, B and C, 11 months after planting. Estimated cane yield was calculated by $ndpr^2L/1000$ (Miller and James 1974, Gravois et al. 1991); where: n = number of stalks.plot⁻¹; d = density at 1.00 g cm⁻³; r = stalk radius (cm) (radius was calculated from the diameter divided by 2); L = stalk height (cm).

In addition, the plants in the unsprayed area A were cut back and allowed to re-grow (ratoon) to maturity, and assessed again for yield. Sucrose and fiber contents were analyzed in a mill room of the South African Sugarcane Research Institute (SASRI) (Schoonees-Muir et al. 2009). All data were analyzed using a One-way ANOVA and Holm-Sidak test ($P < 0.05$).

Acetolactate synthase enzyme and IC_{50} determinations

Leaf ALS enzyme activity (μmol acetoin h^{-1} mg^{-1} protein) was determined colorimetrically (530 nm) by acetoin formed (Yu et al. 2010, Koch et al. 2012) and total protein the method of Bradford (1976).

The concentration of imazapyr required to inhibit ALS activity by 50% (IC_{50}) was determined for the third leaf of Mut1-Mut7 and control N12 from plot A, 5 months after planting. The fresh leaf mass to obtain a maximized initial absorbance at 530 nm for acetoin at 0 μM imazapyr was established per genotype, to correct for basal activity differences. The ALS activity was assayed with 0 - 30 μM imazapyr [PESTANAL® (Sigma-Aldrich)]. The IC_{50} values were calculated from the nonlinear regression analysis of log (inhibitor) vs. response (GraphPad Prism 5.0., GraphPad Software Inc., San Diego, CA, USA). Comparisons of plant IC_{50} values were performed using a One-way analysis of variance (ANOVA) and Holm-Sidak test ($P < 0.05$). Field imazapyr resistance levels in Mut1, Mut6 and N12 control plants were evaluated by ALS assays at 1 and 3, 6, and 12 weeks after imazapyr spraying.

RESULTS AND DISCUSSION

ALS activities and the effect of imazapyr on the enzyme and yield components

The basal ALS activities of Mut1-Mut7 lines and N12 control plants were determined 2 months after planting and prior to herbicide spraying. Mut1 and Mut6 plants had significantly higher ALS activities (190.4 and 179.0 μmoles acetoin h^{-1} mg^{-1} protein, respectively) than control N12 (1.48 and 1.39 times that of N12, respectively) and the other mutants (Table 1). Mutants exhibited 0.77 – 5.36 times greater IC_{50} than N12 (Figure 1). The IC_{50} value of Mut1 was significantly higher than those of Mut2 and control N12; no other significant differences were recorded. By way of comparison, two commercially released imidazolinone-resistant rice mutants have been shown to exhibit IC_{50} values of 13 and 369 times greater than non-mutant rice (Avila et al. 2005). These rice mutants were considered tolerant and resistant, respectively.

After 2 months in the field, plants were sprayed with 0, 312 and 624 g a.i. imazapyr ha^{-1} (areas A, B and C, respectively). Six weeks later, all plants in the untreated area had normal green leaves (Figure 2a), as did those of Mut1, Mut4, Mut5, Mut6, and Mut7 in the sprayed areas B and C. However, the leaves of Mut2, Mut3 and control N12 turned reddish-brown with accumulated 3-deoxyanthocyanidin luteolinidin (spectral identification not shown) (Figure 2b and c), a symptom of IMI herbicide phytotoxicity (Tan et al. 2006). Nine months after herbicide application, only a few Mut2 and Mut3 plants in the 312 g a.i. ha^{-1} treatment were alive; all of the N12, Mut2 and Mut3 plants sprayed with 624 g a.i. ha^{-1} imazapyr had died.

Mut1 and Mut6 were further investigated for their responses to imazapyr over 12 weeks (Figure 3). In the untreated area A, ALS activities were significantly higher ($P < 0.001$) than that of control N12 at week 12 (Figure 3a).

Table 1. Basal ALS activities of mutant (Mut1-Mut7) and control N12 plants. a-b indicate statistically significant differences among genotypes

Genotype	ALS activity (μmoles acetoin h^{-1} mg^{-1} protein)
Mut1	190.4 ± 10.3[b]
Mut2	120.3 ± 4.7[a]
Mut3	138.4 ± 11.5[a]
Mut4	111.4 ± 4.2[a]
Mut5	128.4 ± 10.7[a]
Mut6	179.0 ± 6.9[b]
Mut7	136.9 ± 5.0[a]
N12	128.8 ± 3.8[a]

One-way ANOVA and Holm-Sidak test, $P < 0.05$; n=12, mean ± SE

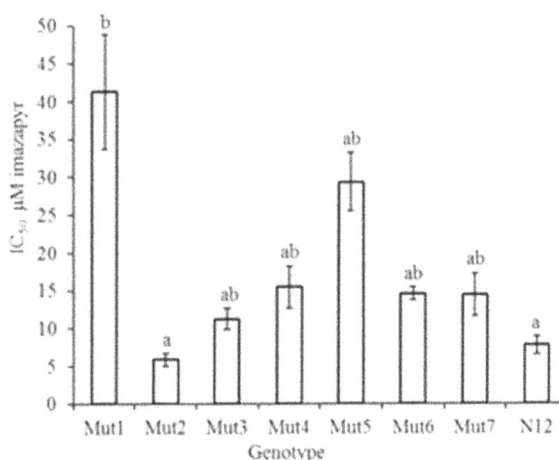

Figure 1. IC_{50} values as a measure of imazapyr resistance in mutant sugarcane genotypes. Plants were tested for ALS activity 5 months after planting. a-b indicate statistically significant differences amongst genotypes. One-way ANOVA and Holm-Sidak test, P < 0.05; n=3, mean ± SE. For analysis, data were log_{10}-transformed; untransformed data are presented.

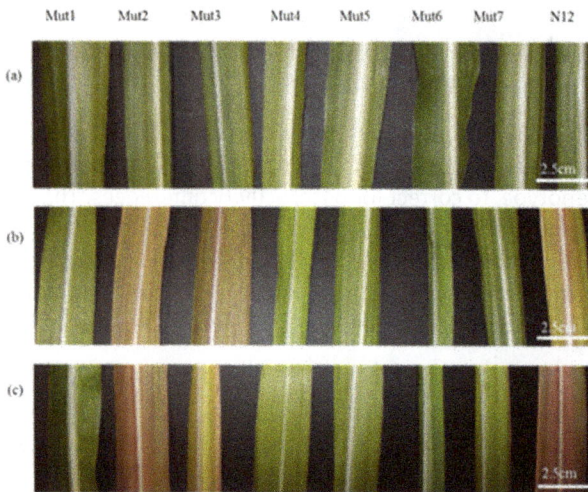

Figure 2. The effect of imazapyr on leaf appearance of plants Mut1-Mut7 and control N12 6 weeks after foliar application. Leaves were collected from (a) untreated; (b) 312 g a.i. ha⁻¹; and (c) 624 g a.i. ha⁻¹ sprayed areas.

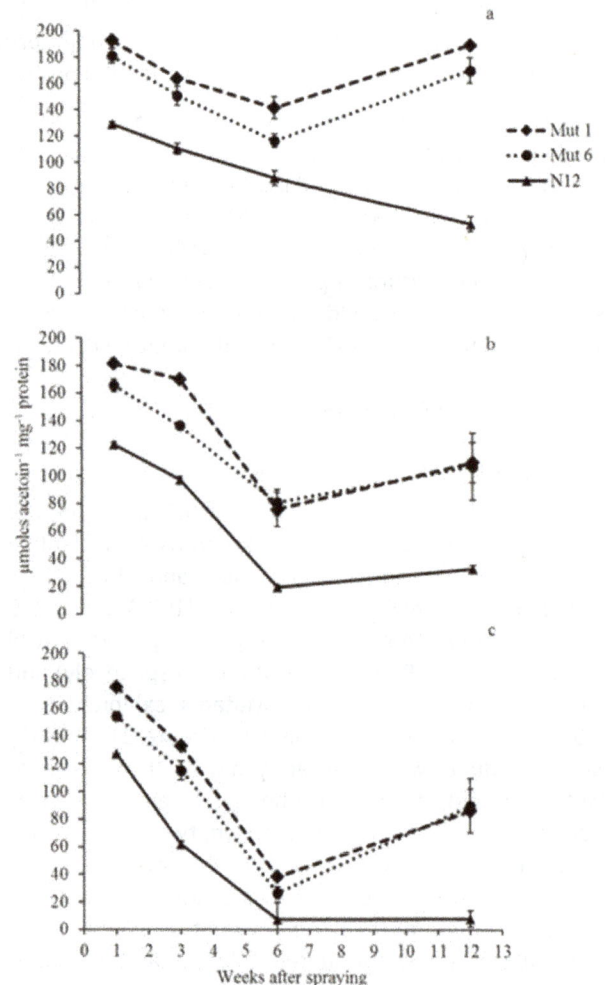

Figure 3. The effect of imazapyr on ALS activity of Mut1, Mut6 and control N12 in field material. Leaf material was collected from: (a) untreated; (b) 312 g a.i. ha⁻¹; (c) 624 g a.i. ha⁻¹ treated areas. (n=3, mean ± SE).

There was a small depression in ALS activity for Mut1 and Mut6 plants at week 6, but an increase at week 12. Over time, there were no significant differences in ALS activity between Mut1 and Mut6 plants, and the ALS activity in N12 control plants decreased significantly. In the treated areas B and C (312 and 624 g a.i. ha⁻¹, respectively) (Figures 3b and c), the ALS activity for the Mut1 and Mut6 plants at week 6 was significantly lower ($P < 0.001$) than that of the unsprayed plants (Figure 3a). The N12 control plants displayed decreased ALS activities at weeks 3, 6 and 12 in the 624 g a.i. ha⁻¹ area. These were significantly lower ($P < 0.001$) than those in corresponding weeks in the untreated area. Sprayed Mut1 and Mut6 exhibited larger depressions in ALS activity at 6 weeks than plants in the untreated area. Again, activity showed recovery at 12 weeks, mirroring that seen in the untreated area. This suggests that the slight depression for Mut1 and Mut6, and the continued decline in activity for control N12 in the untreated area may have been due to herbicide drift from the adjacent treated plots. Eberlein and Guttieri (1994) reported that amounts as small as 1/50th of the normal agricultural imazapyr rate reduced potato yields by two-thirds and Bond et al. (2006) showed that a simulated drift rate of only 8 g a.i. ha⁻¹ could reduce rice yield by 40%.

When sprayed with 624 g a.i. ha⁻¹ imazapyr, ALS activities of Mut1 and Mut6 decreased significantly ($P < 0.001$) from weeks 1 to 6 (Figure 3c). However, at week 12, ALS activities of Mut1 and Mut6 plants were significantly higher ($P < 0.001$) than that of control N12 (Figure 3c). The ALS activity of the N12 control plants was close to zero by week 6, did not recover by week 12 and was always significantly lower ($P < 0.001$) than those of Mut1 and Mut6 (Figure 3c). By harvest, all N12 control plants sprayed with imazapyr had died.

No significant differences in yield components were found between mutants and N12 within the untreated field area in both the plant (Table 2) and the ratoon crops (Table 3). Also, % fibre and % sucrose were not different between mutants and the parent (Table 3). This suggests that the mutation breeding approach did not significantly alter the yield component characteristics of the mutants. Within the sprayed areas, final live stalk numbers and stalk height were significantly reduced in Mut2, Mut3 and in N12 (Table 2).

Imazapyr resistance in plants arising from setts planted in treated soil

In the sprayed half of plot D (1248 g a.i. ha^{-1} imazapyr 3 weeks prior to planting), germination was 7 - 73% of that of the same genotype in the untreated area. Although statistical analysis was not possible due to limited material, germination was higher in Mut1 (61%), Mut4 (73%) and Mut6 (61%) than in the other genotypes (Table 4). By week 12, Mut1 and Mut6 were significantly taller ($P < 0.001$) than the other mutants and control N12 (Table 4). Mut1 and Mut6 were also the least stunted relative to the shoot lengths in the untreated section (75 and 66% of untreated length, respectively). Mut2, Mut3 and control N12 had severely stunted growth compared with their untreated counterparts and the other genotypes.

Potential modes of herbicide resistance in the tested mutant plants

Mut1, and perhaps Mut5, may have mutations in the *als* gene, conferring reduced inhibition of enzyme activity by imazapyr (increased IC$_{50}$; Figure 1). To date, only one such mutation (Ala-559) in the *als* gene of sugarcane has been reported (Khruangchan et al. 2011). Other possible target site alterations could include increased ALS enzyme basal activity due to higher *als* gene transcript levels or gene copies (Boutsalis et al. 1999, Yu et al. 2003). Enhanced enzymatic activity is also possible due to post-transcriptional regulation, increased mRNA stability and/or reduced enzyme degradation (increased half-life) (Yuan et al. 2002). Mut1 and Mut6 showed 1.4-fold increase in basal ALS activities (Table 1). Overproduction of the target enzyme (per unit of protein or fresh weight) increases the number of

Table 2. Yield components and estimated yield of field-grown plants after 11 months. Two months after planting, imazapyr was applied (312 and 624 g a.i. ha^{-1}) to areas B and C; area A was untreated. a-d = statistical difference between each genotype

Field area	Genotype	Live stalk no. plot^{-1}	Live stalk height (cm)	Live stalk diameter (cm)	Estimated yield (kg plot^{-1})
A-Untreated	Mut1	133.67 ± 6.94 [c]	124.65 ± 11.38 [def]	2.02 ± 0.08 [c]	55.04 ± 11.92 [de]
	Mut2	143.67 ± 21.94 [c]	123.58 ± 9.44 [def]	1.70 ± 0.11 [bc]	43.30 ± 12.73 [abcde]
	Mut3	133.67 ± 3.84 [c]	138.97 ± 5.23 [f]	2.03 ± 0.05 [c]	59.79 ± 2.62 [de]
	Mut4	126.67 ± 12.99 [c]	126.93 ± 2.41 [def]	1.96 ± 0.15 [c]	50.01 ± 11.59 [bcde]
	Mut5	160.00 ± 20.00 [c]	136.45 ± 7.88 [ef]	2.10 ± 0.10 [c]	74.39 ± 6.58 [e]
	Mut6	110.33 ± 10.27 [bc]	121.15 ± 11.49 [def]	1.90 ± 0.12 [bc]	40.30 ± 11.24 [abcde]
	Mut 7	109.33 ± 11.20 [bc]	123.08 ± 10.69 [def]	1.90 ± 0.11 [bc]	37.92 ± 4.73 [abcde]
	N12	128.67 ± 5.61 [c]	135.88 ± 3.06 [ef]	1.97 ± 0.04 [c]	53.70 ± 4.54 [de]
	Treatment mean	130.75 B	128.84 C	1.947 C	51.81 B
B-312 g a.i. ha^{-1}	Mut 1	105.67 ± 12.68 [bc]	110.88 ± 11.49 [cdef]	1.70 ± 0.06 [bc]	27.54 ± 6.90 [abcde]
	Mut 2	28.00 ± 16.17 [ab]	46.47 ± 13.69 [ab]	1.30 ± 0.10 [b]	2.80 ± 2.22 [abc]
	Mut 3	11.00 ± 11.00 [a]	55.58 ± 14.04 [abc]	1.57 ± 0.13 [bc]	0.64 ± 0.64 [ab]
	Mut 4	97.33 ± 17.02 [bc]	92.28 ± 9.05 [bcdef]	1.67 ± 0.03 [bc]	20.41 ± 5.60 [abcd]
	Mut 5	142.67 ± 27.63 [c]	110.90 ± 15.78 [cdef]	2.07 ± 0.07 [c]	52.34 ± 12.25 [cde]
	Mut 6	119.33 ± 8.11 [c]	112.45 ± 1.07 [cdef]	2.05 ± 0.10 [c]	43.98 ± 1.07 [abcde]
	Mut 7	120.67 ± 9.13 [c]	115.62 ± 10.41 [def]	1.94 ± 0.03 [c]	41.41 ± 5.60 [abcde]
	N12	0.00 ± 0.00 [a]	0.00 ± 0.00 [a]	0.00 ± 0.00 [a]	0.00 ± 0.00 [a]
	Treatment mean	78.08 A	80.59 B	1.538 B	23.64 A
C-624 g a.i. ha^{-1}	Mut 1	112.33 ± 19.43 [bc]	79.02 ± 15.25 [bcde]	1.78 ± 0.14 [bc]	26.02 ± 11.13 [abcde]
	Mut 2	0.00 ± 0.00 [a]	0.00 ± 0.00 [a]	0.00 ± 0.00 [a]	0.00 ± 0.00 [a]
	Mut 3	0.00 ± 0.00 [a]	0.00 ± 0.00 [a]	0.00 ± 0.00 [a]	0.00 ± 0.00 [a]
	Mut 4	150.00 ± 10.26 [c]	94.02 ± 15.49 [bcdef]	1.86 ± 0.15 [bc]	41.61 ± 13.85 [abcde]
	Mut 5	117.00 ± 34.60 [c]	72.87 ± 13.73 [bcd]	1.78 ± 0.16 [bc]	27.21 ± 12.84 [abcde]
	Mut 6	129.00 ± 11.24 [c]	77.58 ± 14.59 [bcde]	1.90 ± 0.19 [bc]	32.59 ± 13.53 [abcde]
	Mut 7	139.00 ± 22.81 [c]	81.30 ± 13.03 [bcdef]	1.93 ± 0.11 [c]	36.30 ± 15.26 [abcde]
	N12	0.00 ± 0.00 [a]	0.00 ± 0.00 [a]	0.00 ± 0.00 [a]	0.00 ± 0.00 [a]
	Treatment mean	80.92 A	50.60 A	1.155 A	20.47 A

REML analysis and Holm-Sidak test, $P < 0.05$; n=3; mean ± SE

target sites that must be inhibited in order to block amino acid synthesis, diluting the effect of the herbicide (Powles 2010). Alarcón-Reverte et al. (2015) characterized a glyphosate-resistant *Echinochloa* line that lacked known resistance, conferring mutations in the gene of the target enzyme 5-enolpyruvylshikimate-3-phosphate synthase (EPSPS), but they only sequenced part of the gene. That line was similar to susceptible ones in glyphosate absorption, translocation or metabolism, but had a 1.4-fold higher basal EPSPS activity and 5-fold higher LD_{50} than susceptible plants when sprayed with glyphosate. Thus, it may be possible that the increased basal ALS activities seen in Mut1 and Mut6 contribute to their increased imazapyr resistance (Table 1).

Plants also become resistant to herbicides via non-target-site (NTS)-based resistance (Yuan et al. 2007). Non-target-site resistance appears to be controlled by multiple genes, each providing partial quantitative effects, which is difficult to study (Délye 2013). Also, it comprises a range of mechanisms that act to minimize the amount of herbicide reaching the target site, e.g. structural barriers to penetration, physiological exclusion by active transporters and reduced herbicide translocation, and increased metabolic detoxification (Yuan et al. 2007, Powles and Yu 2010). These mechanisms are possible in Mut4 and Mut7, which did not exhibit increased ALS IC_{50} levels or increased ALS basal activity. Similar to Mut1, Mut5 and Mut6, Mut4 and Mut7 were significantly better than the control N12 in terms of stalk survival and height when sprayed with imazapyr (Table 2), and in % germination and shoot height when planted in imazapyr-treated soil (Table 4). Reduced absorption through the cuticle or other physical barrier and reduced translocation are unlikely to be involved as imazapyr resistance was generated in callus cells (Koch et al. 2012). However, in phase I of detoxification, herbicide molecules are activated and functional groups are exposed to phase II conjugation enzymes, and often to oxidation by cytochrome P450 monooxygenases known to participate in IMI metabolism (Manabe et al. 2007). A reduction of herbicide resistance by the P450 monooxygenase inhibitor piperonyl butoxide has been reported for ALS inhibitors in maize and sunflower (Breccia et al. 2012).

The commercial success of IMI resistant mutants (Clearfield®) is partially due to less stringent regulations for mutants than for genetically modified (GM) organisms (Tan et al. 2005), e.g. 'substantial equivalence' requires demonstration

Table 3. An assessment of quality and yield traits of the first ratoon crop of the mutated sugarcane genotypes (Mut1 – Mut7) and the commercial variety N12, for the field area not treated with imazapyr (area A)

Genotype	Stalk no. plot⁻¹	Mass of 12 stalks (kg)	Estimated yield (kg plot⁻¹)	Fiber % (w w⁻¹)	Sucrose % (w w⁻¹)
Mut1	105.33 ± 9.68	7.390 ± 0.41	64.74 ± 6.61	13.18 ± 0.40	11.49 ± 0.29
Mut2	66.33 ± 6.64	5.723 ± 0.44	31.47 ± 3.14	12.80 ± 0.32	12.45 ± 0.31
Mut3	88.67 ± 13.35	6.363 ± 0.73	47.34 ± 9.93	13.34 ± 0.33	11.91 ± 0.46
Mut4	85 ± 7.02	5.680 ± 0.31	40.02 ± 2.84	13.41 ± 0.55	10.09 ± 0.55
Mut5	102.33 ± 24.55	7.777 ± 0.70	69.18 ± 21.97	14.17 ± 0.24	11.36 ± 0.71
Mut6	72.67 ± 13.42	7.167 ± 0.49	43.90 ± 10.15	13.15 ± 0.14	12.22 ± 0.63
Mut7	71 ± 3.46	6.573 ± 0.56	43.90 ± 2.58	13.19 ± 0.27	11.97 ± 0.88
N12	77.33 ± 22.24	7.257 ± 0.17	46.52 ± 13.23	13.83 ± 0.37	11.44 ± 0.84
P value	0.459	0.072	0.256	0.227	0.282

Data were analyzed using a One-Way ANOVA, $P < 0.05$, n=3, mean ± SE. There were no statistically significant differences amongst mutant genotypes and N12.

Table 4. Germination and shoot length in the tested mutants (Mut1-Mut7) and the control N12. The soil in the treated section was sprayed with imazapyr (1248 g a.i. ha⁻¹) 3 weeks prior to planting. a-d denote a statistically significant difference between genotypes

Genotype	Germination after 3 w (as % of that in untreated section)	Shoot length after 12 weeks (as % of shoot length in the untreated section)	Shoot length (mm) after 12 w of geno-types in the treated section
Mut1	61	75	185 ± 8.8[d]
Mut2	27	20	43 ± 3.6[a]
Mut3	38	25	60 ± 6.5[ab]
Mut4	73	44	115 ± 5.2[c]
Mut5	48	34	88 ± 2.7[bc]
Mut6	61	66	179 ± 8.4[d]
Mut7	39	45	124 ± 13.5[c]
N12	7	16	42 ± 5.5[a]

One-way ANOVA and Holm-Sidak test, $P < 0.001$; n=10, mean ± SE

that the GM and the non-GM wild-type lines are similar, except for the transgene (Cellini et al. 2004). However, mutation breeding might cause unintended pleiotropic phenotypes, due to mutation effects on more genes than only the desired one (Manabe et al. 2007). Equivalence between our imazapyr resistant lines and the parent cultivar is therefore being pursued. Thus far, no significant differences in yield and quality components have been identified in imazapyr-untreated lines (Tables 2 and 3). However, yield and quality effects due to mutations will have to be further studied across additional soil types and for an extended crop cycle.

When sprayed with imazapyr, or planted into imazapyr-treated soil, Mut2 and Mut3 are the least resistant (Tables 2, 3 and 4). However, some stalk survival to maturity at the low foliar imazapyr dose (Table 2) and possibly a higher germination rate than that of the control N12 in treated soil (Table 4) in Mut2 and Mut3 indicate a low level resistance to imazapyr by these mutants. Mut2 and Mut3 could be considered 'escapes' from the *in vitro* selection protocol. A more stringent *in vitro* selection protocol could be used in the future to increase the likelihood of obtaining mutants with greater levels of resistance, such as those seen in Clearfield® rice cultivars (Avila et al. 2005).

CONCLUSIONS

Five of the seven mutants tested showed greater imazapyr-resistance in the field than the control N12. Two had a 1.4-fold increase in basal ALS activity. Multiple resistance mechanisms seem to be present across the tested mutants, e.g. point mutations in the *als* gene, increased basal expression, and may include exclusion by transporters and increased detoxification, elements which are being investigated. The demonstrated ability to plant setts of the mutant genotypes during the residual period of imazapyr in the soil will afford farmers improved weed control.

ACKNOWLEDGEMENTS

We thank the National Research Foundation of South Africa (Grants 85573 and 85414) and the University of KwaZulu-Natal for funding and Dr P Campbell, Ms N Sewpersad and the Technical Team of the South African Sugarcane Research Institute (SASRI) for assistance with technical support and field measurements.

REFERENCES

Alarcón-Reverte R, García A., Watson SB, Abdallah I, Sabaté S, Hernández MJ, Dayan FE and Fischer AJ (2015) Concerted action of target-site mutations and high EPSPS activity in glyphosate-resistant junglerice (*Echinochloa colona*) from California. **Pest Management Science 71**: 996-1007.

Avila LA, Lee DJ, Senseman SA, McCauley GN, Chandler JM and Cothren JT (2005) Assessment of acetolactate synthase (ALS) tolerance to imazethapyr in red rice ecotypes (*Oryza* spp) and imidazolinone tolerant/ resistant rice (*Oryza sativa*) varieties. **Pest Management Science 61**: 171-178.

Bond JA, Griffin JL, Ellis JM, Linscombe SD and Williams BJ (2006) Corn and rice response to simulated drift of imazethapyr and imazapyr. **Weed Technology 20**: 113-117.

Boutsalis P, Karotam J and Powles SB (1999) Molecular basis of resistance to acetolactate synthase-inhibiting herbicides in *Sisymbrium orientale* and *Brassica tournefortii*. **Pesticide Science 55**: 507-516.

Bradford MM (1976) A rapid and sensitive method for the quantification of microgram quantities of protein utilizing the principle of protein-dye binding. **Analytical Biochemistry 72**: 248-254.

Breccia G, Gil M, Vega T, Zorzoli R, Picardi L and Nestares G (2012) Effect of cytochrome P450s inhibitors on imidazolinone resistance in sunflower. **Proceedings of the 18th Sunflower Conference, Mar del Plata-Balcarce**, Argentina, p. 507-512.

Campbell PL (2008) Efficacy of glyphosate, alternative post emergence herbicides and tillage for control of *Cynodon dactylon*. **South African Journal of Plant and Soil 25**: 220-228.

Cellini F, Chesson A, Colquhoun I, Constable A, Davies HV, Engel KH, Gatehouse AMR, Kärenlampi S, Kok EJ, Leguay J-J, Lehesranta S, Noteborn HPJM, Pedersen J and Smith M (2004) Unintended effects and their detection in genetically modified crops. **Food and Chemical Toxicology 42**: 1089-1125.

Cockburn JJ, Coetzee HC, Witthöft J, Conlong DE and Van Den Berg J (2012) Exploring the feasibility of push-pull for use in management of *Eldana saccharina* Walker (Lepidoptera, Pyralidae) by small-scale sugarcane growers. **Proceedings of the South African Sugar Technologists' Association 85**: 134.

Délye C (2013) Unravelling the genetic bases of non-target-site-based resistance (NTSR) to herbicides, a major challenge for weed science in the forthcoming decade. **Pest Management Science 69**: 176-187.

Eberlein CV and Guttieri MJ (1994) Potato (Solanum tuberosum) response to simulated drift of imidazolinone herbicides. **Weed Science 42**: 70-75.

Gravois KA, Milligan SB and Martin FA (1991) Indirect selection for increased sucrose yield in early sugarcane testing stages. **Field Crops Research 26**: 67-73.

Khruangchan N, Chatchawankanphanich O and Pornprom T (2011) Acetolactate synthase gene mutation conferring imazapyr tolerance in sugarcane clone. **Journal of the National Research Council of Thailand 41:** 115-127.

Koch AC, Ramgareeb S, Rutherford RS, Snyman SJ and Watt PM (2012) An in vitro mutagenesis protocol for the production of sugarcane tolerant to the herbicide imazapyr. *In vitro* **Cellular and Developmental Biology-Plant 48:** 417-427.

Leibbrandt NB and Snyman SJ (2003) Stability of gene expression and agronomic performance of a transgenic herbicide-resistant sugarcane line in South Africa. **Crop Science 43:** 671-677.

Manabe Y, Tinker N, Colville A and Miki B (2007) CSR1, the sole target of imidazolinone herbicide in *Arabidopsis thaliana*. **Plant Cell Physiology 48:** 1340-1358.

McIntyre RK and Nuss KJ (1998) An evaluation of the variety N12 in field trials. **Proceedings of the South African Sugar Technologists' Association 72:** 28-34.

Miller JD and James NI (1974) The influence of stalk density on cane yield. **Proceedings of the International Society of Sugar Cane Technologists 15:** 177-184.

Millhollon RW (1992) Effect of itchgrass (*Rottboellia cochinchinensis*) interference on growth and yield of sugarcane (*Saccharum* spp. hybrids). **Weed Science 40:** 48-53.

Powles S (2010) Gene amplification delivers glyphosate-resistant weed evolution. **Proceedings of the National Academy of Sciences USA 107:** 955-956.

Powles S and Yu Q (2010) Evolution in action, plants resistant to herbicides. **Annual Review of Plant Biology 61:** 317-347.

Ramgareeb S, Snyman SJ, van Antwerpen T and Rutherford RS (2010) Elimination of virus and rapid propagation of disease-free sugarcane (*Saccharum* spp. cultivar NCo376) using apical meristem culture. **Plant Cell Tissue and Organ Culture 96:** 263-271.

Rutherford RS, Snyman SJ and Watt MP (2014) *In vitro* studies on somaclonal variation and induced mutagenesis, progress and prospects in sugarcane (*Saccharum* spp.) – a review. **Journal of Horticultural Science and Biotechnology 89:** 1-16.

Santos L O, Pinto JJO, Piveta LB, Noldin J A Galon L and Concenço G (2014) Carryover effect of imidazolinone herbicides for crops following rice. **American Journal of Plant Sciences 5:** 1049-1058.

Schoonees-Muir BM, Ronaldson MA, Naidoo G and Schorn PM (2009) SASTA Laboratory Manual including the Official Methods. 5th Edition. ISBN 978-0-620-43586-4.

Tan S, Evans RR and Singh BK (2006) Herbicidal inhibitors of amino acid biosynthesis and herbicide-tolerant crops. **Amino Acids 30:** 195-204.

Tan S, Evans RR, Dahmer ML, Singh BK and Shaner DL (2005) Imidazolinone-tolerant crops, history, current status and future. **Pest Management Science 61:** 246-257.

Yu Q, Han H, Vila-Aiub MM and Powles SB (2010) AHAS herbicide resistance endowing mutations, effect on AHAS functionality and plant growth. **Journal of Experimental Botany 61:** 3925-3934.

Yu Q, Zhang XQ, Hashem A, Walsh MJ and Powles SB (2003) ALS gene proline (197) mutations confer ALS herbicide resistance in eight separated wild radish (*Raphanus raphanistrum*) populations. **Weed Science 51:** 831-838.

Yuan CI, Chaing M and Chen YM (2002) Triple mechanisms of glyphosate-resistance in a naturally occurring glyphosate-resistant plant *Dicliptera chinensis*. **Plant Science 163:** 543-554.

Yuan JS, Tranel PJ and Stewart Jr CN (2007) Non-target-site herbicide resistance: a family business. **Trends in Plant Sciences 12:** 6-13.

25

New strategy for evaluating grain cooking quality of progenies in dry bean breeding programs

Bruna Line Carvalho[1], Magno AntonioPatto Ramalho[1], Indalécio Cunha Vieira Júnior[1] and Ângela de Fátima Barbosa Abreu[2*]

Abstract: *The methodology available for evaluating the cooking quality of dry beans is impractical for assessing a large number of progenies. The aims of this study were to propose a new strategy for evaluating cooking quality of grains and to estimate genetic and phenotypic parameters using a selection index. A total of 256 progenies of the 13th cycle of a recurrent selection program were evaluated at three locations for yield, grain type, and cooked grains. Samples of grains from each progeny were placing in a cooker and the percentage of cooked grains was assessed. The new strategy for evaluating cooking quality was efficient because it allowed a nine-fold increase in the number of progenies evaluated per unit time in comparison to available methods. The absence of association between grain yield and percentage of cooked grains or grain type indicated that it is possible to select high yielding lines with excellent grain aspect and good cooking properties using a selection index.*

Keywords: *Phaseolus vulgaris L., plant breeding, selection index, quantitative genetics, recurrent selection.*

*Corresponding author:
E-mail: angela.abreu@embrapa.br

[1] Federal University of Lavras (UFLA), Department of Biology, CP 3037, 37.200-000, Lavras, MG, Brazil
[2] Embrapa Rice and Beans/UFLA, Rod.GO-462, km 12, Zona Rural, CP 179, 75.375-000, Santo Antônio de Goiás, GO, Brazil

INTRODUCTION

Dry bean (*Phaseolus vulgaris* L.) is considered one of the most important legumes for human consumption (Broughton et al. 2003). The dry bean most consumed in Brazil is the *carioca* bean, which has beige color and brown stripes on the grain. Consumers prefer medium-sized grains (hundred grain weight of 26 g to 27 g), beige color as light as possible, and grains that do not darken rapidly during storage (Araújo et al. 2012).

Recommendation of new cultivars in Brazil is mainly based on their agronomic traits, such as pest and disease resistance and yield (Carbonell et al. 2014, Lima et al. 2014, Melo et al. 2014, Ramalho et al. 2016). Nevertheless, species for human consumption, like dry bean, must receive the approval of producers and also of consumers. In addition to agronomic traits, breeding programs must consider the desired appearance and the time it takes to cook the grains. Incorporating cooking quality in cultivar evaluation would allow researchers to improve cooking and nutritional quality (Wang and Daun 2005, Ribeiro et al. 2014).

Various methods have been used to evaluate cooking time for dry bean grains (Black et al. 1998). Currently, the Mattson cooker method is most used (Proctor 1987, Arruda et al. 2012). The Mattson cooker is used for evaluation of grains from lines before recommending the lines for production. Nevertheless, a great deal of time is spent in this process. When there are a large number of progenies to be evaluated in a short time, as is often the case in dry bean breeding programs, this methodology is not viable. It would be important to explore alternative methods efficient in evaluating cooking parameters of grains from dry bean progenies for researchers to obtain data using a process similar to the one adopted by consumers, that is, cooking the grains in a pressure cooker.

A recurrent selection program for improving the yield and appearance of carioca type grains has been conducted in Brazil since 1990 (Ramalho et al. 2005, Silva et al. 2010). Thirteen selection cycles have been completed so far. It is important to verify if there is variability for cooking time, which has not been used as a selection criterion.

In light of the above, the present study was carried out to propose a new methodology to evaluate the cooking quality of dry bean grains that would make selection for this trait viable when a large number of progenies are to be evaluated, and, in addition, to estimate the genetic and phenotypic parameters to verify the possibility of continued success of recurrent selection in the dry bean crop.

MATERIAL AND METHODS

Experiments were carried out at these three locations in the state of Minas Gerais, Brazil: Lavras (lat 21° 14' S, long 45° 00' W, alt 840 m asl), Lambari (lat 21° 58' S, long 45° 21'W, alt 896 m asl), and Patos de Minas (lat 18° 34' S, long 46° 31' W, alt 832 m asl). The progenies used originated from the 13th cycle of a recurrent selection (RS) program for improving yield and grain type. Details on the RS program are presented in Silva et al. (2010). A total of 252 $S_{0:2}$ progenies and four checks (Carioca, Pérola, BRSMG Talismã, and BRSMG Majestoso cultivars) were evaluated using a 16 x 16 triple lattice design. Plots consisted of two rows of two meters length, with a spacing of 50 cm between rows and 15 seeds/m. Sowing was performed at the beginning of November 2012. Management practices were performed according to recommendations for the crop.

In late summer (February/March2013) the plants were harvested. They were dried in the field and then threshed, and grain yield per plot was calculated. Approximately 30 days after harvest, the progenies were evaluated for grain type. A 60 g sample of each $S_{0:2}$ progeny was placed in a transparent plastic bag (4 x 23 cm), and identified. A 1 to 9 scoring scale was used for classifying grains in regard to general appearance, in which1 represents appearance totally unacceptable according to commercial standards, that is, square shape, dark color, and under- or over-size grains (100 grain weight less than 26 g or higher than 27 g) and 9 represents excellent appearance, that is, oval shape, light beige color, and medium-size grains (100 grain weight between 26 and 27 g). The following aspects were considered in this visual evaluation: shape, size, and grain color, especially in regard to early darkening. Classification was performed by three evaluators, independently.

For evaluation of cooking quality, a new methodology was used. Just after evaluation of grain type, 50-grain samples of each progeny without mechanical damage were placed in voile bags with respective identification, and the bags were tied shut (Figure 1A). Two bags per progeny were prepared, that is, two replications.

Figure 1. Steps of the procedure for evaluation of the percentage of cooked grains. A) Preparation of the samples in voile bags, B) Soaking in distilled water, C) Cooking, D) Removal of samples from the pressure cooker, E) Cooling of grains, F) Grains being arranged in the Mattson cooker, G) Lowering of pegs, H) Counting pegs that perforated the grains.

The bags were placed in distilled water for 90 min (Figure 1B). This time was pre-established in tests conducted by the authors to maximize differentiation among the progenies as quickly as possible. During cooking, the bags were placed at the bottom of an electric pressure cooker with a 3.5 liters capacity to avoid possible variations attributable to positioning. The water level used was ¾ of the volume of the cooker; maintaining the same water in which the bags were soaked. The bags were cooked for 40 min (Figure 1C). After that, the samples were immediately removed from the cooker and the grains were placed on a counter for cooling for 5 min at ambient temperature (Figure 1D and 1E). Sixteen progenies and two controls were included in each cooker, using a 16 x 16 simple lattice design. Evaluation of the percentage of cooked grains was performed with the assistance of the Mattson cooker. Twenty grains per sample, chosen at random, were used (Figure 1F and 1G). The pegs were placed on the grains simultaneously and the number of plungers that immediately perforated the grains completely was recorded (Figure 1H).

Data on yield, percentage of cooked grains, and the grain type score were subjected to analysis of variance by location and across locations. All the effects of the model were considered random, except for the mean, which is a constant, and the location effect. The SAS software (Statistical Analysis System, version 9.3) was used for statistical analysis of the data.

Based on mean square expectations from the analyses of variance per location and joint analysis, variance components and heritability were estimated by the following expressions: genetic variance in location k: $V_{Gk} = \frac{(MSP_k - MSE_k)}{r}$; genetic variance in mean locations: $V_G = \frac{(MSP - MSPxL)}{ra}$; variance of the progeny x location interaction: $V_{GxL} = \frac{(MSPxL - MSE)}{ra}$; phenotypic variance in location k: $V_{pk} = \frac{MSP_k}{r}$; phenotypic variance across locations: $V_p = \frac{MSP}{ra}$; heritability of progenies in location k: $h_k^2 = \frac{(MSP_k - MSE_k)}{MSP_k}$; and heritability of progenies across locations: $h^2 = \frac{(MSP - MSPxL)}{MSP}$ – in which MSP_k is the mean square of progenies in location k; MSE_k is the mean square of error in location k; MSP is the mean square of progenies across locations; $MSPxL$ is the mean square of the progeny x location interaction; MSE is the mean square of error across locations; r is the number of replications; and a is the number of locations. Confidence intervals of heritability were estimated by the expression of Knapp et al. (1985). We also estimated Pearson phenotypic correlations among all pair of traits by location and in the mean across locations and gain from selection for each trait separately, using the mean value of the locations, considering a selection intensity of 5% (Steel et al. 1997, Bernardo 2010). Expected gain from selection for grain yield (Y) and the correlated response in the other traits (X) were also estimated via the following formula presented by Falconer and Mackay (1996):

$$GS_{X(Y)} = ds_{X(Y)} \cdot h_X^2,$$

in which $GS_{X(Y)}$ is the correlated response in trait X (percentage of cooked grains and grain type score) from selection for grain yield (Y); $ds_{X(Y)}$ is obtained from the expression, $M_{sx} - M_{ox}$, with M_{sx} being the mean value of the progenies for trait X and of progenies that were identified as best for grain yield (Y) and M_{ox} being the overall mean of the progenies for trait X; and h_X^2 is the heritability of trait x.

Subsequently, the variables were standardized through the expression:

$$Z_{ij} = \frac{Y_{ij} - \bar{Y}_j}{S_j} \text{ (Steel et al. 1997),}$$

in which Z_{ij} is the standardized variable of treatment i for trait j; Y_{ij} is the observation of the variable of treatment i for trait j; $\bar{Y}_j$ is the general mean value of trait j; and s_j is the phenotypic standard deviation of the variable for trait j. To avoid the occurrence of negative values, the constant 3 was added to each observation Z_{ij}, and then the sum of the standardized variables (ΣZ) was obtained for each location. Using the ΣZ, the selection index involving all three traits, for each location, analysis of variance was carried out, considering each location as a replication. Heritability (h^2) of ΣZ was estimated from the mean squares (MS) of the analysis of variance, as previously described, as well as expected gain from selection, considering the 12 highest-value progenies (that is, selection intensity of 5%) based on grain yield and ΣZ, and the correlated responses expected for each trait separately.

RESULTS

Differences among locations and progenies were evident for the three traits ($p<0.01$). The progeny x location interaction was non-significant only for grain yield ($p>0.05$). The highest mean grain yield was obtained in Lambari, whereas the highest grain score and percentage of cooked grains were obtained in Lavras (Table1). Although the percentages of cooked grains were relatively low, the differentiation among the progenies was the highest. This is why cooking time did not increase. It should be possible to improve cooking time in breeding programs depending on the quality of grains.

The frequency distributions of the mean values showed variability among the progenies for the three traits (Figure 2). The range of variation for grain yield was from 292 to 580 g plot^{-1}, that is, 67.9%, based on the mean of the three locations. In the case of percentage of cooked grains and grain score, variation proportional to the mean was even greater, 134.6 and 130.4%, respectively.

Variation among the progenies can also be observed from the estimates of the components of variance. In the case of grain yield, as there was overlapping of the confidence intervals, it may be inferred that the estimates of genetic variance among progenies (V_G) were similar. Heritability estimates reinforce this observation. The magnitude of the variance component for the progeny x location interaction (V_{GxL}) was small, only 1.7% of the V_G estimate. For this reason, the h^2 estimate for the mean value of the three locations was of greater magnitude than that obtained for individual locations (Table 2).

The estimates for the grain type scores showed greater magnitude of the estimate of heritability among the mean values of the progenies in each location than that obtained for yield and percentage of cooked grains. The V_{GxL} component, which was 72% greater than V_G, showed a considerable progeny x location interaction for this trait (Table 2). The percentage of cooked grains exhibited heritability estimates by location intermediate to heritability estimates obtained for grain yield and grain type scores. The V_{GxL} estimate was 4.6 times greater than the V_G estimate, showing the importance of the progeny x location interaction for this trait (Table 2).

The estimates of the pair-wise correlations among the three traits for the mean value of the locations were low and were highly similar. The correlation between yield and percentage of cooked grains or grain type score was practically nil ($r<0.18$). The correlation between percentage of cooked grains and grain type score was significant, but

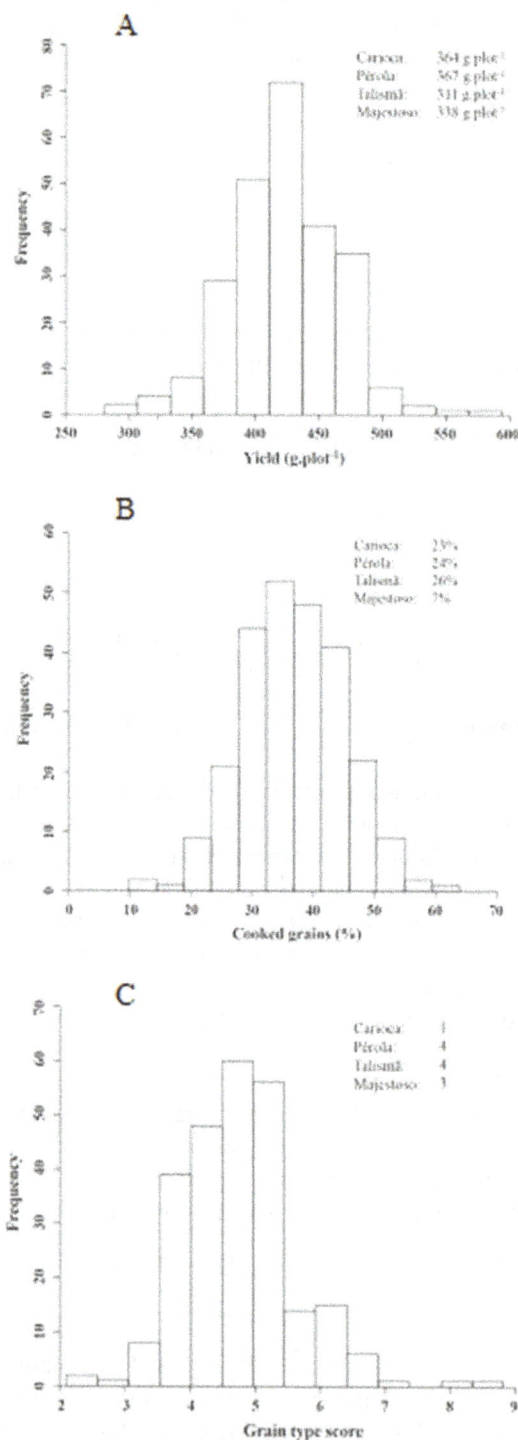

Figure 2. Frequency distribution of the mean values of S$_{0:2}$ progenies evaluated in three locations. A) Grain yield - g plot^{-1}. B) Percentage of cooked grains. C) Grain type score, with 1 being very bad and 9 very good.

Table 1. Mean value of grain yield (g plot^{-1}), grain type (score from 1 to 9), and percentage of cooked grains for each location and across locations, considering progenies and controls separately

Location	Yield	Grain type [+]	Cooked grains
Lavras	423.33	5.16	39.05
Patos de Minas	396.38	4.70	36.55
Lambari	445.59	4.36	33.75
Mean value of the locations			
Progenies	422.87	4.74	36.71
Controls	352.50	3.00	20.00

[+] Grain type score: 1, very bad and 9, very good.

Table 2. Estimates of genetic variance (V_G), phenotypic variance (V_P), progeny x location interaction (V_{GxL}), and heritability (h^2) among $S_{0:2}$ progenies for grain yield (g plot^{-1}), grain type (score from 1 to 9), and percentage of cooked grains. Data obtained per location and across locations

Location	Parameters	Yield	Grain type[+]	Cooked grains
Lavras	V_G	1449.59	1.27	62.09
		(918.11 - 2780.30)[1]	(1.05 - 1.58)	(46.35 - 88.29)
	V_P	4058.71	1.47	104.98
	h^2	35.72	86.39	59.14
		(20.38 - 48.47)[2]	(83.03 - 88.94)	(48.01 - 67.94)
Patos de Minas	V_G	761.22	1.02	60.93
		(414.53 - 1923.53)	(0.83 - 1.30)	(46.87 - 84.07)
	V_P	2841.06	1.29	94.90
	h^2	26.79	79.07	64.20
		(9.33 - 42.31)	(74.64 - 83.47)	(53.76 - 72.24)
Lambari	V_G	665.03	1.23	54.95
		(294.97 - 4553.80)	(1.01 - 1.52)	(41.68 - 77.45)
	V_P	3815.73	1.42	89.80
	h^2	17.43	86.62	61.20
		(-2.27 - 33.81)	(82.87 - 88 84)	(51.94 - 68.89)
Alllocations	V_G	942.64	0.43	10.63
		(677.34 - 1393.73)	(0.32 - 0.61)	(5.48 - 38.59)
	V_P	1819.04	0.75	43.02
	V_{GxL}	15.98	0.74	59.92
	h^2	51.82	57.47	24.72
		(40.52 - 61.28)	(47.49 - 65.82)	(7.05 - 39.50)

[+] Grain type score: 1, very bad and 9, very good. [1] Lower and upper limits of the confidence interval of V_P. [2] Lower and upper limits of the confidence interval of h^2 considering all the S_0 progenies

of relatively small magnitude (r=0.41; p≤0.01). The gain expected from selection of the 12 best progenies was greater than 11% for the three traits when considered separately. Although the magnitude of the estimates of phenotypic correlation between yield and grain score or percentage of cooked grains was low (r<0.18), selection carried out based only on grain yield, as occurs in most breeding programs, contributed to reduced expected gain for both grain score and percentage of cooked grains. The sum of the standardized variables (ΣZ) was also adopted as a selection criterion. The gain expected from selection in relation to the mean was substantial, 31.62%. In addition, this selection led to positive gains for the three traits (Table 3).

DISCUSSION

For application of the new methodology, the first question that arises is if it can replace the method most used for evaluation of dry bean cooking quality, the Mattson method. Thus, we performed preliminary tests using lines (Pádua et al. 2013). In these tests, we were able to demonstrate that the pressure cooker method better differentiated the lines

Table 3. Estimate of gain from selection (GS), considering the three traits simultaneously, using grain yield and the Z index

	Z Index	Yield (g plot^{-1})	Grain type[†]	Grains cooked (%)
		Response of direct selection on each trait		
M_s[1]	-	513.58	6.89	53.72
M_o[2]	-	421.77	4.74	36.46
h^2	-	51.82	57.47	24.72
GS	-	47.58	1.23	4.27
GS/Mo (%)	-	11.28	26.02	11.71
		Correlated response to selection using grain yield		
M_s	-	513.58	4.49	40.64
M_o	-	421.77	4.74	36.46
h^2	-	51.82	57.47	24.72
GS	-	47.58	-0.14	1.03
GS/Mo (%)	-	11.28	-3.06	2.84
		Correlated response to selection using the Z index		
M_s	10.29	474.36	6.47	46.95
M_o	7.00	421.77	4.74	36.46
h^2	67.23	51.82	57.47	24.72
GS	2.21	27.25	0.99	2.59
GS/Mo (%)	31.62*	6.46	20.97	7.12

[†] Grain type score: 1, very bad and 9, very good. [1] Mean of the 5% superior $S_{0:2}$ progenies. [2] Mean considering all the S_0 progenies.

than the classic Mattson Method, since the lines were discriminated via this new method, but not via the former one.

The criterion adopted in the new cooking methodology proposed in this study and the Mattson method is similar, that is, both consider the number of perforated grains using the same device. The difference is that in the Mattson method, the percentage of grains to be perforated is established before conducting the experiment, ranging from 50 to 100%, and the time necessary for achieving this range is recorded. This requires the constant attention of the evaluator until the pre-determined number of pegs perforates the grains. In contrast, the new methodology establishes the cooking time and evaluates the number of perforated grains in each progeny/line. Selection is carried out based on the percentage of cooked grains in a fixed time in an electric pressure cooker. Thus, a pre-established cutoff time may be adopted to achieve the best discrimination among the progenies.

There are some advantages of this method over the Mattson method. The first is in regard to evaluation time. In the Mattson method, it is necessary to soak the grains for 16 hours before performing the test. In the new methodology, soaking the grains is optional. In this study, we established 90 min for soaking, 40 min for cooking, and approximately 30 min for evaluation of 16 samples (quantity placed in each cooker in this study), considering only one operator. To make the process continuous, more samples are placed to soak every 50 min and are afterwards subjected to cooking. Thus, in one workday (eight hours), it is possible to perform nine evaluations, that is, 144 samples. In the conventional method, considering that the time spent on evaluations is an average of 30 min, in eight hours, it would be difficult to evaluate more than 16 samples. Even if the automated system is adopted (Wang and Daun 2005), the amount of time would not change; only the presence/attention of the operator is less necessary. Considering that in a 3.5 liters electric pressure cooker it is possible to evaluate up to 25 samples and not just 16, the advantage is even greater.

In recurrent selection programs, especially for the dry bean crop in which three crop seasons per year can be conducted in Brazil and in other tropical countries, the time between growing seasons is short. Evaluations need to be performed as efficiently as possible and superior progenies need to be identified for the following sowing operation. The cooking quality trait has not been used in these programs as a selection criterion because of the lack of an appropriate methodology to evaluate it. With the use of the methodology proposed in this study, the inclusion of the cooking quality trait as a selection criterion is feasible, because of the possibility of evaluating a relatively large number of samples per day. This is highly desirable in light of the aforementioned consumer demands.

The precision of the experiments in evaluation of this trait was relatively high in all cases, which can be confirmed by the heritability estimates, always greater than 59% (Table 2), a magnitude similar to magnitudes reported in the literature when the Mattson method was used, and the number of progenies/lines evaluated was much lower (Jacinto-Hernandez et al. 2003, Garcia et al. 2012). The mean percentage of cooked grains varied among the locations, with the lowest percentage in Lambari (Table 1). The experiments were planted in late spring (November) and harvest coincided with intense rainfall, especially in Lavras and Patos de Minas. After harvest, the plants were sheltered until the time of threshing. In Lambari, there was no rain at the time of harvest and the plants remained in the field under high temperature for a longer time. There are reports that the climatic conditions, especially high temperatures, at the time of harvest and during drying affect cooking time (Rodrigues et al. 2005, Arruda et al. 2012).

Another aspect that should be noted relates to evaluation of grain appearance. Araújo et al. (2012) observed that for the early darkening of grain, there was no genotype x evaluation time period interaction, and it was already possible to discriminate the progenies 30 days after harvest. Thus, after this minimum period, a grain sample from each progeny, approximately 60 grams, was placed in a transparent plastic bag. This allowed the evaluators to classify the samples in a comparative manner, involving aspects of color (above all, early darkening of grain), grain size, and grain shape. The strategy worked and the accuracy of the evaluations was high, which can be confirmed by the heritability estimates at each location (Table 2).

For grain scores and percentage of cooked grains, the genotype x location interaction was significantly large. The existence of environmental variation at the time of harvest, mentioned above, probably contributed to this interaction. At any rate, it is evident that the response of the progenies in the grain score and cooking quality traits varied across the three locations. In the case of grain yield, reports of the genotype x location interaction are common in the dry bean crop (Lima et al. 2012, Faria et al. 2013), but non-significant interaction estimates have also been reported (Ramalho et al. 2005).

In the literature, there is no consensus in regard to the importance of dominance deviation for yield in the dry bean crop. There are studies that indicate predominance of additive variance (V_A) (Nienhuis and Singh 1988, Corte et al. 2010). Others, however, have indicated that dominance variance (V_D) was also present (Gonçalves-Vidigal et al. 2008). In the case of grain score and percentage of cooked grains, no reports were found regarding variances. Nevertheless, even if V_D was present, the variance among $S_{0:2}$ progenies only captures 1/16 of the estimates of V_D, and so it may be inferred that nearly all the genetic variance was attributable to V_A. Thus, for practical purposes, the heritability estimates obtained in this case may be considered to be in the narrow sense.

The heritability estimates varied among the traits and were smaller for the percentage of cooked grains than for yield and grain type (Table 2). This probably occurred because of the progeny x location interaction, since in the individual analyses, the heritability estimates for this trait were of medium to high magnitude compared to the estimates reported by Elia (2003) and Ribeiro et al. (2006). For grain yield, the heritability estimate of 51.8% was within the limit reported for the same population in cycle 4 (Ramalho et al. 2005) and cycle 8 (Silva et al. 2010). All are within the confidence interval estimated in each cycle. It can be noted that the genetic variability of the population for grain yield has not decreased as a result of the recurrent selection cycles undertaken for yield and grain type. In the case of percentage of cooked grains and grain type score, as these traits were not included in selection in the previous cycles of the program, there is no information on these aspects. But, even so, for the reasons already presented, it may be inferred that there is still enough variability for continued progress in selection.

There was a positive correlation between percentage of cooked grains and grain type (r=0.41; p≤0.01). This may have arisen from pleiotropy and/or linkage of the genes that control these traits (Falconer and Mackay 1996). However, because of the relatively low magnitude of the correlation, it would not be possible to perform indirect selection for cooking quality based only on grain scores.

Correlated response carried out using the ΣZ index led to positive gains for all three traits involved, although of smaller magnitude than when selecting each trait individually (Table 3). However, if the strategy applied in the program in the previous cycles were used, that is, selection with an emphasis on grain yield, the correlated response in cooking quality and grain type would not be significant, and would actually be negative for the grain type score.

We can confirm that it is feasible to implement evaluation of cooking quality in early stages of breeding programs and/or use it in recurrent selection cycles. By doing so, it would be possible to select high yielding lines with better grain quality. In addition, for crops directly consumed by humans, like dry beans, traits associated with appearance and quality have significant importance, and thus it is recommended to apply the ΣZ selection index for selection of the best progenies including all the traits, that is, yield, appearance, and time taken to cook the dry bean. The association between grain yield and percentage of cooked grains or grain type is not complete, in other words, these traits are not controlled by genes with pleiotropic effects, allowing selection of plants with high yield, excellent appearance of grains, and shorter cooking time. It was also shown that recurrent selection was a good strategy for maintaining variability in long-term breeding programs.

ACKNOWLEDGMENTS

The authors are grateful to the Office for Improvement of Personnel in Higher Education (CAPES) and to the National Council for Technological and Scientific Development (CNPq) for financial support.

REFERENCES

Araújo LCA, Ramalho MAP and Abreu AFB (2012) Estimates of genetic parameters of late seed-coat darkening of carioca type dry beans. **Ciência e Agrotecnologia 36**: 156-162.

Arruda B, Guidolin AF, Coimbra JLM and Battilana J (2012) Environment is crucial to the cooking time of beans. **Ciência e Tecnologia de Alimentos 32**: 573-578.

Bernardo R (2010) **Breeding for quantitative traits in plants**. Stemma Press, Woodbury, 390p.

Black RG, Singh U and Meares C (1998) Effect of genotype and pretreatment of field peas (*Pisium sativum*) on their dehulling and cooking quality. **Journal of the Science of Food and Agriculture 77**: 251-258.

Broughton WJ, Hernández G, Blair M, Beebe S, Gepts P and Vanderleyden J (2003) Beans (*Phaseolus* spp.): model food legumes. **Plant and Soil 252**: 55-128.

Carbonell SAM, Chiorato AF, Bolonhezi D, Barros VLNP, Borges WLB, Ticelli M, Gallo PB, Finoto EL and Santos NCB (2014) 'IAC Milênio' - Common bean cultivar with high grain quality. **Crop Breeding and Applied Biotechnology 14**: 273-276.

Corte AD, Moda-Cirino V, Arias CAA, Toledo JFF and Destro D (2010) Genetic analysis of seed morphological traits and its correlations with grain yield in common bean. **Brazilian Archives of Biology and Technology 53**: 27-34.

Elia FM (2003) Heritability of cooking time and water absorption in dry bean (*Phaseolus vulgaris* L.) using a North Carolina design II mating scheme. **Tanzania Journal of Science 29**: 25-34.

Falconer DS and Mackay TFC (1996) **Introduction to quantitative genetics**. Pearson, Malaysia, 464p.

Faria LC, Melo PGS, Pereira HS, Peloso MJD, Brás AJBP, Moreira JAA, Carvalho HWL and Melo LC (2013) Genetic progress during 22 years of improvement of carioca-type common bean in Brazil. **Field Crops Research 142**: 68-74.

Garcia RAV, Rangel PN, Bassinello PZ, Brondani C, Melo LC, Sibov ST and Vianello-Brondani RP (2012) QTL mapping for the cooking time of common beans. **Euphytica 186**: 779-792.

Gonçalves-Vidigal MC, Mora F, Bignotto TS, Munhoz REF and Souza LD (2008) Heritability of quantitative traits in segregating common bean families using a Bayesian approach. **Euphytica 164**: 551-560.

Jacinto-Hernandez C, Azpiroz-Rivero S, Acosta-Gallegos JA, Hernandez-Sanchez H and Bernal-Lugo I (2003) Genetic analysis and random amplified polymorphic DNA markers associated with cooking time in common bean. **Crop Science 43**: 329-332.

Knapp SJ, Stroup WW and Ross WM (1985) Exact confidence intervals for heritability on a progeny mean basis. **Crop Science 25**:192-194.

Lima LK, Ramalho MAP, Abreu AFB, Toledo FHRB and Ferreira RADC (2014) Implications of predictable and unpredictable environmental factors in common bean VCU trials in Minas Gerais. **Crop Breeding and Applied Biotechnology 14**: 146-153.

Lima LK, Ramalho MAP and Abreu AFB (2012) Implications of the progeny x environment interaction in selection index involving characteristics of the common bean. **Genetics and Molecular Research 11**: 4093-4099.

Melo LC, Abreu AFB, Ramalho MAP, Carneiro JES, Paula Júnior TJ, Del Peloso MJ, Pereira HS, Faria LC, Pereira Filho IA, Moreira JAA, Martins M, Vieira RF, Martins FAD, Coelho MAO, Costa JGC, Wendland A, Santos JB, Diaz JLC, Carneiro PCS, Del Giúdice MP and Faria JC (2014) BRSMG Realce: Common bean cultivar with striped grains for the state of Minas Gerais. **Crop Breeding and Applied Biotechnology 14**: 61-64.

Nienhuis J and Singh SP (1988) Genetics of seed yield and its components in common bean (*Phaseolus vulgaris* L.) of Middle-American origin, I.General Combining Ability. **Plant Breeding 101**: 143-154.

Pádua JMV, Ramalho MAP and Abreu AFB (2013) New alternative for assessing cooking time of common bean progenies. **Annual Report of Bean Improvement Cooperative 56**: 17-18.

Proctor JR (1987) Development of a modified Mattson bean cooker procedure based on sensory panel cookability evaluation. **Canadian**

Institute of Food Science and Technology 20: 9-14.

Ramalho MAP, Abreu AFB, Carneiro JES, Melo LC, Paula Júnior TJ, Pereira HS, Del Peloso MJ, Pereira Filho IA, Martins M, Del Giúdice MP and Vieira RF (2016) BRSMG Uai: common bean cultivar with carioca grain type and upright plant architecture. **Crop Breeding and Applied Biotechnology 16**: 261-264.

Ramalho MAP, Abreu AFB and Santos JB (2005) Genetic progress after four cycles of recurrent selection for yield and grain traits in common bean. **Euphytica 144**: 23-29.

Ribeiro ND, Rodrigues JA, Prigol M, Nogueira CW, Storck L and Gruhn EM (2014) Evaluation of special grains bean lines for grain yield, cooking time and mineral concentrations. **Crop Breeding and Applied Biotechnology 14**: 15-22.

Ribeiro SRRP, Ramalho MAP and Abreu AFB (2006) Maternal effect associated to cooking quality of common bean. **Crop Breeding and Applied Biotechnology 6**: 304-310.

Rodrigues JDA, Ribeiro ND, Cargnelutti Filho A, Tretin M and Londero PMG (2005) Cooking quality of common bean grain obtained in differents sowing periods. **Bragantia 64**: 369-376.

Silva GS, Ramalho MAP, Abreu AFB and Nunes JAR (2010) Estimation of genetic progress after eight cycles of recurrent selection for common bean grain yield. **Crop Breeding and Applied Biotechnology 10**: 351-356.

Steel RGD, Torrie JH and Dickey DA (1997) **Principles and procedures of statistics: a biometrical approach.** McGraw-Hill, New York, 666p.

Wang N and Daun JK (2005) Determination of cooking times of pulses using an automated Mattson cooker apparatus. **Journal of the Science of Food and Agriculture 85**: 1631-1635.

26

An improved method for RNA extraction from common bean seeds and validation of reference genes for qPCR

Wendell Jacinto Pereira[1,2]**, Priscila Zaczuk Bassinello**[3]**, Claudio Brondani**[4]** and Rosana Pereira Vianello**[4*]

Abstract: *An RNA extraction method with high integrity and purity as well as the selection of adequate reference genes are prerequisites for gene expression analysis. For common bean seeds, there is no well-defined protocol that can be used in a laboratory routine for gene expression analysis. In this study, an extraction protocol for RNA from common bean seeds, which produced material with good integrity for qPCR (RIN ≥ 6.5), was optimized. In addition, 10 reference genes were evaluated under qPCR standard conditions using different tissue samples of common beans. Gene stabilities were analyzed using the delta-CT method, Bestkeeper, NormFinder and geNorm approaches. The genes β-tubulin and T197 were ranked as the most stable among the sample sets evaluated with different tissue samples, while PvAct and Pv18S were the least stable. To our knowledge, this is the first study evaluating RNA isolation methods and reference gene selection for seeds of Phaseolus vulgaris.*

Key words: *Gene expression; Gene normalization; Grain RNA extraction; Phaseolus vulgaris; Molecular Breeding.*

*Corresponding author:
E-mail: rosana.vianello@embrapa.br

[1] Universidade Federal de Goiás, Instituto de Ciências Biológicas, 74.001-970, Goiânia, GO, Brazil
[2] Universidade de Brasília, Departamento de Biologia Celular, 70910-900, Brasília, DF, Brazil
[3] Embrapa Arroz e Feijão, Laboratório de Grãos e Subprodutos, Rodovia GO-462, km 12, Fazenda Capivara, Zona Rural, 75375-000, Santo Antônio de Goiás, GO, Brazil
[4] Embrapa Arroz e Feijão, Laboratório de Biotecnologia

INTRODUCTION

Experimental validation of gene transcription data is usually performed using real-time quantitative polymerase chain reaction (qPCR), which involves amplifying *in vitro* copies of complementary DNA (cDNA) from an *RNA* template and monitoring the levels of the molecules produced during each cycle in real time (Heid et al. 1996). The advantages of this method include speed, specificity and sensitivity for amplifying the target fragment through relative and absolute transcribed gene quantification (Hu et al. 2014). The qPCR technique has been considered the gold standard for quantifying gene expression since many factors, such as the quality and integrity of the extracted RNA, sample storage, and the correct choice of internal controls, are considered to avoid obtaining biased and/or misleading results (Dheda et al. 2005).

Only a set of RNA that truly represents cell transcription for a particular sample will provide accurate information on the characteristics and expression levels of the transcripts analyzed. Due to variable cell compositions with varying levels of secondary metabolites, polysaccharides, and phenolic and oxidative compounds, a single standardized procedure of RNA extraction for any type of tissue is not viable (Mornkham et al. 2013). Establishment of such procedures is not trivial, and efforts to develop seed RNA extraction protocols have been conducted for different species (Christou et al. 2014, Ma et al. 2015). For the common bean, RNA extraction methods from leaves and roots have been established, which have been based on either commercial kits (Contour-Ansel et al. 2010) or not (Borges et al. 2012). However, a well-defined RNA extraction protocol for common bean grain/seed tissue is not available and, among legumes, there is just one protocol that was created for soybean seeds (Yin et al. 2014).

Because extraction of high-quality RNA and identification of reference genes are among the most important factors for reliable qPCR experiments, we aimed in this paper to establish a suitable RNA extraction method for common bean seeds. Additionally, we evaluated a set of reference genes previously described in the literature, optimized the qPCR conditions and determined the most adequate method for qPCR analysis in different tissues of common beans and under variable experimental conditions.

MATERIAL AND METHODS

Plant samples

Seeds from six cultivars of common bean of the Carioca grain type were used: *BRSMG Madrepérola, BRS Pontal, Pérola*, BRS Estilo, CNFC10467 and Pinto Beans. Plants were grown in a field from August to October 2014, and the plots consisted of 10 rows that were each five meters in length in Santo Antônio de Goiás - GO, Brazil. After harvesting, the seeds were dried at room temperature (21 °C), processed and stored at -80 °C until use. Aiming to define stable reference genes across different types of common bean tissue in addition to seeds, RNA samples were used from the leaves and roots of Pérola and BAT477 that were cultivated under drought stress and the leaves of BRS Realce and BRS Executivo that were inoculated with the fungus *Colletotrichum lindemuthianum*.

Seed total RNA extraction

The total RNA from mature seeds of the six cultivars was extracted using three different protocols: 1) based on TRIzol® RNA Isolation Reagents (Invitrogen™, Carlsbad, California) following the manufacturer's instructions; 2) a commercial kit, PureLink® RNA Mini Kit (Ambion®, Carlsbad, California), which is based on a column extraction method, following the manufacturer's instructions and 3) a protocol proposed by Silva et al. (2011) for coffee seeds, with adaptations as described below. The ground seeds (150 mg, five grains on average) were transferred to 1.5 mL tubes, then 1000 µL of Concert™ Plant RNA Reagent (Invitrogen™, Carlsbad, California) was added and the samples were homogenized by vortexing for 2 min. Subsequently, the tubes were incubated horizontally at room temperature (21 °C) for 5 min. Next, the samples were centrifuged at 11,400 rpm for 2 min at room temperature, and the supernatant was subsequently removed and transferred to fresh tubes. The supernatant was treated with 100 µL of 1.5 mol L^{-1} NaCl and homogenized by inverting the tube eighty times. Thereafter, 300 µL of chloroform was added and homogenized by inverting the tube eighty times. The tubes were then centrifuged for 10 min at 11,400 rpm at 4 °C; the upper phase was recovered and transferred to a new tube. Then, 500 µL of the lysing reagent (Concert™ Plant RNA Reagent) was added to the recovered upper phase, and the initial extraction procedures were repeated. The centrifugation step was repeated, and the final supernatant was extracted; then, 500 µL of isopropanol was added, and the tubes were inverted eighty tubes. The tubes were incubated at room temperature for 10 min. The precipitate was recovered by centrifugation at 11,400 rpm for 10 min at 4 °C followed by washing with 1000 µL of 75% cold ethanol and centrifugation at 11,400 rpm for 1 min at room temperature. To the dried precipitate, 70 µL of autoclaved Diethylpyrocarbonate (DEPC) water was added followed by storage at -80 °C. Three technical replicates were used for each sample. The total RNA extraction from leaves and roots was performed using the PureLink® RNA Mini Kit (Ambion®, Carlsbad, California).

RNA quality control and synthesis of complementary DNA (cDNA)

The quantity and purity of each RNA sample was estimated using a NanoVue™ Plus Spectrophotometer (General Electric Company, GE). Additionally, an RNA integrity number (RIN) was verified with an Agilent 2100 Bioanalyzer (General Electric Company, GE). RNA samples were treated with DNase I (Invitrogen™, Carlsbad, California) following the manufacturer's guidelines. Subsequently, the RNA was transcribed into cDNA using SuperScript® II Reverse Transcriptase (Invitrogen™, Carlsbad, California), as directed by the manufacturer. The cDNA was quantified using a Qubit® ssDNA assay in a Qubit® 2.0 Fluorometer (Life Technologies™, Carlsbad, California).

Optimization of primer concentration for qPCR

The primer sequences corresponding to the target reference genes in different plants were obtained through a literature search (Table 1). Each primer was individually tested for amplification and specificity. The PCR reaction was conducted as described by Müller et al. (2014), and the amplified products were visualized on a 2.0% agarose gel.

For the primer titration, a matrix of reactions was created. A total of 9 reactions were performed varying the concentration of each primer (forward and reverse) independently (50, 300, and 900 nM) to determine the best quantification cycle (C_q) values that avoided primer dimer formation. The reactions were conducted in duplicate using Real Master SYBR ROX Mix (5 Prime, Gaithersburg, Maryland) and 10 ng of cDNA in a final volume of 20 µL, as suggested by the manufacturer. The thermocycling conditions included one cycle at 94 °C for 2 min followed by 40 cycles at 95 °C for 15 sec and 60 °C for 1 min. The qPCR was performed on an ABI7500 Real Time PCR (Life Technologies™).

A dissociation curve based on a melting temperature analysis (Tm) was generated and analyzed using the 7500 software v2.0.5. The Sequence Detection Software (SDS) v.1.3 (Applied Biosystems) was used to calculate the cycle threshold (C_t). Two methods were used to evaluate the amplification efficiencies: 1) a standard curve based on a serial cDNA dilution (128, 64, 32, 16, 8, 4 and 2 ng) and 2) linear regression within the *window-of-linearity by* the LinRegPCR software (Ruijter et al. 2009).

Gene expression stability

The C_q values were used to determine the reference gene stability using a variety of methods: Delta-C_T (Silver et al. 2006), NormFinder (Andersen et al. 2004), geNorm (Vandesompele et al. 2002) and BestKeeper (Pfaffl et al. 2004). The web-based tool RefFinder (Xie et al. 2012) was used to integrate the results of these programs and to generate the final ranking of the tested reference genes.

RESULTS AND DISCUSSION

Several methods to remove polysaccharide and protein contamination from plant RNA have been developed, as described for seeds containing high levels of starch (Li and Trick 2005) and soybean seeds (Yin et al. 2014). In this report,

Table 1. Candidate genes for qPCR analyses with their respective nomenclature and bibliographic citation, GenBank accession number, gene description, primer sequence (forward and reverse) and product size in base pairs (bp)

Gene name	Accession nº	Gene description	Origin of primers	Primer sequences (forward/reverse) 5'- 3'	Size (pb)
Cons 6 (Libault et al. 2008)	CD397253	F-box protein family	soybean	AGATAGGGAAATGGTGCAGGT CTAATGGCAATTGCAGCTCTC	93
Cons 7 (Libault et al. 2008)	AW310136	Insulin-degrading enzyme, Metalloprotease	soybean	ATGAATGACGGTTCCCATGTA GGCATTAAGGCAGCTCACTCT	114
Cons 15 (Libault et al. 2008)	AW396185	CDPK-related protein kinase	soybean	TAAAGAGCACCATGCCTATCC TGGTTATGTGAGCAGATGCAA	97
Pv18S (Santos et al. 2005)	AF207040	18S ribosomal RNA	cocoa	AACGGCTACCACATCCAAGG TCATTACTCCGATCCCGAAG	393
PvAct (Chen et al. 2009)	-----	Actin	bean	GAAGTTCTCTTCCAACCATCC TTTCCTTGCTCATTCTGTCCG	-----
PvUBC9 (Ramírez et al. 2013)	TC34057	Ubiquitin	bean	GCTCTCCATTTGCTCCCTGTT TGAGCAATTTCAGGCACCAA	66
T197 (Thibivilliers et al. 2009)	-----	Guanine nucleotide-binding protein beta subunit-like protein	bean	TGGGCAATTGGACGTTATTAG GCCACGGTCTTGAACATAAAA	80 to 125
Tc127 (Thibivilliers et al. 2009)	Q39257	Ubiquitin	bean	CCAAGGAACTTCAGATTGCTG GTCATCACCATCATCCATTCC	80 to 125
Tc185 (Thibivilliers et al. 2009)	-----	Tubulin beta chain	bean	TTTGGACAACGAGGCTCTCTA GAGATGGTTAAGGTCCCCAAA	80 a 125
ß-tubulin (Eticha et al. 2010)	CV530631	β-tubulin	bean	CCGTTGTGGAGCCTTACAAT GCTTGGGGTCCTGAAACAA	117

three different protocols were used based on their reported impact in the literature. The procedures based on the TRIzol® and column purifications (PureLink®) did not generate RNA from seeds with sufficient purity and integrity for subsequent qPCR analyses (Table 2). The resulting 260/280 ratio (1.8-2.0 desired as an indication of purity) was below 1.5 for the TRIzol® reagent. For the 260/230 ratio (1.8-2.0 desired as an indication of purity), the values obtained were below 0.3 for both TRIzol® and PureLink® reagents, and the RIN values were below 2.6 for all samples.

The adapted protocol from Silva et al. (2011), based on an isolation procedure using the organic solvents phenol and chloroform, in addition to various mixing steps (extremely important for tissue homogenization), allowed the insoluble material to be removed from the samples (as polysaccharides and proteins). The 260/280 ratio obtained was adequate (≥ 2.0). The extraction of high-purity RNA is challenge because of the high protein and carbohydrate content in common bean seeds which might be difficult to remove from samples. The secondary metabolites can either interact with or intercalate in RNA, or taken together during the extraction process due to reagent affinity, which reduce the yield and quality of the RNA. The amounts of these compounds may vary widely among common bean genotypes, reaching values of 32% protein and 23% carbohydrate in the seeds (Brigide et al. 2014). The optimized protocol was efficient at removing these organic compounds. For the 260/230 ratio, low values were obtained (ranging from 0.37 to 0.82) revealing remains of solvents or salts in the RNA solution. These impurities did not compromise the reliability of the qPCR results in this

Table 2. Summary of the common bean seed RNA profile based on the three extraction methods

Extraction method	Seed samples	RIN*	[ ] ng/µL	260/280**	260/230**
Silva et al. (2011) Modified	BRS Estilo	6.93 ± 0.35	2500.0 ± 1078.2	2.16 ± 0.11	0.74 ± 0.21
	BRS Pontal	7.50 ± 0.26	414.4 ± 71.6	2.85 ± 0.09	0.37 ± 0.04
	Madrepérola	7.43 ± 0.12	1333.0 ± 287.1	2.19 ± 0.02	0.61 ± 0.04
	CNFC 10467	8.30 ± 0.44	798.9 ± 280.0	2.26 ± 0.08	0.65 ± 0.10
	Pérola	7.50 ± 0.36	672.3 ± 59.6	2.16 ± 0.06	0.82 ± 0.11
	Pinto Beans	7.77 ± 0.32	1184.0 ± 586.9	2.35 ± 0.18	0.53 ± 0.02
TRIzol® RNA Isolation Reagent	BRS Estilo	1.05 ± 0.07	455.2 ± 35.1	1.30 ± 0.02	0.27 ± 0.00
	BRS Pontal	0.00 ± 0.00	406.0 ± 11.3	0.97 ± 0.05	0.28 ± 0.08
	Madrepérola	0.95 ± 1.34	682.8 ± 72.4	1.48 ± 0.01	0.34 ± 0,18
	CNFC10467	1.75 ± 1.06	376.0 ± 113.7	1.44 ± 0.04	0.27 ± 0.14
	Pérola	1.80 ± 1.13	340.4 ± 127.3	1.45 ± 0.02	0.21 ± 0.07
	Pinto Beans	2.50 ± 0.00	430.0 ± 104.1	1.48 ± 0.00	0.25 ± 0.05
PureLink® RNA Mini Kit	BRS Estilo	0.00 ± 0.00	29.20 ± 0.00	2.09 ± 0.00	0.02 ± 0.00
	BRS Pontal	1.10 ± 1.10	34.00 ± 35.07	2.82 ± 0.97	0.03 ± 0.01
	Madrepérola	1.20 ± 1.20	13.80 ± 0.28	2.06 ± 0.09	0.03 ± 0.00
	CNFC10467	2.40 ± 2.40	0.55 ± 0.46	1.75 ± 0.04	0.55 ± 0.46
	Pérola	2.45 ± 2.45	13.40 ± 1.41	2.00 ± 0.05	0.08 ± 0.05
	Pinto Beans	1.35 ± 1.35	14.30 ± 2.40	2.06 ± 0.25	0.11 ± 0.10

* Values higher than 6.5 were considered adequate
** Desired values ≥ 1.8 and ≤ 2.2

Table 3. Performance evaluation of qPCR for the five selected candidate genes in leaf/root pooled cDNA and seed cDNA

Genes	Tissue	r2 (Standard curve)	Standard curve amplification factor	%Eff (Standard curve)	r2 (LinRegPCR)	LinRegPCR mean Efficiency	%Eff (LinRegPCR)
Pv18S	leaf/root	0.997 ± 0.00	1.64 ± 0.00	63.98 ± 0,51	≥ 0.999	1.55 ± 0.02	55.32 ± 2.63
	seed	0.999 ± 0.00	1.66 ± 0.06	65.72 ± 5,64	≥ 0.999	1.57 ± 0.03	56.96 ± 2.91
ß-tubulin	leaf/root	0.997 ± 0.00	1.97 ± 0.01	96.75 ± 1,38	≥ 0.999	1.85 ± 0.02	85.43 ± 1.88
	seed	0.991 ± 0.01	2.03 ± 0.03	103.02 ± 3,07	≥ 0.999	1.85 ± 0.08	84.76 ± 8.09
PvAct	leaf/root	0.991 ± 0.01	1.96 ± 1.11	95.30 ± 10,45	≥ 0.999	1.67 ± 0.02	66.94 ± 2.22
	seed	0.960 ± 0.03	1.95 ± 0.04	95.22 ± 3,93	≥ 0.999	1.65 ± 0.06	64.93 ± 5.92
Tc127	leaf/root	0.996 ± 0.00	1.96 ± 0.01	95.62 ± 0,45	≥ 0.999	1.83 ± 0.01	82.97 ± 1.26
	seed	0.997 ± 0.00	1.97 ± 0.03	97.01 ± 2,27	≥ 0.999	1.84 ± 0.01	84.01 ± 1.32
T197	leaf/root	0.993 ± 0.00	2.01 ± 0.02	100.67 ± 2,09	≥ 0.999	1.93 ± 0.04	93.33 ± 3.48
	seed	0.993 ± 0.00	1.97 ± 0.03	97.04 ± 3,15	≥ 0.999	1.94 ± 0.03	94.3 ± 2.64

study; this finding is consistent with previous reports in the literature (Cincinnati et al. 2008, Yee et al. 2014). The RIN values ranged from 6.93 to 8.30 (Table 2) and the efficiency of the qPCR reactions was high, as demonstrated by the test of efficiency (Table 3), being used in this study as the criteria for a protocol selection. Most studies have considered the RNA integrity using the RIN algorithm.

Reference genes for qPCR

Concerning the reference genes available for common beans, this work does not intend to confirm information already described in the literature, but rather intends to provide additional guidance on working with common bean reference genes. The purpose is to allow readers access to the most appropriate genes to be used for different plant tissues and the qPCR amplification conditions fundamental to conducting routine laboratory analysis. Important steps of the research process, not available in the published literature, such as primer titrations and a determination of suitability across different tissue samples with strong statistical support, were provided in the present study. Of the genes that were evaluated (Table 1), only the gene *Const15* did not generate an amplified product. It is a constitutive gene of soybeans, and an amplified product may be obtained by redesigning the primer sequences using the reference common bean genome. From the remaining 9 genes, four genes presented the formation of a secondary structure, as demonstrated by the dissociation curve, indicating the need to design new sets of primers for these genes. This optimization step for qPCR conditions, implemented in the present study, was fundamental for the development of a robust assay, which ensures reproducibility between replicates. Of the five remaining primers suitable for evaluation as reference genes (*Pv18S, ß-tubulin, PvAct, Tc127, T197*) of leaf, root and seed tissues, four primers were derived from bean genomic sequences and one from cocoa.

The values of PCR efficiency ranged from 63 to 103% based on the *standard curve* (Table 3). Although the method based on the *standard curve* remains reliable and broadly used (Svec et al. 2015), it often results in unrealistic values greater than 100% for qPCR efficiency. As described by Peirson et al. (2003), several analytical procedures could result in a cumulative error that might lead to efficiency overestimation. To overcome these limitations, mathematical models have been published describing the kinetics of the qPCR reaction and trying to estimate qPCR efficiency from a single reaction (Robledo et al. 2014). The methodology implemented by the LinRegPCR software used in the present study resulted in values of qPCR efficiency ranging from 55 to 94%, which are reduced values compared to the standard curve. The qPCR efficiency was similar, and even higher, for the seed samples compared to the leaf/root samples, using both the standard curve method and LinRegPCR (Table 3). Based on the standard curve method, all genes, except *Pv18S*, showed adequate estimates of efficiencies ranging from 95 to 105%, while for LinRegPCR all genes, except *Pv18S* and *PvAct*, presented values of efficiency higher than 82% (Table 3). For three genes, *ß-tubulin, Tc127*, and *T197*, both methods provided satisfactory levels of efficiency (≥ 82.97%) with the highest value for *T197* (≥ 93.33%). Among all genes, *T197* (encodes a guanine nucleotide-binding protein beta subunit-like protein) was selected from the common bean cDNA libraries; however, in a previous study, it was not recommended for use as a normalizer gene due to the variability of its expression levels (Thibivilliers et al. 2009). As the primer amplifications in this study were carefully adjusted by titration procedure, the potential of each gene and its efficiency as a reference for normalization were evaluated under the adequate qPCR conditions. Failures to amplify a desired target sequence are often assigned to inadequate primer design, and targeting other regions of the gene sequence could be strategic to redesign a new set of primers and obtain a satisfactory amplification.

Expression stability of reference genes

It is well known that gene expression stability is one of the most important criteria for selection of a reference gene. The five selected genes in this study (*Pv18S, ß-tubulin, PvAct, Tc127* and *T197*) showed adequate amplification of cDNA samples from the seeds, leaves and roots of common bean plants. The absolute Cq values individually obtained for each one of the five reference genes are graphically represented on a box plot graphic in Figure 1 where the median raw Cqs are represented by lines. There is a premise that reference genes cannot be regulated by the experimental conditions of the sample set (Robledo et al. 2014). The *geNorm* and *BestKeeper* algorithms are based on the assumption that none of the analyzed genes are co-regulated (Matz et al. 2013). For this reason, the use of more than one algorithm for the validation of reference genes is suggested to give more reliable results (Zyzynska-Granica and Koziak 2012).

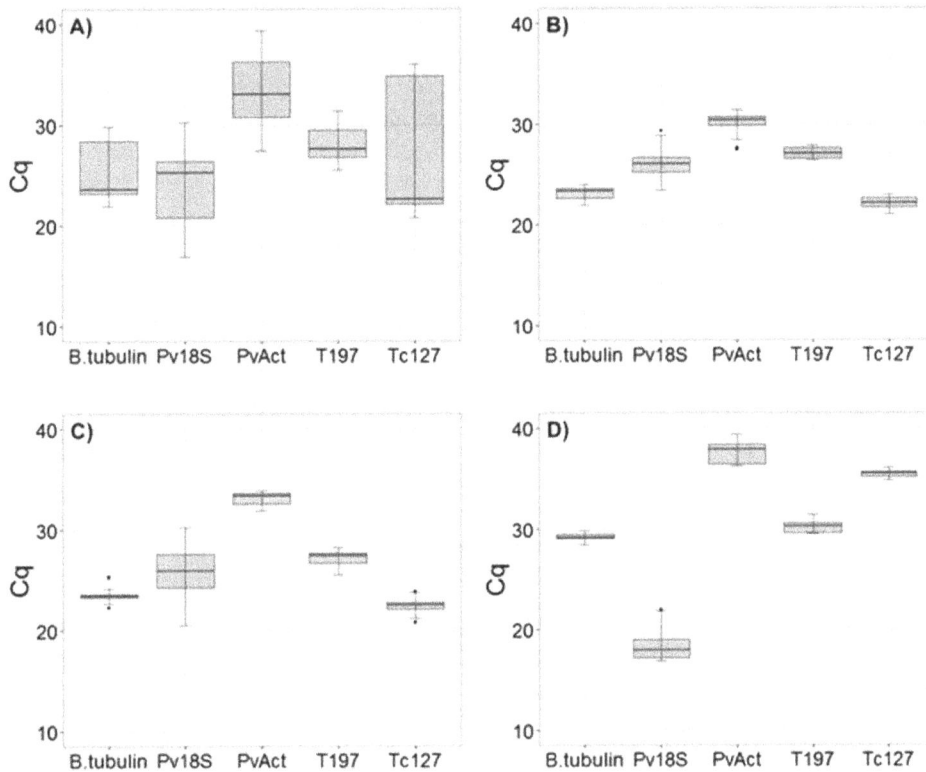

Figure 1. Box plot analysis of Cqs obtained for distinct sample sets of each gene. Each box indicates the middle 50% of the distribution, with a line at the median dividing the box into two parts. A) Cqs distribution values for all samples; B) Cqs distribution for abiotic stress; C) Cqs distribution for biotic stress; D) Cqs distribution for seed samples.

The gene-to-gene correlations were verified by the Pearson correlation coefficient (r) (Pfaffl et al. 2004), and a strong positive correlation for all reference genes ($r \geq 0.80$) was observed (Figure 2), even for *Pv18S*, which presented a negative and strong correlation with the other reference genes evaluated ($r \leq -0.66$). For *Pv18S*, the C_q value tended to be reduced in seeds (Figure 1d) (ranging from 16.92 to 30.23) compared to the other tissues, which is indicative of higher initial copy numbers of this target sequence. The *Pv18S* expression level (C_q value) compared with the remaining genes resulted in statistically significant negative correlations (Figure 2).

The ranking of reference gene stabilities evaluated for the whole set of samples and inter-group variation (abiotic stress, biotic stress and seeds) showed that the genes

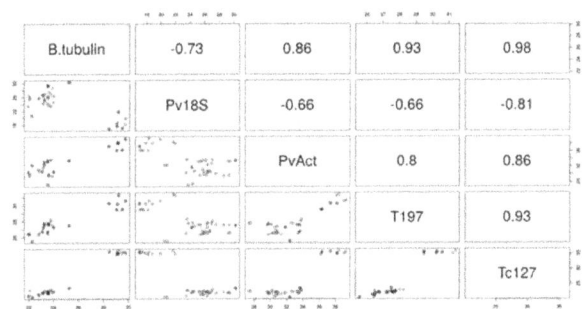

Figure 2. Correlation between Cq values for the reference genes. Values in the boxes: Correlation coefficient (r), p< 0.001.

did not present a high stability of expression across all sample groups (Table 4), as demonstrated by the high standard deviation and by the wide range of C_q values (Figure 1a). For all samples analyzed together, geNorm ranked *T197* and *β-tubulin* as the most stable genes (M = 1.4) with the combined value below the recommended cutoff of 1.5 (Robledo et al. 2014). NormFinder considered the same two genes most stable, with a combined stability value of 0.468. Based on RefFinder and Bestkeeper, the *T197* and *β-tubulin* were also considered the most stable, although the SD (1.26 and

2.37, respectively) and CV (4.52 and 9.50, respectively) determined by Bestkeeper were above the recommended cutoff of 1 (Table 4).

When considering the whole set of samples, the C_q values become more stable and lower (Figures 1b, 1c and 1d). For the samples of leaves and roots submitted to the same experimental conditions (abiotic stress), *Tc127*, *T197*, *β-tubulin* and *PvAct* were ranked as more stable with reduced SD (0.8) by Bestkeeper. The remaining algorithms gave the same result and indicated that the reference genes *T197* and *β-tubulin* were the most stable, with a geNorm value of 0.415. For the samples submitted to biotic stress, the genes *Tc127* and *β-tubulin* were ranked as the best combination by geNorm (M=0.386) and the remaining methods, except for BestKeeper, and the genes ranked with lower SD (below the cutoff of 1) were *β-tubulin* (0.52), *PvAct* (0.58), *Tc127* (0.59) and *T197* (0.64). For the seed samples, geNorm suggested *β-tubulin* and *Tc127* as the most stable (M= 0.605) combination, as well as for the other programs. All genes, except for *Pv18S*, presented SD values lower than one by BestKeeper. For qPCR data normalization, two genes were frequently observed as the optimal number in several experimental plots (Borges et al. 2012, Müller et al. 2014).

In the present study, out of the five genes, *β-tubulin* and *T197*, previously determined for common beans under biotic and abiotic stress, respectively, ranked as the most stable among the sample sets evaluated in different tissue samples (leaves, roots and seeds) and experimental conditions (biotic and abiotic stress). In a previous study, two *β-tubulin* genes (Tub8 and Tub9) were not selected as reference genes (Borges et al. 2012), suggesting that the *β-tubulin* isoforms may have different expression patterns, as seen for other enzymes (Sielski et al. 2014). In addition, the genes *PvAct* and *Pv18S* were the least stable, in accordance with previous studies (Fernandez et al. 2011).

The RNA extraction, optimization and validation of reference genes for expression analysis in different tissues and experimental conditions were aimed at enabling the optimal performance of qPCR experiments in common bean samples. A more adequate method for RNA seed extraction was implemented that produced RNA with high integrity for gene expression studies (RIN ≥ 6.9). Furthermore, a set of reference genes using SYBR Green chemistry were tested and made available for normalization in qPCR experiments for different tissue samples and experimental conditions. This study also provided the validation of the web-based tool RefFinder that integrates a diverse set of methods to compare and rank the best combination of candidate reference genes in common bean samples. The data from the present study strongly support the indication of reference genes for qPCR analysis.

Table 4. Stability rankings obtained with the six determination methods for the individual group of common bean samples

Group	Rank	Comparative Delta-C_T	BestKeeper (SD)	BestKeeper (CV%)	NormFinder	geNorm	RefFinder
all samples	1	*β-tub.*(3.147)	*T197* (1.26)	*T197* (4.52)	*T197* (0.649)	*T197/ β-tub.* (1.403)	*β-tub.*(1.189)
	2	*T197* (3.308)	*β-tub.*(2.37)	*PvAct* (7.67)	*β-tub.*(0.941)		*T197* (1.414)
	3	*PvAct* (3.462)	*PvAct* (2.55)	*β-tub.*(9.50)	*PvAct* (1.517)	*PvAct* (1.712)	*PvAct* (3.000)
	4	*Tc127* (5.215)	*Pv18S* (3.31)	*Pv18S* (13.83)	*Tc127* (4.175)	*Tc127* (2.754)	*Tc127* (4.229)
	5	*Pv18S* (6.87)	*Tc127* (5.20)	*Tc127* (20.11)	*Pv18S* (5.757)	*Pv18S* (4.401)	*Pv18S* (4.729)
abiotic stress (leaves)	1	*T197* (1.017)	*Tc127* (0.46)	*T197* (1.94)	*β-tub.*(0.461)	*T197/ β-tub.* (0.415)	*T197* (1.414)
	2	*β-tub.*(1.018)	*T197* (0.52)	*Tc127* (2.10)	*T197* (0.523)		*β-tub.*(1.565)
	3	*Tc127* (1.062)	*β-tub.*(0.63)	*PvAct* (2.64)	*Tc127* (0.581)	*Tc127* (0.502)	*Tc127* (2.280)
	4	*PvAct* (1.566)	*PvAct* (0.79)	*β-tub.*(2.71)	*PvAct* (1.323)	*PvAct* (1.012)	*PvAct* (4.000)
	5	*Pv18S* (1.626)	*Pv18s* (1.18)	*Pv18S* (4.54)	*Pv18S* (1.415)	*Pv18S* (1.258)	*Pv18S* (5.000)
biotic stress (leaves and roots)	1	*Tc127* (0.991)	*β-tub.*(0.52)	*PvAct* (1.76)	*β-tub.*(0.359)	*β-tub./Tc127* (0.386)	*Tc127* (1.316)
	2	*β-tub.*(1.044)	*PvAct* (0.58)	*β-tub.*(2.22)	*Tc127* (0.390)		*β-tub.*(1.414)
	3	*T197* (1.209)	*Tc127* (0.59)	*T197* (2.36)	*PvAct* (0.657)	*T197* (0.544)	*PvAct* (3.130)
	4	*PvAct* (1.215)	*T197* (0.64)	*Tc127* (2.63)	*T197* (0.816)	*PvAct* (0.644)	*T197* (3.464)
	5	*Pv18S* (2.526)	*Pv18S* (2.25)	*Pv18S* (8.67)	*Pv18S* (1.808)	*Pv18S* (1.397)	*Pv18S* (5.000)
seed	1	*β-tub.*(1.048)	*β-tub.*(0.31)	*Tc127* (0.86)	*β-tub.*(0.302)	*β-tub./Tc127* (0.605)	*β-tub.*(1.189)
	2	*Tc127* (1.071)	*Tc127* (0.31)	*β-tub.*(1.07)	*Tc127* (0.302)		*Tc127* (1.144)
	3	*T197* (1.092)	*T197* (0.50)	*T197* (1.64)	*T197* (0.487)	*T197* (0.684)	*T197* (3.000)
	4	*PvAct* (1.400)	*PvAct* (0.96)	*PvAct* (2.56)	*PvAct* (1.204)	*PvAct* (0.832)	*PvAct* (4.000)
	5	*Pv18S* (2.115)	*Pv18S* (1.36)	*Pv18S* (7.36)	*Pv18S* (2.057)	*Pv18S* (1.345)	*Pv18S* (5.000)

REFERENCES

Andersen CL, Jensen JL and Ørntoft TF (2004) Normalization of real-time quantitative reverse transcription-PCR data: A model-based variance estimation approach to identify genes suited for normalization, applied to bladder and colon cancer data sets. **Cancer Research 64**: 5245-5250.

Borges A, Tsai SM and Caldas DGG (2012) Validation of reference genes for RT-qPCR normalization in common bean during biotic and abiotic stresses. **Plant Cell Reports 31**: 827-838.

Brigide P, Canniatti-Brazaca SG and Silva MO (2014) Nutritional characteristics of biofortified common beans. **Food Science and Technology** 34: 493-500.

Chen JB, Wang SM, Jing RL and Mao XG (2009) Cloning the PvP5CS gene from common bean (*Phaseolus vulgaris*) andits expression patterns under abiotic stresses. **Journal of Plant Physiology 166**: 12-19.

Christou A, Georgiadou EC, Filippou P, Manganaris GA and Fotopoulos V (2014) Establishment of a rapid, inexpensive protocol for extraction of high quality RNA from small amounts of strawberry plant tissues and other recalcitrant fruit crops. **Gene 537**: 169-173.

Cicinnati VR, Shen Q, Sotiropoulos GC, Radtke A, Gerken G and Beckebaum S (2008) Validation of putative reference genes for gene expression studies in human hepatocellular carcinoma using real-time quantitative RT-PCR. **BMC Cancer 8**: 350.

Contour-Ansel D, Torres-Franklin ML, Zuily-Fodil Y and Carvalho MHC (2010) An aspartic acid protease from common bean is expressed 'on call' during water stress and early recovery. **Journal of Plant Physiology 167**: 1606-1612.

Dheda K, Huggett JF, Chang JS, Bustin SA, Johnson MA, Rook GAW and Zumla A (2005) The implications of using an inappropriate reference gene for rela-time reverse transcription PCR data normalization. **Analytical Biochemistry 344**: 141-143.

Eticha D, Zahn M, Bremer M, Yang Z, Rangel AF, Rao IM and Horst WJ (2010) Transcriptomic analysis reveals differential gene expression in response to aluminium in common bean (*Phaseolus vulgaris*) genotypes. **Annals of Botany 105**: 1119-1128.

Fernandez P, Di-Rienzo JA, Moschen S, Dosio GAA, Aguirreza-bal LAN, Hopp HE, Paniego N and Heinz RA (2011) Comparison of predictive methods and biological validation for qPCR reference genes in sunflower leaf senescence transcript analysis. **Plant Cell Reports 30**: 63-74.

Heid CA, Stevens J, Livak K and Williams PM (1996) Real time quantitative PCR. **Genome Research 6**: 986-994.

Hu Y, Chen H, Luo C, Dong L, Zhang S, He X and Huang G (2014) Selection of reference genes for real-time quantitative PCR studies of kumquat in various tissues and under abiotic stress. **Scientia Horticulturae 174**: 207-216.

Li Z and Trick HN (2005) Rapid method for high-quality RNA isolation from seed endosperm containing high levels of starch. **BioTechniques 38**: 872-876.

Libault M, Thibivilliers S, Bilgin DD, Radwan O, Benitez M, Clough SJ and Stacey G (2008) Identification of four soybean reference genes for gene expression normalization. **Plant Genome 1**: 44-54.

Ma Z, Huang B, Xu S, Chen Y, Li S and Lin S (2015) Isolation of high-quality total RNA from chinese fir (*Cunninghamia lanceolata* (Lamb.) Hook). **Plos One 10(6)**: e0130234.

Matz MV, Wright RM and Scott JG (2013) No control genes required: bayesian analysis of qRT-PCR data. **Plos One 8**: 1-12.

Mornkham T, Wangsomnuk PP, Fu Y-B, Wangsomnuk P, Jogloy S and Patanothai A (2013) Extractions of high quality RNA from the seeds of jerusalem artichoke and other plant species with high levels of starch and lipid. **Plants 2**: 302-316.

Müller BS de F, Sakamoto T, Silveira RDD, Zambussi-Carvalho PF, Pereira M, Pappas Jr GJ, Costa MMC, Guimarães CM, Pereira WJ, Brondani C and Vianello-Brondani RP (2014) Differentially expressed genes during flowering and grain filling in common bean (*Phaseolus vulgaris*) grown under drought stress conditions. **Plant Molecular Biology Reporter 32**: 438-451.

Peirson SN, Butler JN and Foster RG (2003) Experimental validation of novel and conventional approaches to quantitative real-time PCR data analysis. **Nucleic Acids Research 31**: e73.

Pfaffl MW, Ales T, Prgomet C and Neuvians TP (2004) Determination of stable housekeeping genes, differentially regulated target genes and sample integrity: BestKeeper – Excel-based tool using pair-wise correlations. **Biotechnology Letters 26**: 509-515.

Ramírez M, Flores-Pacheco G, Reyes JL, Álvarez AL, Drevon JJ, Girard L and Hernández G (2013) Two common bean genotypes with contrasting response to phosphorus deficiency show variations in the microRNA 399-Mediated PvPHO2 regulation within the PvPHR1 signaling pathway. **International Journal of Molecular Sciences 14**: 8328-8344.

Robledo D, Hernádez-Urcera J, Cal MR, Pardo BG, Sánchez L and Martinez P (2014) Analysis of qPCR reference gene stability determination methods and a practical approach for efficiency calculation on a turbot (*Scophthalmus maximus*) gonad dataset. **BMC Genomics 15**: 648.

Ruijter JM, Ramakers C, Hoogaars WMH, Karlen Y, Bakker O, Hoff MJB van den and Moorman AFM (2009) Amplification efficiency: linking baseline and bias in the analysis of quantitative PCR data. **Nucleic Acids Research 37**: e45.

Santos MO, Romano E, Yotoko KSC, Tinoco MLP, Diasa BBA and Aragão FJL (2005) Characterization of the cacao somatic embryogenesis receptor-like kinase (SERK) gene expressed during somatic embryogenesis. **Plant Science 168**: 723-729.

Silva TTA, Abreu LAS, Nascimento VE, Pinho ERVV, Rosa SDVF and Padilha L (2011) Protocolo para extração de RNA em semente de café. In **Proceedings of VII simpósio de pesquisa dos cafés do Brasil**. Available at <http://www.sapc.embrapa.br/arquivos/consorcio/spcb_anais/simposio7/63.pdf> Accessed in Jan 2017.

Sielski NL, Ihnatovych I, Hagen JJ and Hofmann WA (2014) Tissue specific expression of myosin ICisoforms. **BMC Cell Biology 15**: 8.

Silver N, Best S, Jiang J and Thein SL (2006) Selection of housekeeping genes for gene expression studies in human reticulocytes using real-time PCR. **BMC Molecular Biology 7**: 33.

Svec D, Tichopad A, Novosadova V, Pfaffl MW and Kubista M (2015) How good is a PCR efficiency estimate: Recommendations for precise and robust qPCR efficiency assessments. **Biomolecular Detection and Quantification 3**: 9-16.

Thibivilliers S, Joshi T, Campbell KB, Scheffler B, Xu D, Cooper B, Nguyen HT and Stacey G (2009) Generation of *Phaseolus vulgaris* ESTs and investigation of their regulation upon *Uromyces appendiculatus* infection. **BMC Plant Biology 9**: 1-13.

Vandesompele J, Preter KD, Pattyn F, Poppe B, Roy NV, Paepe AD and Speleman F (2002) Accurate normalization of real-time quantitative RT-PCR data by geometric averaging of multiple internal control genes. **Genome Biology 3**: 1-12.

Xie F, Xiao P, Chen D, Xu L, Zhang B (2012) miRDeepFinder: a miRNA analysis tool for deep sequencing of plant small RNAs. **Plant Molecular Biology 80:** 75

Yee JY, Limenta LMG, Rogers K, Rogers SM, Tay VS and Lee EJ (2014) Ensuring good quality RNA for quantitative real-time PCR isolated from renal proximal tubular cells using laser capture microdissection. **BMC Research Notes 7**: 62.

Yin G, Xu H, Liu J, Gao C, Sun J, Yan Y and Hu Y (2014) Screening and identification of soybean seed-specific genes by using integrated bioinformatics of digital differential display, microarray, and RNA-seq data. **Gene 546**: 177-186.

Zyzynska-Granica B and Koziak K (2012) Identification of suitable reference genes for real-time PCR analysis of statin-treated human umbilical vein endothelial cells. **Plos One 7**: e51547.

27

Evaluation of different selection indices combining Pilodyn penetration and growth performance in *Eucalyptus* clones

Andrei Caíque Pires Nunes[1]*, **Marcos Deon Vilela de Resende[2]**, **Glêison Augusto dos Santos[1]**, **Rodrigo Silva Alves[3]**

Abstract: *The present study aimed to evaluate the selection indices efficiency for Pilodyn penetration combined with growth traits in Eucalyptus clones. It was carried out experiments in a randomized block design, with single tree plots and 30 replications. Diameter at breast height (DBH), total height (TH), and Pilodyn penetration as an indicator of basic density (BD) were measured. The volume was estimated. Based on predicted genotypic values, three indices presented the highest accuracies: I_8 (based on partial correlation), I_7 (based on the concept of multivariate BLUP) and I_3 (based on two variables as ratio, which uses a third heritability estimate associated to the ratio DBH/BD, besides the two heritabilities of DBH and BD). Thus, it is possible to optimize the selection by combining properly the variables using their genetic control, precision and the relationships between them. The best options came from using only two no redundant traits DBH and BD.*

Key words: *Selection criteria, accuracy, wood quality, partial correlation, path analysis.*

*Corresponding author:
E-mail: andreicaiquep@gmail.com

[1] Universidade Federal de Viçosa (UFV), Departamento de Engenharia Florestal, Avenida Peter Henry Rolfs, s/n, Campus Universitário, 36.570-900, Viçosa, MG, Brazil
[2] Empresa Brasileira de Pesquisa Agropecuária, Centro Nacional de Pesquisa de Florestas, Estrada da Ribeira, km 111, Bairro Guaraituba, 83.411-000, Colombo, PR, Brazil
[3] UFV, Departamento de Biologia Geral,

INTRODUCTION

Part of the increase in forest production is attributed to breeding programs (Costa et al. 2015). In this scenario, evaluation, characterization and selection of superior genotypes are critical steps in a forest breeding program that aims at maximizing genetic gain (Resende 2002, Bhering et al. 2015). Therefore, it is essential that the breeder carefully sets the goal of the selection, as well as the criteria that will be used to properly manage the breeding population and to generate significant gains in relation to the final commercial product.

Wood quality and volumetric analyses are essential in *Eucalyptus* improvement directed for pulp production (Gomide et al. 2010, Protásio et al. 2014). Although tree volume measurement is commonplace, wood quality studies are costly and time-consuming (Raymond and Apiolaza 2004). Basic density has been considered to be a universal index for assessing the quality of the wood, providing indirect information about other technological traits (Gomide et al. 2010). In spite of basic density importance, its determination is difficult due to high cost and need for tree felling.

Considering these fact, the Pilodyn method has been successfully applied for indirect estimation of wood basic density without to fall the tree (Gouvêa et al. 2011, Couto et al. 2013, Neves et al. 2013). With indirect estimation of basic density and information of tree growth (volume), the genetic selection of superior materials is technically correct.

Thus, in order to adopt more complex criteria to select superior genotypes considering several traits simultaneously, selection index theory can be employed. This theory was first described by Smith (1936), and later by Hazel (1943). According to Freitas et al. (2012), indices allow using a single value to select genotypes, since analysis is carried out by linear combinations of phenotypic data of many characters in study, whose weighting coefficients are estimated in order to maximize the correlation between the index and the true breeding values.

In addition, selection index is more efficient than direct selection, since it enables the distribution of gains between the several traits, in a more homogeneous way and in accordance with the purposes of the breeding program, generating higher total gain (Reis et al. 2011, Freitas et al. 2012, Cruz et al. 2014). Thus, these actions may greatly contribute to maximize the cost/benefit of the breeding program.

Several selection indices have been used for multivariate evaluation of characters of interest in different cultures, particularly *Eucalyptus* (Martins et al. 2006, Reis et al. 2011, 2015). Despite their importance, these indices are based on phenotypic values and do not consider the genotypic correlation and the cause and effect relationships between variables. According to Resende et al. (2014), the use of multivariate mixed models with multiple traits and unstructured covariance matrix is theoretically the most efficient method, and allows considering heterogeneity of variances and covariances. However, in practice, this approach is not used, due to the difficulty in convergence of the iterative analysis and the super parameterization (Resende et al. 2014).

Thus, structural equation modeling is an alternative that allows efficiently representing the standard multi-trait model (Resende et al. 2014). With a functional network of studied traits, it is possible to establish cause and effect relationships between the variables of interest and to compose optimum selection indices using the genotypic values predicted by univariate analysis by the REML/BLUP procedure (Maximum Restricted Likelihood/Best Linear Unbiased Prediction) (Resende et al. 2014, Viana and Resende 2014).

Therefore, the present study investigated the effectiveness of new classes of selection criteria based on partial correlations, direct effect of path analysis, ordinary correlations and heritability for *Eucalyptus*, using genotypic values predicted through the mixed models methodology.

MATERIAL AND METHODS

Experimental network

Experiments were carried out in the areas of CMPC Celulose Riograndense Company, in the municipalities of Minas do Leão (lat 30° 11' S, long 52° 00' W, alt 141 m asl, average temperature 17.5 ºC and annual precipitation 1,422 mm), Encruzilhadas do Sul (lat 30° 27' S, long 52° 39' W, alt 250 m asl, average temperature 17 ºC and annual precipitation 1,368 mm), Dom Feliciano (lat 30° 29' S, long 52° 19' W, alt 378 m asl, average temperature 16 ºC and annual precipitation 1,564 mm) and Vila Nova do Sul (lat 30° 14' S, long 53° 49' W, alt 301 m asl, average temperature 16.8 ºC and annual precipitation 1,133 mm), which are located in the state of Rio Grande do Sul (climate Cfa, according to the climatic classification of Koppen) Brazil. A network of clonal trials with 864 *Eucalyptus* clones was set in 2007. Trees were planted at a spacing of 3.5 x 2.6 m. At each site, it was established an experiment in a randomized block design, with single tree plots and 30 replications.

Data collection

For the indirect estimate of basic density, Pilodyn's method (Greaves et al. 1996) was used when trees were three years old. The measurement with Pilodyn was carried out twice, on each north and south cardinal aspects of the tree. For analysis, the considered number was the mean of the two measurements. It was considered the inverse of Pilodyn penetration depth (mm) as the estimated basic density (BD).

Growth data of the trees were collected at three years of age, as well as the estimated basic density. It was measured the diameter at breast height (DBH), in centimeters (cm), and the total height (TH) of trees, in meters (m). DBH was measured with the aid of a diameter tape, and TH was obtained using a relascope.

To calculate the volume (m^3) without bark (Vol), the model of Leite et al. (1995) was used, as shown below:

$$Vol = 0.000048 \times DBH^{1.720493} \times TH^{1.180736} \times e^{(-3.00555) \times (tx/dbh) \times [1-(d/dbh)^{1+0.228531 \times d}]} + E$$

In which DBH: diameter at 1.3 meters height; TH: total height; tx = 0, for the volume with bark and 1 for volume without bark; d: superior commercial diameter; E = experimental error.

Statistical analysis

The statistical model for analysis of this experimental network in several environments (Resende 2002), with single tree plots is given by $y = Xr + Zg + Hb + Wge + e$, where: y, r, g, h, ge and e are vectors of data, replication effects (fixed), genotypic effects (random), block effects (random), effects of genotype x environment interaction (G x E) (random), and random errors, respectively. In addition, X, Z, H and W are the incidence matrices for r, g, h and ge, respectively. Predicted genotypic values free of interaction, considering all the environments were given by $u + g$, in which u is the mean of all sites. These values predicted for each variable, using univariate analysis, will be used in the selection indices. In addition, it was also obtained the genetic correlations between the analyzed variables. All analyses were carried out using the Selegen-REML/BLUP software (Resende 2016).

Selection Indices

It follows a description of the index used in this paper (Table 1). Details of them and their accuracies are presented by Viana and Resende (2014) and Resende et al. (2014).

Table 1. Description of the indexes and their accuracies used

Concept	Index[1]	Description	Accuracy[1]
Phenotypic index	I_1	$I_1 = \left(\dfrac{DBH}{S_{DBH}}\right)\left(\dfrac{DB}{S_{DB}}\right)$	$r_{g\hat{g}} = \sqrt{\dfrac{n \times h^2}{1 + (n-1) \times h^2}}$
	I_2	$I_2 = \left(\dfrac{Vol}{S_{Vol}}\right)\left(\dfrac{DB}{S_{DB}}\right)$	
Optimum index using a ratio between two variables	I_3	$I_3 = LogDBH - Log(1/BD)$	$r_{g\hat{g}} = \sqrt{\dfrac{n \times h_{y*}^2}{1 + (n-1) \times h_{y*}^2}}$
	I_4	$I_4 = LogVol - Log(1/BD)$	
Genotypic index	I_5	$I_5 = \left(\dfrac{VG_{DBH}}{S_{VG_{DBH}}}\right)\left(\dfrac{VG_{BD}}{S_{VG_{BD}}}\right)$	$Ac_{I5} = \dfrac{sd(I5) \times Ac(I3)}{sd(I3)}$
	I_6	$I_6 = \left(\dfrac{VG_{Vol}}{S_{VG_{Vol}}}\right)\left(\dfrac{VG_{BD}}{S_{VG_{BD}}}\right)$	$Ac_{I6} = \dfrac{sd(I6) \times Ac(I4)}{sd(I4)}$
Multivariate BLUP index	I_7	$I_7 = b_1 g_0 + b_2 g_{a1} + b_3 g_{a2}$	$r_{g\hat{g}} = \sqrt{Var(Index)/\sigma_g^2}$
Partial correlation index	I_8	$I_8 = b_1 g_0 + b_2 g_{a1} + b_3 g_{a2}$	
Ordinary correlation index	I_9	$I_9 = b_1 g_0 + b_2 g_{a1} + b_3 g_{a2}$	$r_{g\hat{g}} = \dfrac{sd(score\ I_x)}{sd(score\ I_{greater})}$
Path analysis index	I_{10}	$I_{10} = b_1 g_0 + b_2 g_{a1} + b_3 g_{a2}$	

[1] Description of the indexes components and accuracies: BD: basic density indirectly estimated by the Pilodyn's method; DBH: diameter at breast height; S_{DBH}: standard deviation of diameter at breast height; S_{Bd}: standard deviation of basic density; Vol: volume of wood without bark; S_{vol}: standard deviation of the volume without bark; Log(1/BD): inverse of the basic density indirectly estimated by the Pilodyn's method on the logarithmic scale; LogDBH: diameter at breast height on the logarithmic scale; LogVol: volume of wood without bark on the logarithmic scale; VG_{BD}: genotypic value of the indirectly estimated basic density; VG_{DBH}: genotypic value of the diameter at breast height; $S_{VG_{DBH}}$: standard deviation of the genotypic value of the diameter at breast height; $S_{VG_{BD}}$: standard deviation of the genotypic value of the basic density; VG_{vol}: genotypic value of volume without bark; $S_{VG_{Vol}}$: standard deviation of the genotypic value of Vol; g_o: standard genotypic value of the objective character (DBHxBD); g_{ai}: is the standard genotypic value of the auxiliary characters (DBH and BD). The weighting coefficients (b_i) of the index are given by (Viana and Resende 2014); $r_{g\hat{g}}$: accuracy of the index; n: number of individuals per clone; h_{y*}^2: heritability of the ratio between two variables given by Resende et al. (2014); h^2: heritability of the phenotypic index; Ac_{I5}: accuracy of the I_5 genotypic index; sd(I5): standard deviation of the I_5 genotypic index score; Ac(I3): accuracy of the I_3 optimum index; sd(I3): standard deviation of the I_3 optimum index score; Ac_{I6}: accuracy of the I_6 genotypic index; sd(I6): standard deviation of the I_6 genotypic index score; Ac(I4): accuracy of the I_4 optimum index; sd(I4): standard deviation of the I_4 optimum index score; $Var(Index)/\sigma_g^2$: ratio between the index and additive variances of the objective trait; sd(score I_x): standard deviation of the I_8, I_9 or I_{10} indexes score; sd(score $I_{greater}$): standard deviation of the score of the index with greater variance among I_7 and I_{10}.

Internal consistency of the indices

As mentioned by Resende et al. (2014), a comparison between alternative selection indices may be carried out by varying the degree of covariance of the variables between each other. Thus, the Cronbach's alpha coefficient (1951) (modified by Resende et al. (2014)) works as an indicator of internal consistency of an index involving n variables. Its formula is given by:

$$\alpha = \frac{n-1}{n}\left(1 - \frac{\Sigma\, v_i^2}{v_t^2}\right)$$

In which $\Sigma\, v_i^2$ = sum of the variances of the n variables; v_t^2 = total variance of the scores of the selection index; n = number of variables.

RESULTS AND DISCUSSION

Genetic parameters and genetic correlations

Genetic parameters of the analyzed traits were estimated (Table 2). Since the present study only aims to evaluate the different selection criteria, genetic parameters related to G × E interactions are not reported here, but the full publication on that can be found in Nunes et al. (2016).

Estimates of individual heritability of the studied characters may be considered low ($h_g^2 = 0.07$ for TH), moderate (from 0.15 for Vol, to 0.23 for I_1), and high (0.59 for BD), Table 2, according to the classification reported by Resende (2002). Elevated heritability value for basic density (0.64) was found by Wei and Borralho (1997) in *Eucalyptus urophylla* S.T. Blake. Muneri and Raymond (2000) and Kube et al. (2001) also reported high values of heritability for Pilodyn penetration and basic density, ranging from 0.60 to 0.70. In spite of high genetic control of basic density, the heritability for growth traits have been reported in literature ranging from 0.10 to 0.22 (Kube et al. 2001), which corroborates with the present work.

Notwithstanding the estimate of broad sense individual heritability, it is observed that the value of this parameter for I_2 (0.16) was similar to the heritability of Vol (0.15), Table 2. Thus, there is the need to develop an index which enables the estimate of a balanced heritability, i.e., that not only resembles to only one of the traits of the index. The same reasoning can be applied to I_1.

Table 2. Estimates of genetic parameters (individual REML) and genotypic correlations (below the genetic parameters) for basic density (BD in kg m⁻³), diameter at breast height (DBH in cm), total height (Th in m), volume (Vol in m³ ha⁻¹ year), phenotypic index DBH×BD (I_1), and phenotypic index Vol×BD (I_2) for *Eucalyptus* clones evaluated in the joint analysis between environments, at three years of age

Parameters[1]	BD	DBH	Th	Vol	I_1	I_2
h_g^2	0.59	0.18	0.07	0.15	0.23	0.16
h_{mg}^2	0.95	0.70	0.54	0.65	0.78	0.67
Acgen	0.97	0.84	0.73	0.80	0.88	0.82
c_{bloc}^2	0.24	0.03	0.08	0.05	0.15	0.07
Overall mean	382.64	13.26	14.72	0.08	6.03	44.92
CVgi (%)	12.39	8.16	5.09	17.89	12.04	19.20
CVe (%)	9.70	15.71	16.50	37.80	20.20	40.08
CVr	1.27	0.51	0.30	0.47	0.59	0.47
	BD	DBH	Th	Vol	I_1	I_2
BD	-	-0.27	-0.01	-0.22	0.67	0.26
DBH		-	0.66	0.97	0.49	0.81
Th			-	0.77	0.48	0.71
Vol				-	0.52	0.86
I_1					-	0.86
I_2						-

[1]Description of genetic parameters: h_g^2: coefficient of individual heritability in the broad-sense (corrected to variance of block), free from interaction; h_{mg}^2: heritability of clone mean; *Acgen*: genetic accuracy in clone selection; c_{bloc}^2: coefficient of determination of block; Overall mean: overall mean of characters between different environments; *CVgi* (%): coefficient of genotypic variation; *CVe* (%): coefficient of experimental variation; *CVr* coefficient of relative variation.

Prediction accuracy of genetic values of the clones was high (Table 2). According to Resende and Duarte (2007), accuracies above 0.70 are sufficient for evaluations in a breeding population, and when the goal is the evaluation of the Value of Cultivation and Use, accuracies must be greater than 0.90. These high accuracy levels justify the great experimental quality, the caution and the technical precision in the establishment and evaluation of experiments. Moreover, the high number of replications (30) enabled obtaining reliable results of clones ranking by their predicted genetic values.

The value of coefficient of environmental variation (CVe) for I_2 (40.08) was twice as higher as the CVe of I_1 (20.20) (Table 2). The prediction accuracy of the breeding values of I_2 (0.82) was relatively lower than the prediction accuracy of I_1 (0.88). These results show that I_1 is more accurate than I_2. This fact corroborates with the highest value of CVe of Vol in relation to the CVe of DBH (Table 2). Considering that the growth traits Vol and DBH make up the indices I_2 and I_1, respectively, by multiplying by BD, the difference in accuracy and in CVe between these two indices is related to greater uncertainty in the estimate of Vol. Thus, since the estimate equation of Vol is composed of DBH and TH, the inclusion of the latter in the estimate of Vol led to higher value of CVe and lower accuracy of I_2, in relation to I_1.

Genotypic correlations between characters was estimated (Table 2) and it was found high correlation value between DBH and Vol (0.97). Nunes et al (2016) reported that is advantageous to perform the indirect selection of Eucalyptus clones aiming at gains in Vol through DBH. Negative values of genetic correlation between BD and DBH, and BD and Vol evidence the need for the study of selection indices involving wood quality and growth characters, simultaneously. Negative values (Kube et al. 2001, Bison et al. 2006) and positive values (Paula et al. 2002, Reis et al. 2011) of genetic correlation between basic density in Eucalyptus and growth characters were reported by different authors. Thus, discrepancies in genetic correlations in each cited work are caused by the genetic variation that exists in the evaluated population and by the different genes that are segregating in relation to the control of growth and wood quality characters (Reis et al. 2011).

Heritability and correlations of optimal indices based on a ratio between two variables

The genetic analyses of a trait using variables as a ratio seems to be unused in forest tree breeding so far. This paper is the first one to evaluate its effectiveness. The results have shown that it is a very promising technique. It was ranked among the three best out of the ten selection indices evaluated.

When comparing the heritability of the phenotypic indices (Table 2), I_1 (0.23) and I_2 (0.16), with the heritability of I_3 and I_4 (Table 3), an increase is observed in the genetic control, and therefore, greater efficiency of these last indices is also observed. For I_3, there was a 37% increase in relation to the phenotypic heritability index, while for I_4, this increase was 18%. These results show that I_3 and I_4 were more efficient than the phenotypic indices in weighing the genotypic values for each variable under study, and thus they were more efficient in weighting the effects of each variable on the index as a whole.

It is observed that the calculated heritability (optimum) of I_4 was lower than the heritability of I_3 (Table 3). This fact can be explained since the elasticity coefficient (K^2) of I_4 (0.53) was lower than the K^2 of I_3 (1.05) (Table 3). Thus, the I_4 had its calculated heritability penalized by the higher variance of volume (Vol), when compared with the variance of the diameter at breast height (DBH). Therefore, I_3 index is more accurate and ideal for selection of superior genotypes in relation to I_4, due to greater accuracy in the measurement of DBH, when compared with the estimate of Vol.

Correlations between indices based on a ratio between two traits and their constituent variables were calculated (Table 3). Genotype correlations of the constituent variables

Table 3. Heritabilities and correlations of a ratio between two variables

Coefficients[1]	I_3	I_4
h^2_{y*}	0.60	0.34
r^g_{y*w*}	0.58	0.65
r^g_{y*x*}	-0.63	-0.61
r^p_{y*w*}	0.30	0.70
r^p_{y*x*}	-0.37	-0.18
k^2	1.05	0.53
k^2_h	1.02	1.77
k	1.03	0.72
k_h	1.01	1.33

[1] Description of the coefficients: h^2_{y*}: heritability of the ratio between two variables (optimum index); r^g_{y*w*}: genotypic correlation of the index with the variable W*, which for I_3 is DBH, and for I_4 is Vol; r^g_{y*x*}: genotypic correlation of the index with variable X*, which for both indices is 1/BD; r^p_{y*w*}: phenotypic correlation of the index with variable W*, which for I_3 is DBH, and for I_4 is Vol; r^p_{y*x*}: phenotypic correlation of the index with the variable X*, which for both indices is 1/BD; k^2: elasticity or relationship between variances, being the phenotypic variance of X (always BD) in the numerator, and the phenotypic variance of W (DBH or Vol) in the denominator; k^2_h: ratio between heritability of variable X (BD) estimated in the original scale and the calculated heritability of the ratio between two variables; k: square root of k^2; k_h: square root of k^2_h. As mentioned in the material and methods, coefficient details can be found in Resende et al. (2014).

of each phenotypic index with these same indices (Table 2) were different from these obtained for I_3 and I_4. The negative correlation between BD and I_3/I_4 is justified, since for calculating the ratio between two variables, it was necessary to carry out analyses considering 1/BD. Therefore, for comparison, it should be noted the magnitude of the correlation, not the negative signal. Thus, it is evident that indices based on the ratio between two variables do not present high correlation with only one of the component character of this index, as found for I_1 and I_2. I_3 and I_4 indices provide a better balance between the two variables that compose it. I_3 presented genotypic correlation with DBH in the order of 0.58 and with 1/BD of -0.63 (Table 3), while I_1 genetic correlation with DBH and BD is 0.49 and 0.67, respectively (Table 2).

Efficiency of selection indices

According to the accuracy values, it can be concluded that the most effective indices are I_8, I_7, I_3, I_5 and I_{10}, while I_6 presents the lowest accuracy value, along with I_9 (Table 4). In general, prediction accuracies were high. This result was obtained due to the high number of replications and consistency in setting up and running the experiment. According to Resende and Duarte (2007), accuracies above 0.90 are considered too high and ensure reliable selection of superior genetic materials, as obtained for I_8, I_7, I_3, I_5, I_{10} and I_4.

This result corroborates the fact that partial correlations and path analyses (I_8 and I_{10}) are more efficient procedures than the ordinary correlations of Pearson (I_9), since they are conditional correlations, unlike the latter (Cruz et al. 2014, Resende et al. 2014). Thus, in the composition of I_8 and I_{10} indices, genotypic values are optimally weighted, and the considered correlations are odd (Resende et al. 2014), and there is no overestimation or underestimation of the index score.

As a report by Resende et al. (2014), for the analysis of a multivariate vector of observations of several traits, the multivariate mixed model is theoretically the most efficient, since it allows considering the complete heterogeneity of variance and covariance. Also, according to these authors, in practice, the use of the multivariate mixed model does not apply, due to the problematic convergence of the iterative analysis and super parametrization. Thus, it is important to use optimum selection indices that incorporate the concept of multivariate BLUP, by global maximization; the use of genotypic correlations and heritabilities; as well as the indices based on the concept of structural equation (Resende et al. 2014). This new approach becomes crucial for the optimization of the selection process, since it generates the same results of the multivariate mixed model, with less effort and high accuracy. In this context, the indices developed in this work can be used for any species and in any situation, in order to optimize the process of selection of superior genetic materials.

According to Resende et al. (2014), the use of structural equations (path analysis) is similar to the use of partial correlation matrices, instead of total correlations. This reduces the complexity of the multivariate mixed model, since it works with clean correlations between each pair of variables, making full rank the covariance matrices (Resende et al. 2014). In the present study, the index of greater efficiency was I_8, which is based on partial correlations between DBH × BD with DBH and BD. However, I_{10}, based on the direct effect of path analysis also showed high accuracy. Thus, it is verified the equivalence of the use of structural equation models and partial correlations, since the path analysis depends on the partial correlations, as reported by Resende et al. (2014).

Among the three best indices, I_3 was slightly higher, since it has greater internal consistency measured by the alpha coefficient. Internal consistency of an index may be studied by the degree of covariance of the variables between each other. Cronbach's alpha coefficient (1951) can be used as an indication of consistency of an index involving these variables (Resende et al. 2014). Thus, the higher the value of the coefficient, the more reliable is the index. I_3, I_5 and I_1 presented the highest internal consistencies (0.50). According to Resende et al. (2014), the lower the specific variance of each variable and the higher the total variance that they produce together, the higher is the alpha coefficient. Thus, when the sum of the variances of the individual variables is reduced, it increases the variance they have in common, that is, the one that ensures the coherence or internal consistency of the index (Resende et al. 2014). Therefore, it is verified that the constituent variables of these indices combine well, i.e., they covary in the index to which they belong (Resende et al. 2014).

The I_5 index presented high accuracy, and was higher than I_1 (Table 4). The index with the use of genotypic values proves to be more efficient, when compared to the phenotypic index. Resende (2002) reports that the use of genotypic

Evaluation of different selection indices combining Pilodyn penetration and growth performance...

179

values will be advantageous when genetic correlations of the variables are close to zero, and when prediction accuracies of genetic values for each character, individually, are high. Such conditions for success in the use of genotypic index are found in the present study, in which the genetic correlation between DBH and BD is null (-0.27) (Table 2), and the prediction accuracies for these two characters are high (0.97 for BD, and 0.84 for DBH) (Table 2). If these conditions are not met, I_8, I_7 or indices based on a ratio between two variables should be used, since they take into account the accuracy of each trait, their correlation and the relationship between variances of the constituent variables of the index.

Genotypic index in function of DBH×BD (I_5), which considers only the heritability of the characters, was as efficient as I_3, which optimally weighs the genotypic values by heritability and correlations between variables. However, since I_3 is optimum by the above mentioned reasons, it was more efficient (Table 4), and should be used especially in experiments in which the constituent variables of the index have non-null genotype correlations. Contrary to what happened to I_5 (0.95 accuracy), I_6 (0.78 accuracy) is the least efficient index. This result can be explained given that the conditions for use of the genotypic indices cited by Resende et al. (2002) are not met in the case of I_6, due to the inherent imprecision of Vol.

Table 4. Accuracies and Cronbach's alpha coefficients (1951) for I_1, I_2, I_3, I_4, I_5, I_6, I_7, I_8, I_9, I_{10} indices, which are based on the concepts of phenotypic index, ratio between two variables (Resende et al. 2014), genotypic index, multivariate blup (Viana and Resende 2014), partial correlation (Viana and Resende 2014), ordinary correlation and direct effects of path analysis (Resende et al. 2014), respectively

Concept	Index	Acurracy	Alfa[4]
Partial Correlelation	I_8	1.00	0.18
Mult Blup[2]	I_7	0.97	0.20
RBTV[1]	I_3	0.96	0.50
Genotypic	I_5	0.95	0.50
Path[3]	I_{10}	0.92	0.20
RBTV[1]	I_4	0.91	-[5]
Phenotypic	I_1	0.88	0.50
Phenotypic	I_2	0.82	-[5]
Ordinary correlation	I_9	0.79	0.21
Genotypic	I_6	0.78	0.49

[1] RBTV: Ratio between two variables; [2] Multi Blup: multivariate Blup; [3] Path: direct effect of path analysis; [4] Cronbach's alpha coefficient (1951) modified by Resende et al. (2014); [5] Negative value.

For the first time in the literature, it was compared so many selection indices, using somehow different concepts such as ratio between two variables, multivariate BLUP, partial correlations and direct effects of path analysis as part of the calculation of the weights. Path analysis is of great importance for the identification of direct and indirect effects of given characters in an objective variable (Cruz et al. 2014). Coefficients of path analysis were estimated based on ordinary genetic correlations of the characters, based on predicted genotypic values calculated by the REML/BLUP approach. Silva et al. (2009) report that path analysis becomes more effective when it is based on predicted genotypic values than when it is applied on phenotypic values. Thus, efficiency of the breeding program increases. Brasileiro et al. (2013), studied the consistency of path analyses using phenotypic and genotypic correlation, and concluded that in unbalanced cases, the use of genetic correlation produces more consistent results. Thus, it is noteworthy the precision of the analysis carried out in this study by the use of genotypic correlations obtained via REML/BLUP.

The five most efficient indices were those based on partial correlation, on the concept of multivariate BLUP, on two variables as a ratio and direct effects from path analysis. Basically, this can be attributed to the use of the following basic quantities: the genetic control of the trait, reliability and precision of the predicted genotypic values, and partial correlations between traits and the breeding objective. For the two variable as a ratio an additional feature is taken into account, the heritability of a third variable which is the own ratio. Thus, the use of indices based on these cited concepts are efficient and effective alternatives in selecting superior *Eucalyptus* genotypes based on several characters, without the complex procedures of multivariate mixed models.

ACKNOWLEDGMENTS

The authors thank the Universidade Federal de Viçosa, CNPq, Capes and the company CMPC Celulose Riograndense.

REFERENCES

Bhering LL, Cruz CD, Peixoto LAP, Rosado AM, Laviola BG and Nascimento M (2015) Application of neural networks to predict volume in eucalyptus. **Crop Breeding and Applied Biotechnology 15**: 125-131.

Bison O, Ramalho MAP, Rezende DSP, Aguiar M and Resende MDV (2006) Comparison Between Open Pollinated Progenies and Hybrids Performance in *Eucalyptus grandis* and *Eucalyptus urophylla*. **Silvae Genetica 55**: 4-5.

Brasileiro BP, Peternelli LA and Barbosa MHP (2013) Consistency of the results of path analysis among sugarcane experiments. **Crop Breeding and Applied Biotechnology 13**: 113-119.

Costa RB, Martinez DT, Silva JC and Almeida BC (2015) Variabilidade e ganhos genéticos com diferentes métodos de seleção em progênies de *Eucalyptus camaldulensis*. **Revista de Ciências Agrárias 58**: 69-74.

Couto AM, Trugilho PF, Neves TA, Protásio TP and Sá VA (2013) Modeling of basic density of wood from *Eucalyptus grandis* and *Eucalyptus urophylla* using nondestructive methods. **Cerne 19**: 27-34.

Cruz CD, Carneiro PCS and Regazzi AJ (2014) **Modelos biométricos aplicados ao melhoramento genético**. Vol 2, Editora UFV, Viçosa, 668p.

Freitas JPX, Oliveira EJ, Jesus ON, Neto AJC and Santos LR (2012) Formação de população base para seleção recorrente em maracujazeiro amarelo com uso de índices de seleção. **Pesquisa Agropecuária Brasileira 47**: 393-401.

Gomide JL, Neto HF and Regazzi AJ (2010) Análise de critérios de qualidade da madeira de eucalipto para produção de celulose kraft. **Revista Árvore 34**: 339-344.

Gouvêa AFG, Trugilho PF, Gomide JL, Silva JRM, Andrade CR and Alves ICN (2011) Determinação da densidade básica da madeira de *Eucalyptus* por diferentes métodos não destrutivos. **Revista Árvore 35**: 349-358.

Greaves BL, Borralho NMG, Raymond CA and Farrington A (1996) Use of a pilodyn for the indirect selection of basic density in *Eucalyptus nitens*. **Canadian Journal of Forest Research 26**: 1643-1650.

Hazel LV (1943) The genetic basis for constructing selection indexes. **Genetics 28**: 476-490.

Kube PD, Raymond CA and Banham PW (2001) Genetic parameters for diameter, basic density, cellulose content and fibre properties for *Eucalyptus nitens*. **Forest Genetics 8**: 285-294.

Leite HG, Guimarães DP and Campos JCC (1995) Descrição e emprego de um modelo para estimar múltiplos volumes de árvores. **Revista Árvore 19**: 75-79.

Martins IS, Martins RCC and Pinho DS (2006) Alternativas de índices de seleção em uma população de *Eucalyptus grandis* Hill ex Maiden. **Cerne 12**: 287-291.

Muneri A and Raymond CA (2000) Genetic parameters and genotype-by-environment interactions for basic density, pilodyn penetration and stem diameter in *Eucalyptus globulus*. **Forest Genetics 7**: 317-328.

Neves AT, Protásio TP, Trugilho PF, Valle MLA, Sousa LC and Vieira CMM (2013) Qualidade da madeira de clones de *Eucalyptus* em diferentes idades para a produção de bioenergia. **Revista de Ciências Agrárias 56**: 139-148.

Nunes ACP, Santos GA, Resende MDV, Silva LD, Higa A and Assis TF (2016) Estabelecimento de zonas de melhoramento para clones de eucalipto no Rio Grande do Sul. **Scientia Forestalis 44**: 00-00.

Paula RC, Pires IE, Borges RCG and Cruz CD (2002) Predição de ganhos genéticos em melhoramento florestal. **Pesquisa Agropecuária Brasileira 37**: 159-165.

Protásio TP, Goulart SL, Neves TA, Trugilho PF, Ramalho FMG and Queiroz LMRSB (2014) Qualidade da madeira e do carvão vegetal oriundos de floresta plantada em Minas Gerais. **Pesquisa Florestal Brasileira 34**: 111-123.

Raymond CA and Apiolaza LA (2004) Incorporating wood quality and deployment traits in *Eucalyptus globulus* and *Eucalyptus nitens*. In Walter C and Carson M (org) **Plantation forest biotechnology for the 21st century**. Research Signpost, Kerala, India, p. 87-99.

Reis CAF, Gonçalves FMA, Ramalho MAP and Rosado AM (2011) Seleção de progênies de eucalipto pelo índice Z por MQM e Blup. **Pesquisa Agropecuária Brasileira 46**: 517-523.

Reis CAF, Gonçalves FMA, Ramalho MAP and Rosado AM (2015) Estratégias na seleção simultânea de vários caracteres no melhoramento do *Eucalyptus*. **Ciência Florestal 25**: 457-467.

Resende MDV (2002) **Genética biométrica e estatística no melhoramento de plantas perenes**. Editora Embrapa, Brasília, 975p.

Resende MDV and Duarte JB (2007) Precisão e controle de qualidade em experimentos de avaliação de cultivares. **Pesquisa Agropecuária Tropical 37**: 182-194.

Resende MDV, Silva FF and Azevedo CF (2014) **Estatística matemática, biométrica e computacional:** modelos mistos, multivariados, categóricos e generalizados (reml/blup), inferência bayesiana, regressão aleatória, seleção genômica, qtl-gwas, estatística espacial e temporal, competição, sobrevivência. Editora Suprema, Viçosa, 881p.

Resende MDV (2016) Software Selegen-REML/BLUP: a useful tool for plant breeding. **Crop Breeding and Applied Biotechnology 16**: 330-339.

Silva FL, Pedrozo CA, Barbosa MHP, Resende MDV, Peternelli LA, Costa PMA and Vieira MS (2009) Análise de trilha para os componentes de produção de cana-de-açúcar via blup. **Ceres 56**: 308-314.

Smith HF (1936) A discriminant function for plant selection. **Annals of Eugenics 7**: 240-250.

Viana AP and Resende MDV (2014) **Genética quantitativa no melhoramento de fruteiras**. Interciência, Rio de Janeiro, 296p.

Wei X and Borralho NMG (1997) genetic control of wood basic density and bark thickness and their relationships with growth traits of *eucalyptus urophylla* in south east china. **Silvae Genetica 46**: 1997.

28

Parents choice and genetic divergence between cambuci fruit tree accessions

Flávio Gabriel Bianchini[1], Rodrigo Vieira Balbi[1], Rafael Pio[1*], Adriano Teodoro Bruzi[1] and Daniel Fernandes da Silva[2]

Abstract: *Fifty-eight cambuci fruit accessions were collected and propagated by seeds. Forty fruits of each accession were collected and evaluated for longitudinal and transverse diameter, fruit weight, number of seeds, seeds weight, total soluble solids, % citric acid, ratio, pH, firmness, vitamin C, and color. The phenotypic correlation between the characters and the relative contribution of the characters for the divergence among accessions were estimated and quantified by the Euclidean genetic distance, and cluster analysis was carried out according to the Neighbour Joining Tree. The significant correlations between the variables allowed the use of indirect selection as an auxiliary tool in the process of domestication and breeding of this species. Weight of 1000 seeds presented the greatest variation and contributed the most with genetic diversity. The expansion of the variability and the association of characters of interest can be promoted by the hybridization of the most divergent accessions, 14 and 43.*

Key words: *Campomanesiaphaea (Berg) Landr.,characters association, variability.*

*Corresponding author:
E-mail: rafaelpio@dag.ufla.br

[1] Universidade Federal de Lavras (UFLA), Departamento de Agricultura, 37.200-000, Lavras, MG, Brazil
[2] UFLA, Departamento de Biologia,

INTRODUCTION

Campomanesia genus belongs to the Myrtaceae family. Native fruit species of the genus *Campomanesia* have potential for commercial cultivation, according to their desirable agronomic characters, such as high yield and high soluble solids (Oliveira et al. 2011). *Campomanesia phaea* (Berg) Landr. is popularly known as cambuci or Cambuci fruit tree (Vallilo et al. 2005), and it spontaneously occurs in the states of São Paulo, Rio de Janeiro and Minas Gerais, at the slope of Serra do Mar, denominated Coastal Atlantic Forest, an endangered vegetation, in small and grouped populations (Maluf and Pisciottano-Ereio 2005). Its *in natura* consumption is limited due to the low carbohydrate content and high acidity. Although its shape is not uniform, it has potential for industrialization for the production of juices, jams and fermented foods, due to its quality attributes, such as high-yield pulp, high acidity, and reasonable concentrations of ascorbic acid (Vallilo et al. 2005).

The large industrial and commercial potential of cambuci fruit tree is the amount of pectin in the pulp. This polysaccharide has high gelling power, a very important property of some proteins used in many industrial foods, such as gelatin gels, candy, textured vegetable protein, jellies, etc (Andrade et al. 2011). In 2006, the Cooperative of Producers of Cambuci and Derivates of Rio Grande da Serra (Cooper Cambucy da Serra) was founded, which brings together 21 members, in order to spread cambuci fruit and demonstrate its cultivation potential and different ways of processing. The cooperative also works together with the Coordination of Integral Technical Assistance (CATI) for the production of new seedlings (Andrade et al. 2011). Much information on the cultivation and management of cambuci fruit tree, and also few reports on the fruit pattern and on the main characters (Bianchini et al. 2015) are available in the literature (Santos et al. 2016).

The dimensions of cambuci fruits can vary, due to their occurrence extension, which goes from mountain regions to areas close to sea level. This fact can cause not only morphological variation, but also variation in the chemical composition of the fruits. Studies on the morphology of fruits and seeds, and chemical characterization of the pulp are common to several species and are carried out to assist in pre-breeding programs of undomesticated species (Moura et al. 2013) and detect the genetic variability between individuals or accessions in a population (Almeida Júnior et al. 2014). In this sense, in order to expand the genetic variability, works that focus on the introduction, exchange, collection, evaluation, documentation and conservation of germplasm have gained attention.

The collection of accessions is essential for the formation of an active germplasm bank at the beginning of a breeding program that aims to increase the variability of a species under study (Chagas et al. 2015). Four types or groups are found in the germplasm collection: a) cultivated species, or obsolete cultivars of varieties derivative from breeding programs or not, which are outdated; b) primitive cultivars or landraces; c) wild relatives of cultivated plants; d) wild species with potential for domestication (Nass 2007).

Studies on the genetic divergence are carried out for the evaluation of the variability between the accessions (Silva et al. 2017). Several methods have been employed to study the diversity, such as the use of molecular markers (Almeida et al. 2011), kinship coefficient (Costa e Silva et al. 2014) and multivariate analysis (Assis et al. 2014). Studies on the genetic diversity provide information on the identification of parents that allow the exploitation of heterotic effect and the obtainment of segregating populations with greater variability in crosses. Moreover, such studies enable the identification of duplicates, and thus reduce costs with the maintenance of germplasm banks.

Two ways of inferring genetic diversity are used: the quantitative and predictive. Diallel analysis is a quantitative way to infer genetic diversity, in which crosses between the parents and their evaluation are necessary. The predictive way is based on morphological differences of nutritional, physiological or molecular nature, quantified in a dissimilarity measure that can express the degree of genetic diversity among parents (Cruz and Carneiro 2003).

The evaluation of Cambuci fruit trees accessions by means of studies on the genetic distance allows obtaining information on the divergence, relating them to possible promising crosses. The accessions are separated into different subgroups, in order to obtain homogeneity and heterogeneity within the subgroups (Cruz et al. 2012).

No studies on the genetics and plant breeding of cambuci fruits are found in the literature. Works related to genetic diversity can assist in the identification of parents for crosses, as well as in pre-breeding programs. The aims of this study were: i) to evaluate the divergence between cambuci fruit tree accessions by means of quantitative and morpho-agronomical characters, and ii) to identify the accessions forcrosses,aiming at the expansion of genetic variability.

MATERIAL AND METHODS

Ripe Cambuci fruitscollected from native plants in different locations of the Atlantic Forest and Serra do Mar, in the states of São Paulo, Minas Gerais and Rio de Janeiro, had their seeds extracted and identified. After drying, seeds were separately sown in 3 liter plastic bags filled with substrate consisting of organic matter. Eight months later, 58 accessions were planted in 2006, spaced 5 x 4 m apart, in an area belonging to the Seedling Production Center of São Bento do Sapucai SP (lat 22° 41' S, long 45°44' W, alt 874 m asl), in a randomized block design, with three blocks. The climate is Cwb type, mesothermal, or high-altitude tropical, with dry winter and rainy summer, according to Köppen.

A sample of 40 ripe fruits was collected in 2014 from the 58 accessions. Fruits were packed in transparent plastic bags, stored in expanded polypropylene box containing ice, and transported to the Federal University of Lavras (UFLA), Lavras-MG. In the Pomology Laboratory of the Fruticulture Sector of UFLA, the dimensions were measured (length and diameter), and mean fruit weight, number of seeds, weight of 1000 seeds, total soluble solids (TSS), percentage of ascorbic acid (acidity), pH, TSS and acidity ratio (ratio), firmness, vitamin C content and color were determined.

The evaluations are described as follows:

- Longitudinal and transverse mean fruit diameter: measured with the aid of a digital caliper (150 mm);

- Mean fruit weight: determined by individual weighing in semianalytical scale;

- Number of seeds: determined by removing and counting the seeds of each fruit of the plot;

- Weight of 1000 seeds: performed by counting the seeds of 10 fruits and weighing them on semianalytical scale, with subsequent ratio for 1000 seeds;

- Total Soluble Solids content (TSS): determined by maceration of fruit pulp samples of each accession in porcelain crucibles, with two readings per sample. The TSS content was determined with the aid of a portable refractometer at 20 ° C, and reading was expressed in ° Brix;

- Percentage of ascorbic acid (titratable acidity - TA): determined by transferring the weight of approximately 10 g to Erlenmeyer flasks, completing the volume of 100 mL with distilled water. Three drops of 1% phenolphthalein indicator were added to the solution, following with the titrations by manual shaking, with 0.05 N NaOH solution, previously standardized with potassium biphthalate. Results were expressed in g citric acid per 100 g pulp;

- Total soluble solids and titratable acidity (ratio): obtained by the ratio between total soluble solids (TSS) and titratable acidity (TA);

- PH: determined in a phmetrer;

- Firmness: measured by the strength necessary for a 3 mm probe coupled with a digital penetrometer to penetrate the fruit, in order to obtain the firmness value;

- Vitamin C: determined by the colorimetric method, using 2,4-dinitrophenylhydrazine. The reading was carried out on a Beckman 640 B spectrophotometer, using a computerized system, and results were expressed in mg of ascorbic acid per 100 g pulp;

- Color: determined at two different points of the fruit, using the colorimeter Minolta CR-400, with determination in the mode CIE L * a * b *. The L* coordinate refers to the luminosity level representing how bright or dark the sample is, with values ranging from 0 (totally black) to 100 (totally white). The a* coordinate may assume values from -80 to +100, in which the extremes correspond to green and red, respectively. Finally, the b* coordinate, with intensity from blue to yellow, can vary from -50 (totally blue) to +70 (totally yellow). Measurements were obtained at two diametrically opposite points in the equatorial zone of the fruit. The color was expressed by the luminosity (L *), which determines the brightness, by the chromaticity (chroma), which determines the intensity of the color, and by the hue angle (°hue), which determines the tonality.

Afterwards, analysis of variance was performed, the genetic parameters and the coefficient of genetic correlation were estimated, and diversity was determined by the Euclidean distance and cluster analysis by the Neighbour Joining Tree (NJT) method (Saitou and Nei 1987). Cluster analyses were carried out using the package "ape" of the R software (R Core Team 2013), while the others were carried out using the Genes software (Cruz 2006).

RESULTS AND DISCUSSION

Analysis of variance of the evaluated fruit characters revealed the existence of genetic variability among the accessions (Table 1). This fact corroborated the estimate of the total variation due to treatments. Most of the observed variation is owing to genetic factors (Table 2).

In general, accessions showed considerable genetic variability values for fruit characters, since the lowest coefficient of genetic variation was 11.47% for diameter. According to Vencovsky (1969) and Resende (2002), this estimate is not considered of low magnitude for this parameter, although the authors state that values greater than 20% should be considered as representative. The study showed coefficient of genetic variation that ranged from 11.47% to 46.86%, which evidences the possibility of selections of contrasting genotypes among the accessions.

Table 1. Summary of the analysis of variance for the characters fruit length, fruit diameter, weight, soluble solids, pH, acidity, firmness, number of seeds per fruit, weight of 1000 seeds, vitamin C, and ratio for 58 cambuci fruit tree accessions

S.V.	df	MS										
		LEN	DIA	WEI	SS	PH	ACI	FIR	NSF	WTS	VTC	RAT
Accessions	57	55.07**	105.11**	479.34**	6.16**	0.30**	0.89**	28.60**	34.34**	3647.65**	1394.76**	6.82**
Blocks	2	8.97	31.36	142.05	0.06	0.15	0.01	0.30	27.92	19.23	31.25	0.27
Error	114	2.83	5.97	27.18	0.14	0.03	0.06	0.40	1.40	3.82	3.69	0.57

** Significantly at 1% level. LEN (fruit length), DIA (fruit diameter), WEI (weight), SS (soluble solids), PH (pH), ACI (acidity), FIR (firmness), NSF (number of seeds per fruit), WTS (weight of 1000 seeds), VTC (vitamin C) and RAT (soluble solids/acidity).

Table 2. Estimates of genetic and phenotypic parameters obtained in the evaluation of 58 Cambuci fruit trees accessions

Traits	CVg(%)	CVg/CVe
Fruit length	11.79	2.48
Fruit Diameter	11.47	2.35
Weight	29.06	2.36
Soluble solids	13.64	3.77
Ph	15.98	1.82
Acidity	22.00	2.20
Firmness	37.81	4.83
Number of seeds per fruit	26.60	2.80
Weight of 1000 seeds	46.86	17.82
Vitamin C	27.00	11.21
Ratio	31.03	1.91

Genetic variance for all characters was superior to the observed environmental variance. This fact is fundamental to this work, since it allows clustering different accessions in more genetic terms. The variation that exists among the accessions can be observed by the variance and by the amplitude of fruit characters (Table 3).

Weight of 1000 seeds presented the greatest variation, with high observed amplitude (166.40). This fact allows inferring that the evaluated accessions present great difference regarding this attribute. Moreover, the percentage of citric acid and the pH were the characters of lower variation amplitude. According to Cruz and Carneiro (2003) and Alexandre et al. (2015), characters that express lower variability are considered of minor importance.

Table 4 shows the estimates of the relative contribution of characters for the divergence. Again, weight of 1000 seeds was the character, which contributed the most to the divergence among the accessions. The second greatest contribution was obtained by vitamin C. Costa e Silva et al. (2014) studied the contribution of different characters for

Table 3. Descriptive statistics obtained for the evaluated cambuci fruit tree accessions

Traits	Estimates				
	Variance	Mean	Maximum	Minimum	Rank
Fruit length	18.36	35.39	46.85	27.24	19.61
Fruit diameter	35.04	50.10	64.94	38.24	26.70
Weight	159.78	42.25	82.33	21.20	61.13
Soluble solids	2.06	10.39	13.14	7.38	5.76
Ph	0.10	1.88	2.87	1.36	1.51
Acidity	0.30	2.39	3.43	0.95	2.48
Firmness	9.53	8.11	19.76	3.33	16.43
Number of seeds per fruit	11.45	12.46	18.43	4.93	13.50
Weight of 1000 seeds	1215.88	74.37	187.38	20.98	166.40
Vitamin C	464.92	79.76	128.8	24.55	104.25
Ratio	2.28	4.65	11.99	2.65	9.34

genetic dissimilarity between peach trees populations and found distribution with lower discrepant values among them, ranging from 3.7% to 12.13%.

Table 5 shows the estimates of the coefficient of phenotypic correlation between the pairs of characters combination. Variation in the magnitude of the estimates was observed, and they were positive or negative. Among the characters, fruit length showed significant high correlations with fruit diameter (0.75) and fruit weight (0.76), and diameter presented significant correlation with fruit weight (0.94), indicating strong relationship between the increase in fruit size with the weight. Good correlation was also observed between the weight of one thousand seeds and length (0.51), diameter (0.43), and fruit weight (0.43), indicating that seed weight influences the increase in fruit size and fruit weight . On the other hand, negative correlation was observed between the number of seeds per fruit with length (-0.19), diameter (-0.29), and fruit weight (-0.25), indicating that fruits with less seeds have smaller dimensions (length and diameter) and lower weight.

Among the chemical characters, soluble solids in the fruits presented significant moderate correlations with pH (0.37) and ratio (0.33), which was expected, since high percentages of soluble solids are found in fruits with higher pH, and in the case of the ratio, the greater the percentage of soluble solids, expressed in ° Brix, the higher is the total soluble solids/acidity ratio. Another expected correlation occurred between acidity and firmness (0.44), since the fruits that did not reach physiological ripe stage have higher acidity contents, and therefore greater firmness. This explains the negative correlation between ratio and firmness (-0.33) and acidity and ratio (-0.82).

Correlations can be useful when the phenotypic evaluations of a particular character are difficult to be obtained. If this character presents significant phenotypic and genotypic correlations with other character of easier measurement, indirect

Table 4. Relative contribution of the traits for the divergence among the evaluated Cambuci fruit tree accessions

Traits[1]	S.j	Value (%)
Fruit length	2883.28	0.15
Fruit diameter	33160.00	1.72
Weight	2152.42	0.11
Soluble solids	42600.52	2.20
Ph	20596.07	1.07
Acidity	18241.25	0.94
Firmness	96658.35	5.00
Number of seeds per fruit	40050.43	2.07
Weight of 1000 seeds	1237836.55	64.00
Vitamin C	435065.13	22.49
Ratio	4917.95	0.25

[1] Calculations were performed using non-standard means.

Table 5. Phenotypic correlation (above) and significance level (below) obtained by the t test among biometric and chemical characters of Cambuci fruit tree fruits

	ACI	SS	LEN	DIA	WEI	NSF	WTS	PH	RAT	FIR	VTC
ACI	-	0.11	0.04	0.13	0.07	0.06	-0.15	0.05	-0.82	0.44	-0.04
SS	0.429	-	-0.51	-0.46	-0.43	0.30	-0.48	0.37	0.33	0.04	-0.14
LEN	0.771	0.001	-	0.75	0.76	-0.19	0.51	-0.07	-0.29	0.07	0.01
DIA	0.328	0.001	0.001	-	0.94	-0.29	0.43	-0.22	-0.29	0.30	-0.29
WEI	0.547	0.001	0.001	0	-	-0.25	0.43	-0.22	-0.24	0.16	-0.01
NSF	0.658	0.024	0.164	0.033	0.066	-	-0.37	0.10	0.14	-0.18	-0.07
WTS	0.276	0.001	0.001	0.001	0.001	0.005	-	-0.38	-0.04	0.08	-0.05
PH	0.669	0.007	0.551	0.114	0.106	0.486	0.004	-	0.07	-0.12	0.19
RAT	7.993	0.014	0.035	0.031	0.086	0.315	0.760	0.581	-	-0.33	-0.06
FIR	0.001	0.770	0.583	0.026	0.235	0.168	0.557	0.366	0.016	-	-0.04
VTC	0.763	0.300	0.458	0.963	0.920	0.611	0.692	0.173	0.649	0.781	-

See code in Table 1.

selection based on the character of easy measurement can be obtained. When two characters present positive and significant correlations, selection in one of these characters results in the improvement of the other. Difficulties arise when two characters have positive and significant correlation and one of them is undesirable, or when the two characters are desirable, but correlation is negative and significant (Nascimento et al. 2014).

In this work, variability among the accessions of the active germplasm bank (BAG) of cambuci fruit tree was observed. The significant correlations between variables allow performing indirect selection as an auxiliary tool in the process of domestication and breeding of this species.

The last step in a study on divergence consists in the cluster analysis, in order to obtain groups of accessions in function of the greater similarity among them. Figure 1 shows that accessions 14 and 43 are the most dissimilar. On the other hand, accessions 14 and 21 are the most similar.

CONCLUSION

Variability for fruit characters among the evaluated cambuci fruit tree accessions was observed. The expansion of genetic variability can be obtained by hybridization among the accessions 14 and 43.

AKNOWLEDGEMENTS

To the Research Support Foundation of the State of Minas Gerais (FAPEMIG) for the financial support of this research.

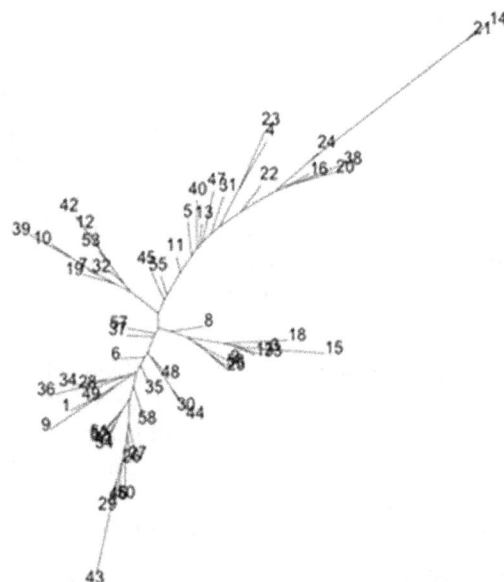

Figure 1. Dispersion diagram according to the Neibour Joining Tree (NJT) method, based on biometric and chemical traits between 58 Cambuci fruit trees accessions.

REFERENCES

Alexandre RS, Chagas K, Marques H IP, Costa P R and Filho JC (2015) Caracterização de frutos de clones de cacaueiros na região litorânea de São Mateus, ES. Revista Brasileira de Engenharia Agrícola e Ambienta l8: 785-790.

Almeida Júnior EB, Chaves LJ and Soares TN (2014) Genetic characterization of a germplasm collection of cagaiteira, a species native to the cerrado. Bragantia 73: 246-252.

Almeida MCC, Chiari L, Jank L and Valle CB (2011) Diversidade genética molecular entre cultivares e híbridos de Brachiaria spp. e Panicum maximum. Ciência Rural 11: 1998-2003.

Andrade BAGF, Fonseca PYG and Lemos F (2011) Cambuci: o fruto, o bairro, a rota: história, cultura, sustentabilidade e gastronomia. Ourivesaria da Palavra, São Paulo, 175p.

Assis GML, Santos CF, Flores OS and Valle CB (2014) Genetic divergence among Brachiara humidicola (Rendle) Schweick hybrids evaluated in the Western Brazilian Amazon. Crop Breeding and Applied Biotechnology 14: 224-231.

Bianchini FG, Balbi RV, Pio R, Silva DF, Pasqual M and Vilas Boas EVB (2015)

Caracterização morfológica e química de frutos de cambucizeiro. Bragantia 75: 10-18.

Chagas EA, Lozano RMB, Chagas PC, Bacelar-Lima CG, Garcia MIR, Oliveira JV, Souza OM, Morais BS and Araújo MCR (2015) Intraspecific variability of camu-camu fruit in native populations of northern Amazonia. Crop Breeding and Applied Biotechnology 15: 265-271.

Costa e Silva JO, Cremasco JPG, Matias RGP, Silva DFB, Salazar AH and Bruckner CH (2014) Divergência genética entre populações de pessegueiro baseada em características da planta e do fruto. Ciência Rural 10: 1770-1775.

Cruz CD (2006) Programa GENES: biometria. Editora UFV, Viçosa, 382p.

Cruz CD and Carneiro PCS (2003) Modelos biométricos aplicados ao melhoramento. Editora UFV, Viçosa, 585p.

Cruz CD, Regazzi AJ and Carneiro PCS (2012) **Modelos Biométricos Aplicados ao Melhoramento Genético**. Editora UFV, Viçosa, 516p.

Maluf AM and Pisciottano-Ereio WA (2005) Secagem e armazenamento de sementes de cambuci. **Pesquisa Agropecuária Brasileira 40:** 707-714.

Moura NF, Chaves LJ and Naves RV (2013) Caracterização física de frutos de pequizeiro (*Caryocarbrasiliense*Camb.) do cerrado. **Revista Árvore**

37: 905-912.

Nascimento WMO, Gurgel FL, Bhering LL and Ribeiro OD (2014) Pré-melhoramento do camucamuzeiro: estudo de parâmetros genéticos e dissimilaridade. **Revista Ceres 61**: 538-543.

Nass LL (2007) **Recursos genéticos vegetais**. Embrapa Recursos Genéticos e Biotecnologia, Brasília, 858p.

Oliveira MC, Santana DG and Santos CM (2011) Biometria de frutos e sementes e emergência de plântulas de duas espécies frutíferas do gênero *Campomanesia*. **Revista Brasileira de Fruticultura 33**: 446-455.

R Core Team (2013) **R: A language and environment for statistical computing**. R Foundation for StatisticalComputing, Vienna.

Resende MDV (2002) **Genética biométrica e estatística no melhoramento de plantas perenes**. Embrapa Informação Tecnológica, Brasília, 975p.

Saitou N and Nei M (1987) The Neighbor-joining method: A new method for reconstructing phylogenetic trees. **Molecular Biology and Evolution 4**: 406-425.

Santos DN, Nunes CF, Setotaw TA, Pio R, Pasqual M and Cançado GMA (2016) Molecular characterization and population structure study of cambuci: strategy for conservation and genetic improvement. **Genetics and Molecular Research15**: 1-13.

Silva VA, Machado JL, Resende JC, Oliveira AL, Figueiredo UJ, Carvalho GR, Ferrão MAG and Guimarães RJ (2017) Adaptability, stability, and genetic divergence of conilon coffee in Alto Suaçuí, Minas Gerais, Brazil. **Crop Breeding and Applied Biotechnology 17**: 25-31.

Vallilo MI, Garbelott ML, Oliveira E and Lamardo LCA (2005) Características físicas e químicas dos frutos do cambucizeiro (*Campomanesiaphaea*). **Revista Brasileira de Fruticultura 27**: 241-244.

Vencovsky R (1969) Genética quantitativa, I. In Kerr WE **Melhoramento e genética**. Edições Melhoramento, São Paulo, p.17-38.

29

Evaluation of total flavonoid content and analysis of related EST-SSR in Chinese peanut germplasm

Mingyu Hou[1,2], Guojun Mu[1], Yongjiang Zhang[3], Shunli Cui[1], Xinlei Yang[1] and Lifeng Liu[1*]

Abstract: *As important antioxidants and secondary metabolites in peanut seeds, flavonoids have great nutritive value. In this study, total flavonoid contents (TFC) were determined in seeds of 57 peanut accessions from the province of Hebei, China. A variation of 0.39 to 4.53 mg RT g^{-1} FW was found, and eight germplasm samples containing more than 3.5 mg RT g^{-1} FW. The TFC of seed embryos ranged from 0.14 to 0.77mg RT g^{-1} FW. With a view to breeding high-quality peanut varieties with high yields and high TFC, we analyzed the correlations between TFC and plant and pod characteristics. The results of correlation analysis indicated that TFC was significantly negatively correlated with pod number per plant (P/P) and soluble protein content (SPC). We used 251 pairs of expressed sequence tag - simple sequence repeat (EST-SSRs) primers to sequence all germplasm samples and found four EST-SSR markers that were significantly related to TFC.*

Key words: *Peanut (Arachis hypogaea L.), flavonoid, germplasm, EST-SSR, correlation analysis.*

*Corresponding author:
E-mail: lifengliu@126.com

[1] Agricultural University of Hebei, Laboratory of Crop Germplasm Resources for Northern China, Ministry of Education/Key Laboratory of Crop Germplasm Resources of Hebei Province/College of Agronomy, Baoding 071001, China
[2] Agricultural University of Hebei, College of Life Science, Baoding 071001, China
[3] Agricultural University of Hebei, College of Agronomy, Baoding 071001, China

INTRODUCTION

Flavonoids are secondary metabolites contained in a wide variety of plants, and their pharmacological effects include anti-inflammatory, anti-hyperlipidemia, anti-cancer, immunity-promoting, and obesity-preventing effects (Shao et al. 2011, Destro et al. 2013). Flavonoid content is used as an important index of crop quality in breeding, where screening of high-flavonoid germplasm in crops such as rice and wheat has been reported. Additional reports have found a correlation between flavonoid content and other characters, including grain shape and seed-coat color, as well as related quantitative trait loci (QTLs) (Zhang et al. 2010, Shao et al. 2011, Wang et al. 2015).

Peanut seeds contain abundant flavonoid secondary metabolites (Wang et al. 2008), are a popular food source, have preventive effects on metabolic diseases such as diabetes and hypertension, and can also be used in food production as an antioxidant of edible oil (Chukwumah et al. 2009, Lopes et al. 2011). Therefore, the flavonoid content of peanut seeds has already aroused the attention of many scholars. Wang et al. (2013) found that the flavonoid contents of a mini-core collections of American peanut ranged from 1.32 to 136.73 μg g-1 and also identified non-significant correlations between flavonoid content and both seed-coat color and fatty acid content. In 27 peanut cultivars, Chukwumah et al. (2009) assessed the TFC (0.28–1.40 mg CE g^{-1} FW; CE: catechin equivalent;

FW: fresh weight), but found no correlation between TFC and seed-coat color. Studies on peanut flavonoids, however, have so far focused on the flavonoid types found in seeds (Wang et al. 2008), the variation in the range of flavonoid contents, and the relationship between flavonoid content and seed-coat color (Wang et al. 2013, Chukwumah et al. 2009, Yoav et al. 2012). High quality is important for cultivar breeding, at the same time yield and resistant traits are also major factors in breeding research. Godoy et al. (2014) developed some peanut cultivars with high oleic and high yield. And some peanut cultivars with leaf spot resistance and high yield were bred by Suassuna et al. (2015). So it will be significant to study the relationships between flavonoid content and other biological characteristics.

Flavonoid content can be considered a quantitative characteristic, as it varies according to each peanut accession (Chukwumah et al. 2009, Wang et al. 2013). The development of molecular markers associated with flavonoid content could greatly improve the efficiency of screening germplasm for high flavonoid content. At present, some studies based on molecular markers associated with peanut traits have addressed the contents of protein, fat, and fatty acid, but few those of flavonoids. Sarvamangala et al. (2011) and Pandey et al. (2014) successfully identified QTLs for peanut protein, fat, and fatty acid content using an SSR genetic map constructed with a population of recombinant inbred lines (RIL). In addition, transgenic can also promote peanut breeding. Hassan et al. (2016) transformed chitinase gene into peanut to enhance resistance to leaf spot. A RIL population was also used for the initial location of QTLs for peanut TFC by Mondal et al. (2015), but further research on this topic is required.

China has the largest peanut–producing area in the world. The province of Hebei ranks third in total production and fourth in planted area in China. In this study, we determined the TFC of 57 peanut landraces and assessed correlations between TFC and botanical yield as well as nutritional quality traits. In addition, we screened EST-SSR markers to detect associations with TFC. The results of this study not only defined the distribution of TFC in peanut germplasm in more detail, but also provide important guidance for molecular breeding of peanut with high yield, high quality, and high flavonoid content.

MATERIAL AND METHODS

The 57 peanut landraces from the province of Hebei used in this study (Table 1) were provided by the Crop Germplasm Resource Laboratory, of the Agricultural University of Hebei. The experiments were carried out at the experimental station of Baoding of this university (lat 38° 81' N, long 115° 67' E, alt 19 m asl). The above peanut landraces were planted on May 10, 2013, routinely managed, and harvested at maturity in September. Three plants per variety were harvested and dried, and the seeds within the browned outer shells were stored at 4 °C.

The peanut seeds were ground with a universal high-speed smashing machine, passed through a 20-mesh sieve, and weighed. The resulting extract was analyzed to evaluate TFC, fat content (FC), SPC and total sugar content (TSC). The FC was calculated by the Chinese National Standard Method GB/T 14772-2008. The SPC and TSC were determined as proposed by Wang et al. (2009).

The TFC of the peanut seeds in the shell was determined according to the protocol of Chukwumah et al. (2009). The alumina colorimetric method was used, and the absorbance value of reactants determined at 510 nm. Rutin (RT) purchased from Shanghai Yuanye Bio-Technology Co., Ltd. was used as a standard. After removing the seed coat, the TFC of the peanut embryo was determined by the above method. Results were expressed as rutin equivalent on a fresh weight basis (mg RT g^{-1} FW).

Three individual plants with varietal characteristics were selected during the harvest for each landrace. Plant characteristics were determined, including plant height (PH), branch length (BL), and branching number (BN). A single plant was harvested and dried naturally, and we subsequently measured pod traits, such as P/P, 100-pod weight (100 PW), 100-seed weight (100 SW), pod kernel thickness (PKT), and reticulation depth (RD); the latter two were measured with a Vernier caliper.

Peanut leaves were sampled from plants 15 days after sprouting, and the SDS method was used for total DNA extraction (Santos et al. 2014). We used 215 pairs of EST-SSR primers to detect genetic polymorphisms in the 57 accessions, including 95 pairs of primers of Wei et al. (2011), and 123 pairs of the GM series (http://marker.kazusa.or.jp/Peanut). The sequencing protocols of PCR (Polymerase Chain Reaction) of Santos et al. (2014) were applied. The primers were

Table 1. Peanut landraces for research

No.	Landrace	Origin	Botanical type	No.	Landrace	Origin	Botanical type
1	Zunhualiyang	Zunhua city	*hypogaea*	30	Gaoyibanman	Gaoyi county	*hypogaea*
2	Xinhedayiwohou	Xinhe county	*hypogaea*	31	Dahongpao	Baoding city	*hypogaea*
3	Hejianbanpaman	Hejian city	*hypogaea*	32	Nanqinghuasheng	Nanqing county	*hypogaea*
4	Xinnongyiwohou	Julu county	hypogaea	33	Wenanpaman	Wenan county	*hypogaea*
5	Luanpingdahuasheng	Luanping county	*hypogaea*	34	Funingjiubilou	Funing county	*hypogaea*
6	Xinledabacha	Xinle city	hypogaea	35	Zhuoxianyibajiu	Zhuo county	*hypogaea*
7	Jiaoheyiwohou	Jiaohe county	*hypogaea*	36	Wuyihuasheng	Wuyi county	*hypogaea*
8	Baxianhuasheng	Ba county	*hypogaea*	37	Yuanshibanmanguo	Yuanshi county	*hypogaea*
9	Yixianyiwohou	Yi county	*hypogaea*	38	Renqiuyiwohou	Renqiu city	*hypogaea*
10	Wuqingyiwohou	Wuqing county	*hypogaea*	39	Dingxiandahuasheng	Dingzhou city	*hypogaea*
11	Daminglianhua	Daming county	*hypogaea*	40	Xianxianhuawo	Xianxian county	*hypogaea*
12	Renqiutieba	Renqiu city	*hypogaea*	41	Funinghuasheng	Funing county	*hypogaea*
13	Suninghuasheng	Suning county	*hypogaea*	42	Qianxiliyang	Qianxi county	*hypogaea*
14	Yaoyanghuasheng	Yaoyang county	*hypogaea*	43	Guangzongsaojiao	Guangzong county	*hypogaea*
15	Funingliyang	Funing county	*hypogaea*	44	Shulihuasheng	Xinji city	*hypogaea*
16	Funingdali	Funing county	*hypogaea*	45	Qinhuangdaoliyang	Qinhuangdao city	*hypogaea*
17	Qiananhuasheng	Qianan city	*hypogaea*	46	Yanshanbansaman	Yanshan county	*hypogaea*
18	Longyaoyiwohou	Longyao county	*hypogaea*	47	Qingyuanyiwohou	Qingyuan county	*hypogaea*
19	Neiqiuyiwohou	Neiqiu county	*hypogaea*	48	Baodinghong	Baoding city	*hypogaea*
20	Fengrundahuasheng	Fengrun county	*hypogaea*	49	Suning xiaobaguo	Suning county	*vulgaris*
21	Hebeidalidun	Tangshang city	*hypogaea*	50	Lulongxiaohuasheng	Lulong county	*vulgaris*
22	Damingyibajiu	Daming county	*hypogaea*	51	Shenxiansilihong	Shenzhou city	*fastigiata*
23	Xiangheyiwohou	Xianghe county	*hypogaea*	52	Shenxianxiaobaguo	Shenzhou city	*vulgaris*
24	Yanshanwuyawo	Yanshan county	hypogaea	53	Juluxiaohuasheng	Julu county	*vulgaris*
25	Gaoyiyiwohou	Gaoyi county	*hypogaea*	54	Hengshuiyiwohou	Hengshui city	*vulgaris*
26	Yangmuhuasheng	Tangshan city	*hypogaea*	55	Hejianxiaobaguo	Hejian city	*vulgaris*
27	Quyangbanman	Quyang county	*hypogaea*	56	Yuanshiyiwohou	Yuanshi county	*vulgaris*
28	Luanxianliyang	Luanping county	*hypogaea*	57	Pingshanzhongli	Pingshan county	*vulgaris*
29	Damingdayanghuasheng	Daming county	*hypogaea*				

synthesized by Beijing TsingKe Co., Ltd. The functions of correlated markers were determined by BLAST searches of the NCBI GenBank database. The SAS 9.1.3 software package was used for correlation analysis.

RESULTS AND DISCUSSION

The TFC in seeds of the 57 peanut accessions varied between 0.39 and 4.53 mg RT g^{-1} FW, with a 11.6 times higher maximum than minimum value. The TFC of 35 accessions were between 1.5 and 3.5 mg RT g^{-1} FW (Figure 1). The TFC of the Yanshanwuyawo landrace was highest (4.53 mg RT g^{-1} FW), followed by Renqiuyiwohou (4.32 mg RT g^{-1} FW), Longyaoyiwohou (4.04 mg RT g^{-1} FW), Xinnongyiwohou (3.92 mg RT g^{-1} FW), Xiangheyiwohou (3.87 mg RT g^{-1} FW), Luanpingdahuasheng (3.84 mg RT g^{-1} FW), Damingyibajiu (3.79 mg RT g^{-1} FW), and Lulongxiaohuasheng (3.61 mg RT g^{-1} FW). The highest TFC among the 27 peanut cultivars in the study by Chukwumah et al. (2009) was 1.40 mg CE g^{-1} FW, while the average TFC determined in this study was higher than 1.40 mg CE g^{-1} FW. Both rutin and catechin are standards for flavonoid analysis. This discrepancy may be due to the different germplasm used in each study. The highest isoflavone content in wild soybean germplasm found in the province of Heilongjiang was 6.81 mg g^{-1} FW (Lin et al. 2005). Germplasm sources of resistance to leaf spot were more abundant in wild peanut species and landraces than in commercial cultivars (Suassuna et al. 2015). For being members of the Leguminosae family, wild peanut and peanut landraces are likely to contain germplasm with higher flavonoid contents, which is a yet untapped potential. The eight high-flavonoid accessions identified in this experiment will be useful references for the genetic breeding of high-flavonoid peanut.

Table 2. Phenotypic values of plant, pod and quality traits and their correlation analysis with TFC in peanut germplasm

Traits	TFC (mg RT g⁻¹ FW)	PH (cm)	BN	BL (cm)	P/P	100 PW (g)	100 SW (g)	RD (mm)	PKT (mm)	FC (%)	SPC (%)	TSC (%)
TFC in embryo (mg RT g⁻¹ FW)	0.40											
TFC (mg RT g⁻¹ FW)	2.33				*						*	
PH (cm)		50.27		**	–	–	–	–	–	–	–	–
BN			13.47		–	–	–	–	–	–	–	–
BL (cm)				56.95	–	–	–	–	–	–	–	–
P/P					13.69					–	–	–
100 PW (g)						143.01	**		*	–	–	–
100 SW (g)							63.57		*	–	–	–
RD (mm)								0.27	**	–	–	–
PKT (mm)									0.87	–	–	–
FC (%)										51.48	–	**
SPC (%)											2.33	
TSC (%)												11.61

* and **: significant correlation at the 0.05 and the 0.01 probability levels, respectively. –: not detected.
TFC: total flavonoid content; PH: plant height; BN: branching number; BL: branch length; P/P: pod number per plant; 100 PW: 100-pod weight; 100 SW: 100-seed weight; RD: reticulation depth; PKT: pod kernel thickness; FC: fat content; SPC: soluble protein content; TSC: total sugar content.

The TFC of the 57 peanut germplasm embryos varied between 0.14 and 0.77 mg RT g⁻¹ FW, of which 42 accessions had contents between 0.3 and 0.6 mg RT g⁻¹ FW (Figure 1). The TFC of two landraces exceeded 0.60 mg RT g⁻¹ FW: Pingshanzhongli (0.67 mg RT g⁻¹ FW) and Shenxianxiaobaguo (0.77 mg RT g⁻¹ FW). The amount of flavonoids distributed in the seed coat and embryo varied according to each peanut accession. In 48 accessions, the embryo accounted for less than 30% of the TFC of the seeds (Figure 2) and in only four accessions, the embryo contained more than 50% of the seed TFC, namely Qingyuanyiwohou (0.42 mg RT g⁻¹ FW in embryos, accounting for 59.25%), Juluxiaohuasheng (0.26 mg RT g⁻¹ FW, 66.98%), Yuanshibanmanguo (0.41 mg RT g⁻¹ FW, 54.84%), and Yuanshiyiwohou (0.28 mg RT g⁻¹ FW, 68.39%).

Flavonoid secondary metabolites are abundant in plant seed coats. Research on soybean and rice has shown that the flavonoid content in the seed coat is higher than in the embryo per unit of seeds (Bordiga et al. 2014, Min et al. 2015). Research has also indicated that the flavonol content in rice embryos is much higher than that in seed coats (Galland et al. 2014); however, few studies have analyzed the flavonoid content in peanut seed embryos. It has been hypothesized that the types and amounts of flavonoids distributed in the seed coat and embryo may be related to the plant genotype. As an index of edible quality, the flavonoid content in the embryo has gained particular attention in peanut breeding. While the TFC in the embryos of the above four accessions (Qingyuanyiwohou, etc.) were not the absolute highest among the tested germplasm,

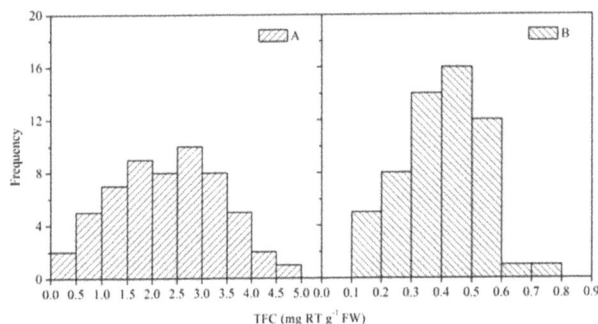

Figure 1. Histogram for the frequency distribution of TFC of seed (A) and embryo (B) in peanut germplasm

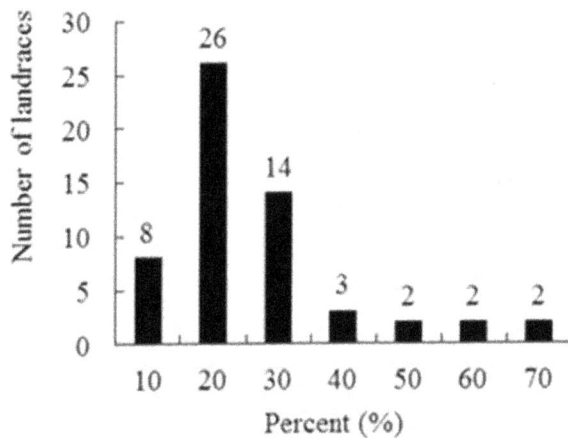

Figure 2. The percentage of TFC contained in the embryos of peanut seeds

the TFC in the embryos were higher than those in the seed coats, indicating that there is a pathway for concentrating flavonoids in the embryo. The types of flavonoids distributed in the seed coat and embryo, as well as the regulation mechanisms of flavonoid concentration in the seed coat and embryo still require further study.

Among the 57 peanut accessions there was abundant genetic variation for each assessed plant characteristic, including the traits related to pod, plant, and quality (Table 2). Among plant and pod characteristics, 100 PW demonstrated the highest variation with a standard deviation (SD) of 28.50, while the range was largest for RD, with a 36.5-fold difference between the maximum and minimum values. In terms of quality characteristics, the SPC ranged from 0.17% (Dahongpao) to 4.42% (Xianxianhuawo), while the TSC ranged from 7.81% (Fengrundahuasheng) to 18.01% (Luanpingdahuasheng). The FC of landrace Shenxiansilihong was the highest (58.97%) and that of Funingdali the lowest (45.16%).

Adequate vegetative growth of plants can generate a sufficient photosynthetic output to provide raw materials for seeds and improve their yield and quality. The yield level of legumes influences their fat and protein contents. For example, 100-seed weight is positively correlated with oil content in *Ricinus communis* (Severino et al. 2015). In this study, the TFC in peanut was negatively correlated with plant and pod characteristics. Only the negative correlation between TFC and P/P was significant (Pearson correlation co-efficient = -0.244, P = 0.034). In a future study, the genetic mechanisms underlying this relationship should be investigated to establish a theoretical basis for the breeding of peanut varieties with high yields and flavonoid contents.

As a secondary metabolite, flavonoids are synthesized via the phenylpropanoid pathway, while the initiator phenylalanine is generated in the shikimate pathway. The aromatic amino acids in plant proteins are also derived from the shikimate pathway, and the synthesis of fatty acids begins with the decarboxylation of the glycolytic intermediate pyruvic acid. Therefore, the flavonoids, proteins and fats stored in seeds originate from and compete for the same glycolytic intermediate pyruvic acid. It is hypothesized, therefore, that flavonoids would be negatively correlated with proteins and fats and positively correlated with sugar. In soybean seeds, however, the oil content is negatively correlated with protein content, and soybean isoflavone is negatively correlated with fat and positively correlated with protein content (Charron et al. 2005). In the above-mentioned mini-core collection of American peanut studied by Wang et al. (2013), no significant correlation between the fatty acid content and the contents of flavonoids such as quercetin and genistein was detected. In this study, we analyzed the correlation between TFC in peanut and several primary storage substances. Our results were similar to those found for soybean by Charron et al. (2005). Peanut TFC was negatively correlated with FC and SPC, though the correlation was only significant for SPC (Pearson correlation coefficient: -0.254, P = 0.028). TFC also exhibited a weak positive correlation with TSC (Table 2). The Luanpingdahuasheng landrace exhibited the highest TSC (18.01%) and a high TFC (3.84 mg RT g^{-1} FW).

Compared to other secondary metabolites such as salicylic acid and lignin, the heritability of flavonoid content is higher (Caseys et al. 2015). Germplasm with high flavonoid contents can therefore be selected by breeding. Molecular markers could accelerate genetic breeding for high flavonoid content (Sakiyama et al. 2014). Rhodes et al. (2014)

Table 3. The marker loci associated with flavonoid of peanut seed

EST-SSR	GM2284	GM2156	GM2067	GM1878
Pearson (Sig)	-0.256 0.027	0.22 0.05	-0.256 0.027	-0.253 0.029
Gene bank number of EST	CL1Contig14875	CL11377Contig1	CL1Contig2585	CL1Contig4088
Functions of EST	Heat stress transcription factor B-4	Tic20, a central, membrane-embedded component of the precursor protein translocon of the inner envelope of chloroplasts (TIC)	*Phaseolus vulgaris* agglutinin, hexaprenyldihydroxybenzoate methyltransferase, alkylated DNA repair protein	Mitochondrial outer membrane protein porin
Identity	74-76%	80-86%	73-77%	77-89%
Gene bank number of identity fragment	XM_003603010.1 NM_001254675.2 XM_003550286.2 XM_003528246.2	XM_003519583.2 XM_011097236.1 XM_002267833.3 XM_004148215.2 XM_008450812.1	XM_007142180.1 XM_003545029.2 XM_004292604.2	XM_008356496.1 XM_004288200.2 XM_009356339.1 XM_008239788.1 XM_006466369.1 XM_003611539.1

identified a QTL for flavonoid synthesis genes in sorghum in a genome-wide association study (GWAS). On the basis of 50 extant QTLs for isoflavone in soybean, Wang et al. (2015) located 21 new QTLs using a RIL population, including many sites related to the environment. In peanut, Mondal et al. (2015) located five QTLs for flavonoid content using a RIL population, including pPGPseq_12A07–UBC 841, GM 633–Ah242, RM14E11–TE 389, ARS 796–SSR_HO115593, and S89–TC11D09, with a minimum distance between markers of 21.0 cM. In this study, 215 pairs of EST-SSR markers were used to determine genetic differences in the 57 accessions. As a result, we discovered polymorphisms in 57 EST-SSRs. By correlation analysis, we found that four EST-SSRs were significantly correlated with TFC (Table 3): GM2284, GM2156, GM2067, and GM1878.

The functions of these four TFC-related EST-SSR markers were analyzed and each EST had more than 70% identity with gene sequences of other species (Table 3). The outer membrane protein porin is essential for transmembrane transport and signal transduction, while the heat-shock transcription factor promotes the transcription of genes when the plant is under stress. Lectins respond to the secondary metabolites generated under biotic stress to help plants adapt to the environment. As flavonoids are secondary metabolism products of plants under external stress, theoretically, the expressions of these genes should be positively correlated with flavonoid content in plants, however, we found that all markers except GM2156 were negatively correlated with TFC. The distance of these markers to flavonoid QTLs and the regulation mechanism of these ESTs need to be corroborated to confirm their functions in flavonoid synthesis in peanut seeds.

ACKNOWLEDGEMENTS

The authors thank the National Natural Science Foundation of China (31471523) and Introduction of International Advanced Agriculture Science and Technology (2013-Z65) for financial support.

REFERENCES

Bordiga M, Gomez-Alonso S, Locatelli M, Travaglia F, Coïsson J, Hermosin-Gutierrez I and Arlorio M (2014) Phenolics characterization and antioxidant activity of six different pigmented *Oryza sativa* L. cultivars grown in Piedmont (Italy). **Food Research International 65**: 282-290.

Caseys C, Stritt C, Glauser G, Blanchard T and Lexer C (2015) Effects of hybridization and evolutionary constraints on secondary metabolites: The genetic architecture of phenylpropanoids in European populus species. **PLOS ONE 26**: e0128200. doi: 10.1371/journal.pone.0128200.

Charron CS, Allen FL, Johnson RD, Pantalone VR and Sams CE (2005) Correlations of oil and protein with isoflavone concentration in soybean [*Glycine max* (L.) Merr.]. **Journal of Agricultural and Food Chemistry 53**: 7128-7135.

Chukwumah Y, Walker LT and Verghese M (2009) Peanut skin color: a biomarker for total polyphenolic content and antioxidative capacities of peanut cultivars. **International Journal of Molecular Science 10**: 4941-4952.

Destro D, Faria AP, Destro TM, Faria RT, Gonçalves LSA and Lima WF (2013) Food type soybean cooking time: a review. **Crop Breeding and Applied Biotechnology 13**: 194-199.

Galland M, Boutet-Mercey S, Lounifi I, Godin B, Balzergue S, Grandjean O, Morin H, Perreau F, Debeaujon I and Rajjou L (2014) Compartmentation and dynamics of flavone metabolism in dry and germinated rice seeds. **Plant Cell Physiology 55**: 1646-1659.

Godoy IJ, Santos JF, Carvalho CRL, Michelotto MD, Bolonhezi D, Freitas RS, Kasai FS, Ticelli M, Finoto EL and Martins ALM (2014) IAC OL 3 and IAC OL 4: new Brazilian peanut cultivars with the high oleic trait. **Crop Breeding and Applied Biotechnology 14**: 200-203.

Hassan M, Akram Z, Ali S, Ali GM, Zafar Y, Shah ZH and Alghabari F (2016) Whisker-mediated transformation of peanut with chitinase gene enhances resistance to leaf spot disease. **Crop Breeding and Applied Biotechnology 16**: 108-114.

Lin H, Lai YC, Qi N, Li H, Zhang XB and Yang XF (2005) Screening of germplasm with high content of isoflavones in wild and cultivated soybean in Heilongjiang. **Journal of Plant Genetic Resources 6**: 53-55.

Lopes RM, da Silveria Angostini-Costa T, Gimenes MA and Silveria D (2011) Chemical composition and biological activities of *Arachis* species. **Journal of Agriculture and Food Chemistry 59**: 4321-4330.

Min HK, Kim SM, Baek SY, Woo JW, Park JS, Cho ML, Lee J, Kwok SK, Kim SW and Park S (2015) Anthocyanin extracted from black soybean seed coats prevents autoimmune arthritis by suppressing the development of Th17 cells and synthesis of proinflammatory cytokines by such cells, via Inhibition of NF-kB. **PLOS ONE 10**: e0138201.

Mondal S, Phadke RR and Badigannavar AM (2015) Genetic variability for total phenolics, flavonoids and antioxidant activity of testaless seeds of a peanut recombinant inbred line population and identification of their controlling QTLs. **Euphytica 204**: 311-321.

Pandey MK, Wang ML, Qiao L, Feng S, Khera P, Wang H, Tonnis B, Barkley NA, Wang J, Holbrook CC, Culbreath AK, Varshney RK and Guo B (2014) Identification of QTLs associated with oil content and mapping FAD2 genes and their relative contribution to oil quality in peanut (*Arachis hypogaea* L.). **BMC Genetics 15**: 133.

Rhodes DH, Hoffmann Jr L, Rooney WL, Ramu P, Morris GP and Kresovich S (2014) Genome-wide association study of grain polyphenol concentrations in global Sorghum [*Sorghum bicolor* (L.) Moench] germplasm. **Journal of Agriculture and Food Chemistry 62**: 10916-10927.

Sakiyama NS, Ramos HCC, Caixeta ET and Pereira MG (2014) Plant breeding with marker-assisted selection in Brazil. **Crop Breeding and Applied Biotechnology 14**: 54-60.

Santos JM, Barbosa GVS, Neto CER and Almeida C (2014) Efficiency of biparental crossing in sugarcane analyzed by SSR markers. **Crop Breeding and Applied Biotechnology 14**: 102-107.

Sarvamangala C, Gowdaa MVC and Varshney RK (2011) Identification of quantitative trait loci for protein content, oil content and oil quality for groundnut (*Arachis hypogaea* L.). **Field Crops Research 122**: 49-59.

Severino LS, Mendesa BSS and Lima GS (2015) Seed coat specific weight and endosperm composition define the oil content of castor seed. **Industrial Crops and Products 75**: 14-19.

Shao YF, Jin L, Zhang G, Lu Y, Shen Y and Bao JS (2011) Association mapping of grain color, phenolic content, flavonoid content and antioxidant capacity in dehulled rice. **Theoretical and Applied Genetics 122**: 1005-1016.

Suassuna TMF, Suassuna ND, Moretzsohn MC, Bertioli SCML, Bertioli DJ and Medeiros EP (2015) Yield, market quality, and leaf spots partial resistance of interspecific peanut progenies. **Crop Breeding and Applied Biotechnology 15**: 175-180.

Wang DM, Lv SX and Wang JS (2009) **Laboratory course of biochemistry**. Science Press, Peking, 92p.

Wang ML, Chen CY, Tonnis B, Barkley NA, Pinnow DL, Pittman RN, Davis J, Holbrook CC, Stalker HT and Pederson GA (2013) Oil, fatty acid, flavonoid, and resveratrol content variability and FAD2A functional SNP genotypes in the U.S. peanut mini-core collection. **Journal of Agriculture and Food Chemistry 61**: 2875-82.

Wang ML, Gillaspie AG, Morris JB, Pittman RN, Davis J and Pederson GA (2008) Flavonoid content in different legume germplasm seeds quantified by HPLC. **Plant Genetic Resources 6**: 62-69.

Wang Y, Han YP, Zhao X, Li YG, Teng WL, Li DM, Zhan Y and Li WB (2015) Mapping isoflavone QTL with main, epistatic and QTL × Environment effects in recombinant inbred lines of soybean. **PLOS ONE 10**: e0118447.

Wei XY, Liu L F, Cui S L, Chen HY and Zhang JJ (2011) Development of EST-SSR markers in peanut (*Arachis hypogaea* L.). **Frontier Agriculture of China 5**: 268-273.

Yoav Shem-Tov, Hana B, Aharon S, Ilan H, Shmuel G and Ran H (2012) Determination of total polyphenol, flavonoid and anthocyanin contents and antioxidant capacities of skins from peanut (*Arachis hypogaea*) lines with different skin colors. **Journal of Food Biochemistry 36**: 301-308.

Zhang MW, Zhang RF, Zhang FX and Liu RH (2010) Phenolic profiles and antioxidant activity of black rice bran of different commercially available varieties. **Journal of Agriculture and Food Chemistry 58**: 7580-7587.

30

Reaction of common bean lines to *Xanthomonas axonopodis pv. phaseoli and Curtobacterium flaccumfaciens pv. flaccumfaciens*

Tamires Ribeiro[1*], Cleber Vinicius Giaretta Azevedo[1], Jose Antonio de Fatima Esteves[1], Sérgio Augusto Morais Carbonell[1], Margarida Fumiko Ito[2] and Alisson Fernando Chiorato[1]

Abstract: *The aim of this study was to evaluate the resistance of 58 common bean lines against common bacterial blight (Xanthomonas axonopodis pv. phaseoli) and bacterial wilt (Curtobacterium flaccumfaciens pv. flaccumfaciens). The experimental design consisted of completely randomized blocks, with four replications per pathogen. The results were subjected to variance analysis by the F test at 1% probability. Significant differences between the treatments indicated different resistance levels among the lines against both pathogens. According to the Scott-Knott test, six lines were resistant to Xanthomonas axonopodis pv. phaseoli, 14 moderately resistant, and 38 susceptible. To Curtobacterium flaccumfaciens pv. flaccumfaciens, 11 lines were resistant, 26 moderately resistant and 21 susceptible. Among these, the lines Pr10-3-4/1, Pr10-5-2/1 and Pr10-5-2/2 of the black bean group and C10-2-4/2 of the Carioca group were resistant to both major bacterial diseases affecting common bean in Brazil.*

Key words: *Phaseolus vulgaris L., plant breeding, common bacterial blight, bacterial wilt.*

*Corresponding author:
E-mail: tamires_r1@yahoo.com.br

[1] Instituto Agronômico, Centro de Análise e Pesquisa Tecnológica do Agronegócio dos Grãos e Fibras, Avenida Barão de Itapura, 1481, 13.020-902, Botafogo, Campinas, São Paulo, Brazil
[2] Instituto Agronômico, Centro de Fitossanidade

INTRODUCTION

Common bean (*Phaseolus vulgaris* L.) is a major source of vegetable protein for direct human consumption. In addition, it also contains carbohydrates, dietary fiber, B-complex vitamins, iron, calcium, and other minerals, playing an important role in the diet of the Brazilian population (Vieira et al. 2006).

According to data of CONAB (2015), the mean grain yield in Brazil is about 1.095 kg ha^{-1}, well below the productive potential of a crop which, under appropriate conditions, can yield more than 4.000 kg ha^{-1}. This low productivity can be attributed to the incidence of pests and diseases, adverse environmental conditions, low-yielding cultivars, and sowing outside the agricultural zones (Oliveira et al. 2005).

Among the main diseases affecting common bean are common bacterial blight, caused by *Xanthomonas axonopodis* pv. *phaseoli* (Smith) Dye and bacterial wilt, caused by *Curtobacterium flaccumfaciens* pv. *flaccumfaciens* (Hedges) Collins & Jones. These pathogens are widespread in the producing regions, causing yield losses, especially when stimulated by favorable environmental conditions such as high temperatures (Theodoro 2004)

The first pathogen symptoms of common bacterial blight appear on the shoot, consisting primarily of small water-soaked areas in the leaves, evolving to necrosis and imperfections in the seeds such as discoloration of the hilum, yellow spots, and wrinkling of the seed coat, which can reduce yields by 10 to 70% (Diaz et al. 2001, Bianchini et al. 2005). The inheritance of resistance to this pathogen is genetically complex, described by several authors as oligogenic or polygenic (Kelly et al. 2003, Santos et al. 2003, Manzanera et al. 2005). According to Zapata et al. (2010), Ferreira and Grattapaglia (2003), and Marquez et al. (2007), the number of genes, degrees and interactions involved in the expression of this trait may vary. Thus, the strong environmental influence in the evaluation period of the genotypes, can explain the low heritability observed in studies focused on the introgression of resistance into segregating common bean populations. Another factor that hampers the development of resistant genotypes for breeding programs is the genetic diversity of the pathogen (Mkandawire et al. 2004).

The pathogen symptoms of bacterial wilt begin with the colonization of vascular tissues, leading to the drying of apical leaflets, yellowing and gradual wilting of leaves, yellowish areas and necrosis of the parenchyma, as well as to yield drop (Maringoni 2002). According to Valentini et al. (2011), the resistance inheritance of this pathogen is polygenic and, according to Souza et al. (2006b), Wendland et al. (2008), and Torres et al. (2009b), the occurrence of genetic diversity and widespread dissemination in the producing regions of Brazil, makes the development of resistance sources even more difficult. For this reason, techniques have been developed to identify this bacterium in common bean crops (Maringoni 2002, Hsieh et al. 2005, Herbes et al. 2008), as well as to evaluate the resistance of genotypes and lines, with a view to the development of new resistant cultivars (Maringoni 2002, Souza et al. 2006a, Theodoro et al. 2007).

Both bacterial diseases are controllable by phytotechnical treatments such as crop rotation, elimination of crop residues and sowing of healthy seeds, whereas the use of resistant cultivars is the most efficient method to minimize production costs, avoiding significant yield and grain quality losses (Hsieh et al. 2005, Souza et al. 2006a, Huang et al. 2007b).

Moderate resistance to common bacterial blight or bacterial wilt was identified in the genotypes IAC Pyatã, IAC Diplomata, CNFC 10408, L 185633, IAPAR 16, UTF 6, PB 4, BRS Campeiro, IPR Chopim, XAN 159, LP 99-79, LP 93-23, L 64-5132, LP 01-51, PI 2072620, SCS 202-GUARÁ, IAPAR 81, L 264219, Iapar 80, LH 11, BRS Radiante, SM 9906, UTF 4, Iapar 20, IPR Uirapuru, and IAPAR 3 by Maringoni (2002), Rava et al. (2003), Souza et al. (2006a), Theodoro and Maringoni (2006), Costa et al. (2008), Silva et al. (2009), and Maringoni et al. (2015). These authors emphasized the importance of obtaining resistance sources to both bacterial diseases. Thus, the purpose of this study was to evaluate the resistance reaction of 58 advanced common bean lines to *X. axonopodis* pv. *phaseoli* and *C. flaccumfaciens* pv. *flaccumfaciens*.

MATERIAL AND METHODS

The experiments were conducted at the Experimental Center of Farm Santa Elisa, Instituto Agronômico-IAC, in Campinas-SP (lat 22º 54' S and long 47º 03' W, and alt 854 m asl), from January 02 to June 30, 2014. The experimental design was arranged in completely randomized blocks, with four replications for each evaluated pathogen. Each repetition consisted of one pot with two plants. Inside the greenhouse, the temperature varied from 28 °C to 32 °C during the experimental period.

Fifty-eight advanced common bean lines derived from the following crosses were evaluated: IPR Colibri x P5-4-4-1; Gen C2-1-1 x IAC Alvorada; IAC Alvorada x C6-9-10-1; Gen C4-8-2- 2 x IPR Colibri; LP02-02 x IAC Alvorada; IAC Alvorada x IAC Ybaté; Branquinho x IAC Imperador, Pr15-3-4-1 x Acesso Argentino; Pr15-5-15-1 x LP04-72; IAC Diplomata x LP04-72; IPR - Uirapuru x (IAC Una x XAN 251); IAC Una x LP04-72; P11-5-9-1 x Una IAC; IAC Diplomata x LP04-72; IAC Diplomata x (IAC Una x Acesso Argentino); (IAC Una x Acesso Argentino) x IAC Diplomata; P5-3-9-2 x IPR Colibri; IPR Colibri x IAC Imperador; LP04-72 x Pr13-3-4-1; (IAC Diplomata x LP04-72) x IAC Una; and P12-1-11-1 x LP04-72.

To evaluate the reactions of the lines to *X. axonopodis* pv. *phaseoli* (Common bacterial blight), seeds of 58 lines and the susceptible control (Rosinha G2) were disinfected with 70% ethanol and then with 1.25% sodium hypochlorite for 5 min. Subsequently, they were spread on paper sheets for germination and placed in BOD at 28 ºC for three days. After this period, the seedlings were transplanted into pots containing 500 g substrate (organic compound and soil, 1:1) and placed in the greenhouse.

The isolate 11.280 of *X. axonopodis* pv. *phaseoli*, from the Plant Health Center of the Instituto Agronômico-IAC, Campinas, SP, was used. The isolate was multiplied on PDA (potato, dextrose, agar) and incubated at 28°C for 48 hours. Thereafter, inoculum was prepared by addition of distilled water and sterilization in the bacterial colony, scraping with a glass slide, and concentration adjustment to 108 CFU mL^{-1}.

Plants in the V_2 stage were inoculated by the technique of multiple needles, according to Pompeu and Crowder (1973). The primary leaves were perforated with light pressure to allow the pathogen to enter the plant. Then the pots were placed in a moist chamber for 48 hours, at temperatures between 25 °C and 28 °C, and then transferred to the greenhouse.

Ten days after inoculation, the plants were evaluated on a 1 - 9 scale as follows: 1 to 2 - plants free of disease symptoms; 3 to 6 - small water-soaked areas; and 7 to 9 - plant tissue necrosis (Rava and Sartorato (1994). The resistance of genotypes was determined as follows: resistant lines had mean scores between 1 and 2; moderately resistant, between 2.1 and 5; and susceptible, between 5.1 and 9.

To evaluate reactions to *C. flaccumfaciens* pv. *flaccumfaciens,* seeds of the 58 lines and the pathogen-susceptible control (Rosinha G$_2$) were pre-germinated under laboratory conditions, as described above and transplanted into pots in the greenhouse.

The isolate used in this study was Feij-14627, provided by the Faculdade de Ciências Agronômicas, UNESP, in Botucatu. The isolate was multiplied in NA (Nutrient Sucrose Agar) culture medium and incubated at 28 °C for 72 hours. Inoculation was carried out in a greenhouse when the plants reached the V_3 developmental stage, by drilling two holes into the stem between the cotyledons and the primary leaves, using an entomological needle after dipping into the bacterial colony (Maringoni 2002).

Thirty days after inoculation, the plants were evaluated on a 1 - 9 scale adapted by Maringoni (2002) as follows: 0, plants without disease symptoms; 1 - mosaic symptoms on the leaves; 2 - 10% withered leaves; 5 - 25% of wilting and yellowing leaves; 7 - 50% withered leaves, yellowing and necrosis; and 9 - 75% withered leaves, yellowing and necrosis. The resistant genotypes were determined as follows: resistant lines scored between 1 and 2; moderately resistant, between 2.1 and 5; and susceptible, between 5.1 and 9.

The experiments were conducted separately. The results were subjected to analysis of variance (ANOVA) using the statistical software Genes (Cruz 2013), and differences between means were compared by the Scott-Knott test at 5% probability.

RESULTS AND DISCUSSION

The results of this study show the importance of knowing the resistance reaction of common bean genotypes to common bacterial blight and bacterial wilt. The evaluation of these genotypes is useful in breeding programs, to choose continuous sources of disease resistance, coupled with important agronomic traits, such as early cycle, high yield, upright growth, and resistance to grain darkening, with a view to develop superior genotypes for the productive sector.

The data of the evaluations of the 58 common bean lines regarding resistance to *Xanthomonas axonopodis* pv. *phaseoli* and *C. flaccumfaciens* pv. *flaccumfaciens* were subjected to analysis of variance by the F test at 1% probability. Table 1

Table 1. Summary of analysis of variance of 58 common bean lines inoculated with common bacterial blight (*Xanthomonas axonopodis* pv. *phaseoli*) and bacterial wilt *(Curtobacterium flaccumfaciens* pv. *flaccumfaciens)*

Sources of variation	df	Mean square	
		CBB	BW
Treatments	57	236.5848**	0.9849**
Error	174	5.0785	0.02369
Total	231		
CV (%)		13.65	7.40
Mean		16.50	2.07

** Significant at 1% probability by the F test; CBB= Common Bacterial Blight and BW= Bacterial Wilt.

shows significant differences between treatments, indicating different resistance levels of the common bean lines to the two studied pathogens. The experimental precision, with coefficients of variation of 7.40% and 13.65%, indicated low environmental influence during the experiments, ensuring reliability of the results.

Differential reactions of the lines to the *X. axonopodis* pv. *phaseoli* isolate were shown by the Scott-Knott test. Of the 58 lines, 6 were resistant to the pathogen (10%), 21 moderately resistant (36.20%), and 31 were susceptible (53.44%) (Table 2).

The low percentage of resistant genotypes to common bacterial blight can be explained by the occurrence of additive and non-additive effects, resulting in complex inheritance, as reported by Marquez et al. (2007). Six QTLs in F_3 plants resulting from the BAC-6 and HAB-52 cross were identified by Santos et al. (2003). Five of these QTLs were associated with resistance of leaves and one of pods, with a phenotypic variation from 12.7 to 68.7% for leaf and 12.9% for pod resistance. These results highlight the complexity of the trait, where the genes that control leaf resistance are not the same as those that control pod resistance, indicating the occurrence of oligo- or polygenic interaction, reinforcing the complex nature of resistance to the pathogen.

Among 56 evaluated cultivars, Silva (2009) identified 21 as resistant to common bacterial blight and among 61

Table 2. Resistance of 58 common bean lines to common bacterial blight (*Xanthomonas axonopodis* pv. *phaseoli*) and bacterial wilt (*Curtobacterium flaccumfaciens* pv. *flaccumfaciens*)

| Line | Mean | | Line | Mean | |
	Common Bacterial Blight	Bacterial Wilt		Common Bacterial Blight	Bacterial Wilt
1. Pr10-4-4/11	1.00 aR	3.50 bMR	30. C10-2-4/36	7.00 fS	5.00 dMR
2. Pr10-5-2/1	1.00 aR	2.00 aR	31. C10-2-5/8	7.00 fS	5.00 dMR
3. Pr10-5-2/2	1.00 aR	1.50 aR	32. P10-1-2/13	7.00 fS	3.25 bMR
4. C10-2-4/2	2.00 bR	1.00 aR	33. Pr10-4-3/13	7.00 fS	4.50 cMR
5. Pr10-3-4/1	2.00 bR	1.75 aR	34. Pr10-4-4/14	7.00 fS	7.50 fS
6. Pr10-3-5/10	2.00 bR	7.50 fS	35. C10-2-16/1	7.50 fS	7.00 fS
7. C10-2-4/57	3.75 cMR	2.00 aR	36. P10-1-3/1	7.50 fS	6.00 eS
8. P10-1-3/16	3.75 cMR	5.50 eS	37. P10-1-4/2	7.50 fS	7.00 fS
9. P10-1-1/12	4.50 dMR	2.00 aR	38. Pr10-5-1/14	7.50 fS	3.75 cMR
10. P10-1-9/38	4.50 dMR	2.00 aR	39. Pr10-5-1/2	7.50 fS	3.00 bMR
11. Pr10-8-3/2	4.50 dMR	7.50 fS	40. Pr10-5-2/4	7.50 fS	7.00 fS
12. Pr10-4-4/4	4.5 dMR	6.00 eS	41. Pr10-7-1/3	7.50 fS	4.00 cMR
13. Pr10-4-4/5	4.75 dMR	4.00 cMR	42. C10-2-16/5	8.00 gS	3.00 bMR
14. C10-2-17/1	5.00 dMR	3.00 bMR	43. C10-2-16/9	8.00 gS	3.00 bMR
15. C10-2-4/35	5.00 dMR	2.00 aR	44. P10-1-1/19	8.00 gS	4.50 cMR
16. C10-2-4/41	5.00 dMR	6.50 eS	45. Pr10-3-4/2	8.00 gS	6.00 eS
17. P10-1-4/23	5.00 dMR	5.00 dMR	46. Pr10-4-4/27	8.00 gS	3.00 bMR
18. Pr10-4-2/10	5.00 dMR	3.50 bMR	47. C10-2-17/4	8.50 gS	5.50 eS
19. Pr10-5-2/3	5.00 dMR	2.00 aR	48. Pr10-3-2/35	8.50 gS	8.50 gS
20. Pr10-7-1/6	5.00 dMR	8.50 gS	49. Pr10-4-3/14	8.50 gS	5.00 dMR
21. Pr10-3-3/8	5.25 eS	3.25 bMR	50. Pr10-7-1/16	8.50 gS	5.50 eS
22. C10-2-16/8	6.00 eS	3.25 bMR	51. C10-2-16/7	9.00 hS	6.50 eS
23. C10-2-17/3	6.00 eS	2.00 aR	52. P10-1-3/17	9.00 hS	4.00 cMR
24. C10-2-4/12	6.00 eS	5.50 eS	53. Pr10-3-3/9	9.00 hS	4.00 cMR
25. C10-6-2/11	6.00 eS	4.00 cMR	54. Pr10-3-3/10	9.00 hS	8.00 gS
26. P10-1-1/8	6.00 eS	4.50 cMR	55. Pr10-4-4/19	9.00 hS	3.25 bMR
27. Pr10-3-5/35	6.00 eS	3.75 cMR	56. Pr10-4-4/39	9.00 hS	8.00 gS
28. P10-1-9/39	6.75 fS	2.00 aR	57. Pr10-5-1/15	9.00 hS	3.25 bMR
29. C10-2-17/2	7.00 fS	7.50 fS	58. Pr10-8-3/1	9.00 hS	8.00 gS

Means values followed by different lowercase letters are significantly different between lines by the Scott-Knott test, at 5% probability and different uppercase letters indicate reactions of the lines to the common bacterial blight and bacterial wilt isolates (R= resistant: scores between 1.00 and 2.00; MR= moderately resistant: scores between 2.10 and 6,00 S= susceptible: scores between 6.10 and 9.00).

genotypes, Costa et al. (2008) identified the cultivars Magnifico, Radiante and BRS Pontal as resistant. In our study, six lines were selected, one of which belongs to the Carioca group (C10-2-4/2) and five to the black bean group (Pr10-4-4/11, Pr10-5-2/1, Pr10-5-2/2, Pr10-3-4/1 and Pr10-3-5/10). These six resulted from the respective crosses: IAC Alvorada x C6-9-10-1; IAC Una x LP04-72; IPR-Uirapuru x (IAC Una x XAN 251); IPR-Uirapuru x (IAC Una x XAN 251); (IAC Diplomata x LP04 -72) x IAC Una and (IAC Una x Acesso Argentino) x IAC Diplomata (Table 3).

The resistant lines of the black bean group (Pr10-4-4/11, Pr10-5-2/1, Pr10-5-2/2, Pr10-3-4/1 and Pr10-3-5/10), were derived from the parents IPR-Uirapuru, IAC Una or IAC Diplomata. The former two were classified, respectively, as resistant and moderately resistant by Silva et al. (2009), while IAC Diplomata was classified as susceptible by Azevedo et al. (2015). The resistant line of the Carioca group C10-2-4/2 was derived from the parent IAC Alvorada, classified as susceptible to common bacterial blight by Azevedo et al. (2015).

The difficulty in the development of resistant genotypes motivated several authors to approach this problem by identifying bacteria in seeds. However, Denardin and Agostini (2013) and Silva et al. (2013) described the complexity of pathogen identification on seeds and highlighted the importance of finding resistance sources, due to the wide dissemination of the pathogen in the producing areas of common bean.

The Scott-Knott test showed a differential reactions among the lines to *C. flaccumfaciens* pv. *flaccumfaciens*. Among the 58 lines, 11 were resistant to the pathogen (18.96%), 26 moderately resistant (44.82%) and 21 were susceptible (36.20%) (Table 2).

Of 333 tested genotypes, Souza et al. (2006a) found 18% to be resistant, which is consistent with our results. The low percentage of genotypes resistant to bacterial wilt was mentioned by Theodoro and Maringoni (2006). These authors evaluated 73 lines and found only two resistant cultivars (Mouro Piratuba and Vagem Amarela). According to Valentini et al. (2011), the low rate of resistant genotypes can be explained by the occurrence of additive and non-additive effects in the inheritance of bacterial wilt resistance. These authors identified more than three resistance genes in two populations resulting from the crosses IAC Carioca Aruã x SCS Guará and IAC Carioca Pyatã x Perola.

In this study, 11 lines were classified as bacterial wilt resistant, 7 of which belong to the Carioca group (C10-2-4/2, C10-2-4/57, P10-1-1/12, P10-1-9/38, C10-2-4/35, C10-2-17/3, and P10-1-9/39) and 4 to the black bean group (Pr10-5-2/1,

Table 3. Common bean lines and their respective original crosses resistant to common bacterial blight (*Xanthomonas axonopodis* pv. *phaseoli*) and bacterial wilt (*Curtobacterium flaccumfaciens* pv. *flaccumfaciens*)

Common bacterial blight		
Line	Cross	Market class
1. C10-2-4/2	IAC Alvorada x C6-9-10-1	Carioca
2. Pr10-3-5/10	(IAC Una x Acesso Argentino) x IAC Diplomata	Black
3. Pr10-3-4/1	(IAC Diplomata x LP04-72) x IAC Una	Black
4. Pr10-5-2/1	IPR - Uirapuru x (IAC Una x XAN 251)	Black
5. Pr10-5-2/2	IPR - Uirapuru x (IAC Una x XAN 251)	Black
6. Pr10-4-4/11	IAC Una x LP04-72	Black
Bacterial Wilt		
Line	Cross	Market class
1. C10-2-4/2	IAC Alvorada x C6-9-10-1	Carioca
2. C10-2-4/57	IAC Alvorada x C6-9-10-1	Carioca
3. P10-1-1/12	IPR Colibri x P5-4-4-1	Carioca
4. P10-1-9/38	P5-3-9-2 x IPR Colibri	Carioca
5. C10-2-4/35	IAC Alvorada x C6-9-10-1	Carioca
6. C10-2-17/3	Gen C2-1-1 x IAC Alvorada	Carioca
7. P10-1-9/39	P5-3-9-2 x IPR Colibri	Carioca
8. Pr10-5-2/1	IPR - Uirapuru x (IAC Una x XAN 251)	Black
9. Pr10-5-2/2	IPR - Uirapuru x (IAC Una x XAN 251)	Black
10. Pr10-3-4/1	(IAC Diplomata x LP04-72) x IAC Una	Black
11. Pr10-5-2/3	(IAC Diplomata x LP04-72) x IAC Una	Black

Pr10-5-2/2, Pr10-3-4/1, and Pr10-5-2/3), resulting from the respective crosses IAC Alvorada x C6-9-10-1, IAC Alvorada x C6-9-10-1, IPR Colibri x P5-4-4-1, P5-3-9-2 x IPR Colibri, IAC Alvorada x C6-9-10-1, Gen C2-1-1 x IAC Alvorada, P5-3-9-2 x IPR Colibri, IPR-Uirapuru x (IAC Una x XAN 251), IPR-Uirapuru x (IAC Una x XAN 251), (IAC Diplomata x LP04-72) x IAC Una, and (IAC Diplomata x LP04-72) x IAC Una (Table 3).

Of these 11 lines, C10-2-4/57, Pr10-3-4/1, Pr10-5-2/1, Pr10-5-2/2, C10-2-4/2, C10 2-17/3, and C10-2-4/35 were originated from the parents IAC Diplomata, IAC Alvorada, IAC Una, or IPR-Uirapuru. IAC Diplomata and IAC Alvorada were classified as resistant, while IAC Una was classified as susceptible by Maringoni et al. (2015). Parent IPR-Uirapuru was classified as susceptible to *C. flaccumfaciens* pv. *flaccumfaciens* by Theodoro et al. (2007) and Maringoni et al. (2015).

According to Souza and Maringoni (2008), resistant genotypes involve the pathogen by protoplasmic projections, preventing its installation in the xylem vessels, while in susceptible genotypes the water transport is obstructed by the presence of bacterial cells. These results are related to the disease symptoms, e.g., plant wilting, yellowing, underdevelopment, and death, observed at different levels of aggressiveness in the 58 lines evaluated in this study.

The lines C10-2-4/2, Pr10-3-4/1, Pr10-5-2/1, and Pr10-5-2/2 (Table 3) were resistant to common bacterial blight and bacterial wilt, the two major bacterial diseases affecting common bean in Brazil. The development of these cultivars is extremely important to maintain the yield and grain quality of common bean, given the lack of resistant cultivars in the productive sector, the physiological variability and wide dissemination of the pathogens in crop areas.

ACKNOWLEDGEMENTS

The authors thank the Fundação de Pesquisa do Estado de São Paulo (FAPESP) for the financial support.

REFERENCES

Azevedo CVG, Ribeiro T, Silva DA, Carbonell SAM and Chiorato AF (2015) Adaptabilidade, estabilidade e resistência a patógenos em genótipos de feijoeiro. **Pesquisa Agropecuária Brasileira 50**: 912-922.

Bianchini A, Maringoni AC and Carneiro SMPG (2005) Doenças do feijoeiro (*Phaseolus vulgaris* L.). In Kimati H, Amorim L, Rezende JAM, Bergamin Filho A and Camargo LEA. **Manual de fitopatologia**. 2ⁿᵈ edn, Editora Ceres, São Paulo, 333-349.

CONAB - Companhia Nacional de Abastecimento (2015) Available at <http://www.conab.gov.br/OlalaCMS/uploads/arquivos/15_12_11_11_02_58_boletim_graos_dezembro_2015.pdf>. Accessed on 25 Dec, 2015.

Costa JGC, Rava CA, Puríssimo JD, Peloso MJD, Melo LC and Faria LC (2008) Reação de genótipos de feijoeiro comum ao crestamento bacteriano comum e à murcha de curtobacterium. **Revista Ceres 55**: 93-395.

Cruz CD (2013) GENES - a software package for analysis in experimental statistics and quantitative genetics. **Acta Scientiarum Agronomy 35**: 271-276.

Denardin ND'A and Agostini VA (2013) Detection and quantification of *Xanthomonas axonopodis* pv. *phaseoli* and its variant fuscans in common bean seeds. **Journal of Seed Science 35**: 428-434.

Diaz CG, Bassanezi RB, Godoy CV, Lopes DB and Bergamin Filho A (2001) Quantificação do efeito do crestamento bacteriano comum na eficiência fotossintética e na produção do feijoeiro. **Fitopatologia Brasileira 26**: 71-76.

Ferreira ME and Grattapaglia D (2003) Introdução ao uso de marcadores moleculares em análise genética. **Embrapa-Cenargen 3**: 220.

Herbes DH, Theodoro GF, Maringoni AC, Dal Piva CA and Abreu L (2008) Detecção de *Curtobacterium flaccumfaciens* pv. *flaccumfaciens* em sementes de feijoeiro produzidas em Santa Catarina. **Tropical Plant Pathology 33**: 53-156.

Hsieh TF, Huang HC, Mündel HH, Conner RL, Erickson RS and Balasubramanian PM (2005) Resistence of common bean (*Phaseolus vulgaris*) to bacterial wilt caused by *Curtobacterium flaccumfaciens* pv. *flaccumfaciens*. **Phytopathology 153**: 245-249.

Huang HC, Erickson RS and Hsieh TF (2007b) Control of bacterial wilt of bean (*Curtobacterium flaccumfaciens* pv. *flaccumfaciens*) by seed treatment with *Rhizobium leguminosarum*. **Crop Protection 26**: 1055-1061.

Kelly JD, Gepts P, Miklas PN and Coyne DP (2003) Tagging and mapping of genes and QTL and molecular marker-assisted selection for traits of economic importance in bean and cowpea. **Field Crops Research 82**: 135-154.

Manzanera MAS, Asensio C and Singh SP (2005) Gamete selection for resistance to common and halo bacterial blights in dry bean intergene pool populations. **Crop Science 46**: 131-135.

Maringoni AC (2002) Comportamento de cultivares de feijoeiro comum à murcha-de- curtobacterium. **Fitopatologia Brasileira 27**: 157-166.

Maringoni AC, Ishiszuka MS, Silva AP, Soman JM, Moura MF, Santos RL, Júnior TAFS, Chiorato AF, Carbonell SAM and Júnior NSF (2015) Reaction and colonization of common bean genotypes by *Curtobacterium flaccumfaciens* pv. *flaccumfaciens*. **Crop Breeding and Applied Biotechnology 15**: 87-93.

Marquez ML, Terán H and Singh SP (2007) Selecting common bean with genes of different evolutionary origins

Reaction of common bean lines to Xanthomonas axonopodis pv. phaseoli and Curtobacterium...

201

for resistance to *Xanthomonas campestris* pv. *phaseoli*. **Crop Science 47**: 1367-1374.

Mkandawire ABC, Mabagala RB, Guzman P, Gepts P and Gilbertson RL (2004) Genetic and Pathogenic variation of common blight bacteria (*Xanthomonas axonopodis* pv. *phaseoli* and *X. axonopodis* pv. *phaseoli var. fuscans*). **Phytopathology 94**: 593-603.

Oliveira AD, Fernandes EJ and Rodrigues TJD (2005) Condutância estomática como indicador de estresse hídrico em feijão. **Engenharia Agrícola 25**: 86-95.

Pompeu AS and Crowder LV (1973) Métodos de inoculação e concentrações bacterianas de *Xanthomonas phaseoli*, para a herança da reação a doença em *Phaseolus vulgaris* sob condições de câmara de crescimento. **Ciência e Cultura 25**: 1078-1081.

Rava CA and Sartorato A (1994) Crestamento bacteriano comum. Principais doenças do feijoeiro comum e seu controle. Embrapa, Brasília (CNPAF 300).

Rava CA, Costa JGC, Fonseca JR and Salgado AL (2003) Fontes de resistência à antracnose, crestamento-bacteriano-comum e murcha-de-curtobacterium em coletas de feijoeiro comum. **Revista Ceres 50**: 797-802.

Santos AS, Bressan-Smith RE, Pereira MG, Rodrigues R and Ferreira CF (2003) Genetic Linkage Map of *Phaseolus vulgaris* L. and identification of QTLs responsible for resistance to *Xanthomonas axonopodis* pv. *phaseoli*. **Fitopatologia Brasileira 28**: 5-10.

Silva A, Santos I, Balbinot AL, Matei G and Oliveira PH (2009) Reação de genótipos de feijão ao crestamento bacteriano comum, avaliado por dois métodos de inoculação. **Ciência Agrotécnica 33**: 2019-2024.

Silva FC, Souza RM, Zacaroni AB, Lelis FMV and Figueira AR (2013) Otimização da técnica de PCR para a detecção de *Xanthomonas axonopodis* pv. *phaseoli* em sementes de feijão. **Summa Phytopathologica 39**: 45-50.

Souza VL and Maringoni AC (2008) Análise ultraestrutural da interação de *Curtobacterium flaccumfaciens* pv. *flaccumfaciens* em genótipos de feijoeiro. **Summa Phytopathologica 34**: 318-320.

Souza VL, Maringoni AC and Krause-Sakate R (2006b) Variabilidade genética em isolados de *Curtobacterium flaccumfaciens*. **Summa Phytopathologica 32**: 170-176.

Souza VL, Maringoni AC, Carbonell SAM and Ito MF (2006a) Resistência genética em genótipos de feijoeiro a *Curtobacterium flaccumfaciens* pv. *flaccumfaciens*. **Summa Phytopathologica 32**: 339-344.

Theodoro GF (2004) Reação de cultivares locais de feijão a *Xanthomonas axonopodis* pv. *phaseoli*, em condições de campo. **Revista Brasileira Agrociência 10**: 373-375.

Theodoro GF and Maringoni AC (2006) Effect of potassium levels in the severity of bacterial wilt in common bean cultivars. **Summa Phytopathologica 32**: 139-146.

Theodoro GF, Herbes DH and Maringoni AC (2007) Fontes de resistência à murcha-de-curtobacterium em cultivares locais de feijoeiro, coletadas em Santa Catarina. **Ciência e Agrotecnologia 31**: 333-339.

Torres JP, Silva Júnior TAF and Maringoni AC (2009b) Detecção de *Xanthomonas axonopodis* pv. *phaseoli* em sementes de feijoeiro provenientes do Estado do Paraná, Brasil. **Summa Phytopathologica 35**: 136-139.

Valentini G, Baldissera JNC, Morais PPP, Stähelin D, Heidemann JC, Stenger F, Elias HT, Guidolin AF and Coimbra JLM (2011) Herança da resistência em feijão à murcha causada por *Curtobacterium flaccumfaciens* pv. *flaccumfaciens*. **Pesquisa Agropecuária Brasileira 46**: 1045-1052.

Vieira C, Paula Júnior TJ and Borém A (2006) **Feijão**. Editora UFV, Viçosa, 600p.

Wendland A, Alencar NA, Melo LC, Costa JGC, Del Peloso MJ, Pereira HS, Faria LC, Côrtes MVCB and Brondani RPV (2008) Padrão de sintomas de isolados de *Curtobacterium flaccumfaciens* pv. *flaccumfaciens* em dois genótipos de feijoeiro. **Boletim de Pesquisa e Desenvolvimento Embrapa Arroz e Feijão 33**: 19.

Zapata M, Beaver JS and Porch TG (2010) Dominant gene for common bean resistance to common bacterial blight caused by *Xanthomonas axonopodis* pv. *phaseoli*. **Euphytica 179**: 373-382.

PERMISSIONS

LIST OF CONTRIBUTORS

Dayana Rotili Nunes Picolotto and Daly Roxana Castro Padilha
Universidade Federal de Mato Grosso do Sul, Campus de Chapadão do Sul, CP 112, 79.560-000, Chapadão do Sul, MS, Brazil

Vespasiano Borges de Paiva Neto
Universidade Federal do Vale do São Francisco, Campus de Ciências Agrárias, Rodovia BR 407, km 119, Lote 543, PSNC, s/n, C1, 56.300-990, Petrolina, PE, Brazil

Fábio de Barros
Instituto de Botânica, Núcleo de Pesquisa Orquidário do Estado, CP 68041, 04.045-972, São Paulo, SP, Brasil

Ana Cláudia Ferreira da Cruz and Wagner Campos Otoni
Universidade Federal de Viçosa, Instituto de Biotecnologia Aplicada à Agropecuária (BIOAGRO), Laboratório de Cultura de Tecidos, Campus Universitário, Avenida Peter Henry Rolfs, s/n, 36.570-900, Viçosa, MG, Brazil

Aloka Kumari, Ponnusamy Baskaran and Johannes Van Staden
University of KwaZulu-Natal Pietermaritzburg, Research Centre for Plant Growth and Development, School of Life Sciences, Scottsville 3209, South Africa

Aleksandra Dimitrijević, Ivana Imerovski, Dragana Miladinović, Sandra Cvejić, Siniša Jocić, Tijana Zeremski and Zvonimir Sakač
Institute of Field and Vegetable Crops, Maksima Gorkog 30, 21000 Novi Sad, Serbia

Otávio Luiz Gomes Carneiro and Danilo Hottis Lyra
Universidade de São Paulo (USP), Escola Superior de Agricultura "Luiz de Queiroz" (ESALQ), Departamento de Genética, Avenida Pádua Dias, 11, 13.418-900, Piracicaba, SP, Brazil

Silvia Regina Rodrigues de Paula Ribeiro, Marcio Lisboa Guedes and César Augusto Brasil Pereira Pinto
Universidade Federal de Lavras (UFLA), Campus UFLA, Departamento de Biologia, CP 3037, 37.200 000, Lavras, MG, Brazil

Carolina Mariane Moreira
Instituto Federal de Educação, Ciência e Tecnologia do Sul de Minas Gerais, Campus Poços de Caldas, Avenida Dirce Pereira Rosa, 300, 37.713-100, Poços de Caldas, MG, Brazil

Paulo Eduardo Teodoro, Leonardo de Azevedo Peixoto and Leonardo Lopes Bhering
Universidade Federal de Viçosa, Departamento de Biologia Geral, 36.571-000, Viçosa, MG, Brazil

Erina Vitório Rodrigues and Bruno Galvêas Laviola
Embrapa Agroenergia, 70297-400, Brasília, DF, Brazil

Rubens Marschalek, Jose Alberto Noldin, Ester Wickert, Klaus Konrad Scheuermann, Moacir Antonio Schiocchet, Domingos Savio Eberhardt, Ronaldir Knoblauch, Eduardo Hickel, Gabriela Neves Martins and Alexander de Andrade
Epagri, EEI, CP 277, 88301-970, Itajaí, SC, Brazil

Juliana Vieira Raimondi
Avantis, Av. Marginal Leste, n. 3600, 88339- 125, BC, SC, Brazil

Gladyston Rodrigues Carvalho and Juliana Costa de Rezende and Cesar Elias Botelho
Empresa de Pesquisa Agropecuária de Minas Gerais (EPAMIG), Unidade Sul, Campus da Universidade Federal de Lavras (UFLA), CP 176, 37.200-000, Lavras, MG, Brazil

Gabriel Ferreira Bartholo and Antônio Carlos Baião de Oliveira
Embrapa Café, Parque Estação Biológica - PqEB., 70.770-901, Brasília, DF, Brazil

Antônio Alves Pereira
EPAMIG, Unidade Sudeste, Vila Gianetti, 46/47, Campus da UFV, 36.570-000, Viçosa, MG, Brazil

Felipe Lopes da Silva
Universidade Federal de Viçosa (UFV), Departamento de Fitotecnia, Avenida P.H. Rolfs, 36.570-000, Viçosa, MG, Brazil

Rafael Augusto Vieira, Renato da Rocha and Carlos Alberto Scapim
Universidade Estadual de Maringá, Departamento de Agronomia, 87.020-900, Maringá, PR, Brazil

Antonio Teixeira do Amaral Junior
Universidade Estadual do Norte Fluminense Darcy Ribeiro, Laboratório de Melhoramento Genético Vegetal, 28.013-602, Campos dos Goytacazes, RJ, Brazil

Telma Nair Santana Pereira, Ingrid Gaspar da Costa Geronimo and Messias Gonzaga Pereira
Universidade Estadual Norte Fluminense Darcy Ribeiro, Laboratório de Melhoramento Genético Vegetal, Avenida Alberto Lamego, 2000, Horto, 28.013-602, Campos dos Goytacazes, RJ, Brazil

Aparecida Bandini Rossi
Universidade do Estado de Mato Grosso, Departamento de Ciências Biológicas, Fundação Universidade do Estado de Mato Grosso, Campus Universitário de Alta Floresta, Residencial Flaboyant, 78.580-000, Alta Floresta, MT, Brazil

Carlos T. Bainotti, Enrique Alberione, Nicolás Salines, Dionisio Gomez, Jorge Fraschina, José Salines, María B. Formica, Guillermo Donaire, Lucio Lombardo, María M. Nisi, Martha B. Cuniberti, Leticia Mir, María B. Conde and Marcelo Helguera
INTA EEA Marcos Juárez, Ruta 12, km 3, (2580) Marcos Juárez, Córdoba, Argentina

Silvina Lewis
Instituto de Recursos Biológicos, INTA, (1686) Hurlingham, Buenos Aires, Argentina

Pablo Campos
INTA EEA Bordenave, Zona Rural, (8187) Bordenave, Buenos Aires, Argentina

Leonardo S. Vanzetti
INTA EEA Marcos Juárez, Ruta 12, km 3, (2580) Marcos Juárez, Córdoba, Argentina
Instituto de Recursos Biológicos, INTA, (1686) Hurlingham, Buenos Aires, Argentina
INTA EEA Bordenave, Zona Rural, (8187) Bordenave, Buenos Aires, Argentina
CONICET, Av. Rivadavia 1917 (C1033AAJ) CABA, Argentina

Ignácio José de Godoy, João Francisco dos Santos, Andrea Rocha Almeida de Moraes, Rogério Soares de Freitas and Cassia Regina Limonta de Carvalho
Instituto Agronômico (IAC)/Apta, Av. Barão de Itapura, 1481, CP 28, 13.020-902, Campinas, SP, Brazil

Marcos Doniseti Michelotto, Denizart Bolonhezi, Everton Luis Finoto and Antonio Lúcio Melo Martins
Apta, Departamento de Descentralização do Desenvolvimento (DDD), Av. Barão de Itapura, 1481, CP 28, 13.020-902, Campinas, SP, Brazil

Pedro Luiz Scheeren, Eduardo Caierão, Márcio Só e Silva, Luiz Eichelberger, Alfredo do Nascimento Júnior, Eliana Maria Guarienti, Martha Zavariz de Miranda, Ricardo Lima de Castro, Leila Costamilan, Flávio Martins Santana, João Leodato Nunes Maciel, Maria Imaculada Pontes Moreira Lima, João Leonardo Fernandes Pires, Douglas Lau, Paulo Roberto Valle da Silva Pereira and Gilberto Rocca da Cunha
Embrapa Trigo, Rodovia BR 285, km 294, CP 451, 99.001-970, Passo Fundo, RS, Brazil

Gabrielen de Maria Gomes Dias
Universidade da Integração da Lusofonia Afro-Brasileira (UNILAB), Instituto de Desenvolvimento Rural, Avenida da Abolição, 3, 62.790-000, Redenção, CE, Brazil

Joyce Dória Rodrigues Soares, Adalvan Daniel Martins and Moacir Pasqual
Universidade Federal de Lavras (UFLA), Departamento de Agricultura, CP 3037, 37.200-000, Lavras, MG, Brazil

Suelen Francisca Ribeiro
UFLA, Departamento de Biologia

Eduardo Alves
UFLA, Departamento de Fitopatologia

Thiago Vincenzi Conrado and Daniel Furtado Ferreira
Universidade Federal de Lavras (UFLA), Departamento de Biologia, Campus, CP 3037, 37.200-000, Lavras, MG, Brazil

Carlos Alberto Scapim
Universidade Estadual de Maringá (UEM), Agronomia, 87.080-000, Maringá, PR, Brazil

Wilson Roberto Maluf
UFLA, Departamento de Agricultura

João Luís da Silva Filho, Camilo de Lelis Morello, Nelson Dias Suassuna, Francisco José Correia Farias and Taís de Moraes Falleiro Suassuna
Embrapa Algodão, CP 147, 58.428-095, Campina Grande, PB, Brazil

Fernando Mendes Lamas
Embrapa Agropecuária Oeste, CP 661, 79.804-970, Dourados, MS, Brazil

Murilo Barros Pedrosa
Fundação Bahia, Rodovia BR 020/242, S/N, km 50,7, 47.850-000, Zona Rural, Luís Eduardo Magalhães, BA, Brazil

José Lopes Ribeiro
Embrapa Meio-Norte, CP 01, 64.006-220, Teresina, PI, Brazil

Marcelo Vivas and Rogério Figueiredo Daher
Universidade Estadual do Norte Fluminense Darcy Ribeiro (UENF), Centro de Ciências e Tecnologias Agropecuárias, Laboratório de Engenharia Agrícola, 28.013-602, Campos dos Goytacazes, RJ, Brazil

Silvaldo Felipe da Silveira, Janieli Maganha Silva Vivas, Pedro Henrique Dias dos Santos and Beatriz Murizini Carvalho
UENF, Laboratório de Entomologia e Fitopatologia

Antonio Teixeira do Amaral Júnior and Messias Gonzaga Pereira
UENF, Laboratório de Melhoramento Genético Vegetal

Ariano Martins de Magalhães Júnior, Paulo Ricardo Reis Fagundes, Daniel Fernandes Franco, Eduardo Anibele Streck, Gabriel Almeida Aguiar and Paulo Henrique Karling Facchinello
Embrapa Clima Temperado, Rodovia BR-392, km 78, 9° Distrito, Monte Bonito, CP 321, 96.010 971, Pelotas, RS, Brazil

Orlando Peixoto de Morais
Embrapa Arroz e Feijão, Rodovia GO-462, km 12, Fazenda Capivara, Zona Rural, CP 179, 75.375-000, Santo Antônio de Goiás, GO, Brazil

Félix Gonçalves de Siqueira
Embrapa Agroenergia, Parque Estação Biológica (pqEB), PqEB s/n°, CP 40.315, 70.770- 901, Brasília, DF, Brazil

Edelclaiton Daros, Ricardo Augusto de Oliveira, José Luis Camargo Zambon, João Carlos Bespalhok Filho, Bruno Portela Brasileiro, Oswaldo Teruyo Ido, Lucimeris Ruaro and Heroldo Weber
Universidade Federal do Paraná (UFPR), Departamento de Fitotecnia e Fitossanitarismo, 80.035-050, Curitiba, PR, Brazil

Suresh Kumar
ICAR - Indian Grassland and Fodder Research Institute, Division of Crop Improvement, Jhansi-284003, India
ICAR - Indian Agricultural Research Institute, Division of Biochemistry, Pusa Campus, New Delhi, Delhi-110012, India

Sheena Saxena and Madan G. Gupta
ICAR - Indian Grassland and Fodder Research Institute, Division of Crop Improvement, Jhansi-284003, India

Priscilla Neves de Santana
Faculdade Integrada Aparício Carvalho (FIMCA), 78.912-640, Porto Velho, RO, Brazil

Américo José dos Santos Reis and Lázaro José Chaves
Universidade Federal de Goiás, Escola de Agronomia, Campus Samambaia, 74.690-900, Goiânia, GO, Brazil

Antônio Vander Pereira, Francisco José da Silva Lédo and Juarez Campolina Machado
Embrapa Gado de Leite, Rua Eugênio do Nascimento, 610, Dom Bosco, 36.038-330, Juiz de Fora, MG, Brazil

Leonardo Cunha Melo, Helton Santos Pereira, Luís Cláudio de Faria, Thiago Lívio Pessoa Oliveira de Souza, Adriane Wendland, José Luis Cabrera Díaz, Mariana Cruzick de Souza Magaldi and Joaquim Geraldo Cáprio da Costa
Embrapa Arroz e Feijão, Rod. GO 462, km 12, CP 179, 75.375-000, Santo Antônio de Goiás, GO, Brazil

Hélio Wilson Lemos de Carvalho
Embrapa Tabuleiros Costeiros, Avenida beira Mar, 3250, Bairro Jardins, CP 44, 49.025-040, Aracaju, SE, Brazil

Carlos Lásaro Pereira de Melo
Embrapa Agropecuária Oeste, Rod. BR 163, km 253,6, CP 449, 79.804-790, Dourados, MS, Brazil

Antônio Félix da Costa
Instituto Agronômico de Pernambuco, Avenida General San Martin, 1371, Bairro Bongi, 50.761-000, Recife, PE, Brazil

Ariano Martins de Magalhães Júnior, Paulo Ricardo Reis Fagundes, Daniel Fernandes Franco Eduardo Anibele Streck, Gabriel Almeida Aguiar and Paulo Henrique Karling Facchinello
Embrapa Clima Temperado, Rodovia BR-392, km 78, 9° Distrito, Monte Bonito, CP 321, 96.010-971, Pelotas, RS, Brazil

Orlando Peixoto de Morais, José Manoel Colombari Filho, Paulo Hideo Nakano Rangel and Francisco Pereira Moura Neto
Embrapa Arroz e Feijão, Rodovia GO-462, km 12, Fazenda Capivara, Zona Rural, CP 179, 75.375-000, Santo Antônio de Goiás, GO, Brazil

Antônio Carlos Centeno Cordeiro
Embrapa Roraima, Rodovia BR 174, km 8, Distrito Industrial, CP 133, 69.301-970, Boa Vista, RR, Brazil

José Almeida Pereira
Embrapa Meio Norte, Av. Duque de Caxias, 5.650, Bairro Buenos Aires, CP 001, 64.006- 220, Teresina, PI, Brazil

RS Rutherford, KZ Maphalala, AC Koch and SJ Snyman
School of Life Sciences, University of KwaZulu-Natal, Westville campus, Private Bag X54001, Durban, 4000, South Africa
South African Sugarcane Research Institute, Private Bag X02, Mount Edgecombe, KwaZulu- Natal, 4300, South Africa

MP Watt
South African Sugarcane Research Institute, Private Bag X02, Mount Edgecombe, KwaZulu- Natal, 4300, South Africa

Bruna Line Carvalho, Magno AntonioPatto Ramalho and Indalécio Cunha Vieira Júnior
Federal University of Lavras (UFLA), Department of Biology, CP 3037, 37.200-000,Lavras, MG, Brazil

Ângela de Fátima Barbosa Abreu
Embrapa Rice and Beans/UFLA, Rod.GO-462, km 12, Zona Rural, CP 179, 75.375-000, Santo Antônio de Goiás, GO,Brazil

Wendell Jacinto Pereira
Universidade Federal de Goiás, Instituto de Ciências Biológicas, 74.001-970, Goiânia, GO, Brazil
Universidade de Brasília, Departamento de Biologia Celular, 70910-900, Brasília, DF, Brazil

Priscila Zaczuk Bassinello
Embrapa Arroz e Feijão, Laboratório de Grãos e Subprodutos, Rodovia GO-462, km 12, Fazenda Capivara, Zona Rural, 75375-000, Santo Antônio de Goiás, GO, Brazil

Claudio Brondani and Rosana Pereira Vianello
Embrapa Arroz e Feijão, Laboratório de Biotecnologia

Andrei Caíque Pires Nunes and Glêison Augusto dos Santos
Universidade Federal de Viçosa (UFV), Departamento de Engenharia Florestal, Avenida Peter Henry Rolfs, s/n, Campus Universitário, 36.570-900, Viçosa, MG, Brazil

Marcos Deon Vilela de Resende
Empresa Brasileira de Pesquisa Agropecuária, Centro Nacional de Pesquisa de Florestas, Estrada da Ribeira, km 111, Bairro Guaraituba, 83.411-000, Colombo, PR, Brazil

Rodrigo Silva Alves
UFV, Departamento de Biologia Geral

Flávio Gabriel Bianchini, Rodrigo Vieira Balbi, Rafael Pio and Adriano Teodoro Bruzi
Universidade Federal de Lavras (UFLA), Departamento de Agricultura, 37.200-000, Lavras, MG, Brazil

Daniel Fernandes da Silva
UFLA, Departamento de Biologia

Mingyu Hou
Agricultural University of Hebei, Laboratory of Crop Germplasm Resources for Northern China, Ministry of Education/Key Laboratory of Crop Germplasm Resources of Hebei Province/College of Agronomy, Baoding 071001, China
Agricultural University of Hebei, College of Life Science, Baoding 071001, China

Guojun Mu, Shunli Cui, Xinlei Yang and Lifeng Liu
Agricultural University of Hebei, Laboratory of Crop Germplasm Resources for Northern China, Ministry of Education/Key Laboratory of Crop Germplasm Resources of Hebei Province/College of Agronomy, Baoding 071001, China

Yongjiang Zhang
Agricultural University of Hebei, College of Agronomy, Baoding 071001, China

Tamires Ribeiro, Cleber Vinicius Giaretta Azevedo, Jose Antonio de Fatima Esteves, Sérgio Augusto Morais Carbonell and Alisson Fernando Chiorato
Instituto Agronômico, Centro de Análise e Pesquisa Tecnológica do Agronegócio dos Grãos e Fibras, Avenida Barão de Itapura, 1481, 13.020-902, Botafogo, Campinas, São Paulo, Brazil

Margarida Fumiko Ito
Instituto Agronômico, Centro de Fitossanidade

Index